NO PLACE FOR A PURITAN

NO PLACE FOR A PURITAN

The Literature of the California Deserts

Edited by Ruth Nolan

Heyday Books, Berkeley, California
Santa Clara University, Santa Clara, California
Inlandia Institute, Riverside, California

This book was made possible in part by a generous grant from The James Irvine Foundation.

Library of Congress Cataloging-in-Publication Data

No place for a Puritan : the literature of the California deserts / edited by Ruth Nolan.
    p. cm. -- (A California legacy book)
  Includes index.
  ISBN 978-1-59714-098-0 (pbk. : alk. paper)
  1. Deserts--California--Literary collections. 2. California--Literary collections. I. Nolan, Ruth.
  PS571.C2N62 2009
  810.9'3581540794--dc22
                   2009013994

Cover image created from original photo by Jeff Jensen.
Book Design: Lorraine Rath
Printing and Binding: Thomson Shore, Dexter, MI

This California Legacy book was copublished by Santa Clara University, Inlandia Institute, and Heyday Books. Orders, inquiries, and correspondence should be addressed to:
    Heyday Books
    P. O. Box 9145, Berkeley, CA 94709
    (510) 549-3564, Fax (510) 549-1889
    www.heydaybooks.com

10 9 8 7 6 5 4 3 2 1

To the next desert generation: my daughter Tarah and
my nephews Mikhael and James Dean

CONTENTS

Introduction

I was ten years old in 1973 when my father first drove me in his old Volkswagen bug from my hometown of San Bernardino, embedded in the smog of southern California sixty miles east of Los Angeles, up the long, steep grade of Interstate 15 and over the four-thousand-foot lip of Cajon Pass. I held my breath as we reached the top and I saw, for the first time in my life, a land that was as wide and vast as the sea. There, at the edge of the Mojave Desert, a long necklace of headlights stretched east for forty miles; toward the west, the sky was lit with rose and orange hues. We descended toward the small town of Victorville, racing past Joshua trees whose thick-needled fists etched the sky gracefully and fiercely against the sunset. I knew then and there that I'd found my place, my calling, my landscape. I stuck my head out the window and looked up: there was the evening star, a slice of moon alongside it. I was instantly and forever smitten.

This was an empty and imposing land, rife with promise of danger and thrill. I sensed that an entirely new adventure lay in wait for our family there, where we intended to relocate to be near my father's new job. My intuitions were confirmed when my mother later opened a kitchen drawer to find a baby Mojave green rattlesnake; when I went to bed serenaded by a symphony of coyotes every night; when my brother went to the hospital with dehydration after climbing a harsh rock peak near our house on an August afternoon. The desert was as silent as a church during a funeral and as wide open and empty as a schoolyard on a Sunday, but it was never, ever boring.

Little did I know, on that first drive to the high desert, that the road we drove overlaid an older route, an Indian trail used for thousands of years by different tribes to traverse California's desert from the coast to the Colorado River and other inland areas, from waterhole to waterhole. The trail in part follows the one hundred and fifty miles of the Mojave River, which for centuries has flowed north—in some places above ground and through dense shoulders of cottonwoods, in other stretches underground beneath a vast and arid floodplain covered with deep sand—to its final resting place at Soda Dry Lake. By the seventeenth century, Spanish priests and early explorers were using this same route; in the nineteenth and twentieth centuries, westbound settlers used the Mojave Road, or the Mormon Trail (same route, different name), to safely cross the desert to and from California's densely populated, climate-friendly coast and the interior regions of the country.

As a child I perceived that this is a place of wildness and possibility, of rock hunting and tortoise sightings, of flash floods and years when rain never falls. But perhaps more significantly, this is also a peopled place. It is a land rife with stories of courage, despair, and resignation, where hopes are fulfilled and dreams dissolved. The desert is not, as it's been stereotyped, a waterless void: thousands of springs and waterholes grace it. They are often hidden, detectable only to the longtime desert resident who spots a lone cottonwood or sprig of weed in the Mojave, or a cluster of native Washingtonian fan palms tucked into deep canyons of the western Colorado. Just as the desert has been stereotyped, people who have made the desert home have been reduced to two-dimensional caricatures in the annals of American history and in a literary canon that favors cities, farms, and forests. With a little digging, one discovers that the literature of California deserts is every bit as exuberant, varied, and charming as the literature of its more populated and gentrified sister regions.

The average reader is certainly familiar with stories of the rugged desert survivalist, the consummate "desert rat" or gold miner, grizzled and worn by sun, who wears a rattlesnake-skin headband and roams the desert with a bag of tools. However, the stories of the earliest people, California's desert Indians, have gone largely untold. Their creation stories and songs are rich contributions to literature and depict an active relationship between the landscape and an ancient culture, a relationship that thrives to this day. In stock desert literature, stories of rugged western settlers, gunslingers, and stagecoach riders who brave the desert's harsh expanses and pray they'll make it to water have been greatly emphasized. The true story of California's Mojave and western Colorado deserts is as rich and textured as their vast geography, which covers twenty-five hundred miles and parts of seven of the state's counties.

When I was ten years old, I knew nothing of the history of people in the place I found so entrancing. By the time I reached adulthood and began to teach desert literature courses at College of the Desert, I realized that I had underestimated the breadth and depth of the literature of California's desert. I'd never seen a collection that aptly reflected my own experiences and gave coherent meaning to the threads of "desert" woven into the works I'd read. In my own studies, any mention of the California desert was always overshadowed and left unexplained. It seemed to me that the desert was a literary underdog, employed as a fearful setting or as a metaphor for triumph over adversity but never depicted in a broader, fuller sense. Along with other inhabitants and enthusiasts who wanted a substantial, honest literary exploration of California's deserts, I needed and deserved more.

It is my hope that, in this collection, I've given readers meaningful access to the history and culture of the desert, where the landscape is so diverse that it seems to resist classification as a single region. For example, the Mojave River flows northward across an otherwise arid stretch of land, defying the stereotype of a waterless geography. In the southern portion of western Colorado, centuries-old sand dunes struggle to survive alongside golf courses. The Santa Rosa Mountains, one of the rockiest, most barren ranges in the West, overlook the Salton Sea, a vast, landlocked body of water situated at more than two hundred feet below sea level. Likewise, this collection contains pieces that seem to describe vastly different places but in fact describe facets of the same dynamic geography.

This anthology is organized thematically: these themes include dangers, crossings, refuge and exile, the lure of the desert, making a home there, changes, and conservation and protection. The order follows the arc of outsiders' changing perceptions of the desert: from the idea that the desert is a terrifying wasteland to either be avoided or hurried across—at the risk of one's life—to the belief that the desert is a place of spiritual renewal and mystery; from an exotic and forbidding place to visit to a landscape tamed by irrigation and development, a permanent home for thousands of residents. All of these attitudes toward the desert exist today, but the last two sections of the book explore them in a contemporary context. Environmental awareness has dawned, and we've discovered that the desert, far from being the disposable wasteland it was once thought to be, is in fact a fragile, overcrowded, overused, and intensely threatened landscape.

At the heart of this collection is some of the best writing found in the American literary canon. There are stories, poems, journal entries, and news stories that incorporate unique icons of the desert: the roadrunner, the remote homesteading cabin, the mirage. There are stories that thrill, frighten, sadden, and inspire: a man foolishly and arrogantly collecting live rattlesnakes; a lone woman striving to make a home in a remote desert canyon; Asian American farmers in the Imperial Valley suffering unbearable personal loss; and a family coping with incarceration in a World War II concentration camp. There are meditations on how the desert landscape parallels the human spirit, and tales of ethnically diverse people carving communities out of the farthest corners of the California desert. People from the region's diverse Indian tribes—the Timbisha Shoshone, the Cahuilla, the Serrano, the Chemehuevi, the Mojave, and others—have participated in this project, and an essay commemorating the passage of the historic California Desert Protection Act in 1994 is also included. In this collection, anything can happen, and often does: familiar voices are included alongside literature that has been obscurely published, is just arriving on the scene, or has been long out of print.

Decades have passed since the desert first took my breath away, and much has changed. I now live in the area of the desert near Palm Springs, where golf courses and posh resorts crowd the horizon, and the endangered bighorn sheep is commemorated in decorative statues in nearby shopping malls. The entire California desert is threatened with overpopulation, pollution, and other social and climactic ills facing contemporary society. The population of Victorville has exploded to more than one hundred thousand people, and smog now fills the easy expanses once billed by real estate flyers as "the land of the champagne climate." A new threat to the desert is the rush to install solar- and wind-power facilities throughout vast tracts of the remaining open desert spaces as our nation turns to alternate energy sources. What remains of the open spaces I saw as a child from the summit of Cajon Pass will likely soon be transformed into massive power grids, fed by acre upon acre of windmills and solar farms. Areas that once provided abundant food to the region's native people are now all but barren of plant life.

The desert has come to seem much smaller to me now, but the literary legacy seems larger. This is a land of people, of struggle, loss, and success, and more recently, a region of politically charged and intensely debated land-use issues. The California desert has long been—and continues to be—far more than a mere wasteland waiting for people to carelessly exploit or briefly endure it. In the stories in this collection, the desert sings. It hums with the pulse of overlapping human lives, a river of sound that sometimes overflows its shores, and at other times travels quietly underground.

Acknowledgments

First and foremost, my warmest thanks to The James Irvine Foundation and Inlandia Institute for the generous grant that made this book possible, as well as to Terry Beers of Santa Clara University's California Legacy Project, for his enthusiastic support of this book.

I am deeply grateful to the amazing leaders at Heyday Books: Malcolm Margolin for his ready belief in and support of this book, as well as for providing me the opportunity to work on this challenging and rewarding anthology; Gayle Wattawa for her steady presence and guidance every step of the way; and David Chu, proofreader and support team member.

Many other people have helped make this project a reality, and my sincere thanks go to each of them. First, I would like to thank Jeff Green for his hard work in helping me find selections and decide what to include, and for endless hours of permissions-seeking, scanning, photocopying, organizing, and serving the role of local and irreplaceable support staff. I also received wonderful assistance from Claudia Derum and Jon Fernald, librarians at College of the Desert, and I thank them for the ready access I was given to the campus desert collection, as well as the City of Palm Desert library's desert collection. For their generosity in sharing their cultural knowledge and resources to make this book possible, I'd like to thank Dr. Katherine Siva Sauvel; the Malki Museum and its staff; the Agua Caliente Band of Cahuilla Indians; Pauline Esteves; Theresa Mike; Dr. Clifford Trafzer; Dr. Lowell Bean; the Native American Land Conservancy; and Phil Klasky. I'd like to thank Marion Mitchell-Wilson of Inlandia Institute; B. H. Fairchild, Professor Emeritus, California State University, San Bernardino; Rob Fulton, Director of the Desert Studies Center; Julie Warren of the Palm Springs Library; the Palm Springs Art Museum staff; Cliff Walker, Mojave Desert historian; Betsy Knaak of the Anza-Borrego Desert Natural Historical Association; Diana Lindsay of Sunbelt Books; Tom Budlong; the Mojave River Valley Museum staff; Joan Brooks; Hal Rover of the Palm Desert Historical Society; Les Snodgrass and other staff of the 29 Palms Historical Society; Peter Wild; Ann Japenga; Lawrence Hogue; the staff of the Coachella Valley Historical Association in Indio, California; Harry Quinn; Joshua Tree National Park Ranger Caryn Davidson; and several of my colleagues at College of the Desert, including Dr. Jerry Patton, Dr. Ellen Hardy, Dr. Kathlyn Enciso, Steve Acree, Linda Lawliss, Jean Waggoner, and Denise Toland. Special thanks go to those who assisted me in tracking down difficult permissions, including Keith and Corless W. Eldred; Kenny Paul; Cathy Brant; Susan Zwinger; and Ambar Tovar and the Cesar E. Chavez Foundation.

ACKNOWLEDGMENTS

I also received ideas and help from many others, as well as inspiration and, at times, much needed support and belief in the project. Thanks go to: Rob Roberge; Tod Goldberg; Allison Hedgecoke; Juan Felipe Herrera; Margarita Luna Robles; Karen Wilson; Mary Sojourner; Ethan Applegarth; Reggie Woollery; Ching-In Chen; Mary Curtin; Allison Johnson; Celeste de Blasis; Alaska Whelan; Kath Abela Wilson; Deborah P. Kolodji; Armi Atil; Brian Brown; Gayle Brandeis; Joel Lamore; Michael Cluff; Rowena Silver; Howard Wilshire; Jane Nielsen; Judy Kronenfeld; Maureen Alsop; Cati Porter; Lavina Blossom; Lucia Galloway; Chrystine Julian; Kathryn Jordan; Linda McMillin Pyle; Philip Helland; Mike Cipra; Swami Ramananda; Sri Sri Anandamayi Ma; and my many students at College of the Desert, whose enthusiasm for learning and literature have been an inspiration for this book.

I would like to thank my highly supportive family: my parents, Beverly and Joseph Nolan; my daughter, Tarah Fenelon; my brothers, John, Jerry, and Patrick; my aunts Eileen De Vito and Jeanne Bruce; my cousin Beth Pinkerton; and many other extended family members whose love and support gave me courage and focus throughout this project.

I would especially like to thank my parents for moving our family to California's Mojave Desert, a fortuitous relocation that gave me an instant and enduring love for the California desert.

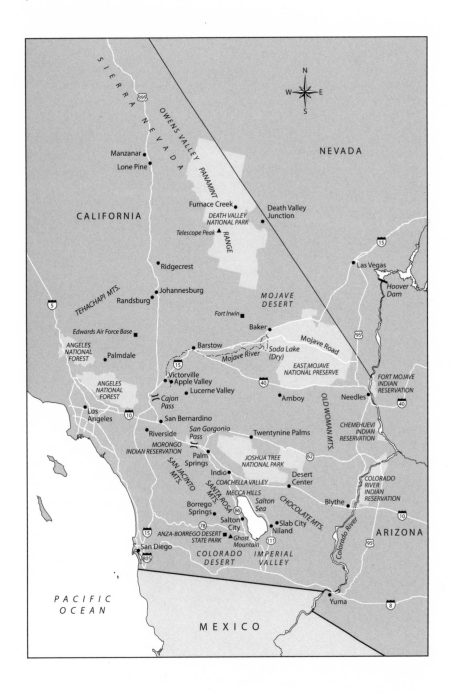

DANGERS OF THE DESERT

MARY AUSTIN

*Born in 1868 in Illinois, Mary Austin moved to California in 1888. As a
writer she is best known for her landmark book* The Land of Little Rain. *First
published in 1903, it provides an aesthetic and compelling description of the
people and landscape of the Mojave Desert. Austin also studied and wrote widely
on Native American culture. She died in Santa Fe, New Mexico, in 1934.*

The following excerpt from The Land of Little Rain *focuses on the land,
plants, animals, and people of Death Valley, Owens Valley, and Inyo County in
the early twentieth century.*

from *The Land of Little Rain*

East away from the Sierras, south from Panamint and Amargosa, east and south
many an uncounted mile, is the Country of Lost Borders.

Ute, Paiute, Mojave, and Shoshone inhabit its frontiers, and as far into the
heart of it as a man dare go. Not the law, but the land sets the limit. Desert is
the name it wears upon the maps, but the Indian's is the better word. Desert
is a loose term to indicate land that supports no man; whether the land can
be bitted and broken to that purpose is not proven. Void of life it never is,
however dry the air and villainous the soil.

This is the nature of that country. There are hills, rounded, blunt, burned,
squeezed up out of chaos, chrome and vermilion painted, aspiring to the
snowline. Between the hills lie high level-looking plains full of intolerable sun
glare, or narrow valleys drowned in a blue haze. The hill surface is streaked with
ash drift and black, unweathered lava flows. After rains water accumulates in the
hollows of small closed valleys, and, evaporating, leaves hard dry levels of pure
desertness that get the local name of dry lakes. Where the mountains are steep
and the rains heavy, the pool is never quite dry, but dark and bitter, rimmed
about with the efflorescence of alkaline deposits. A thin crust of it lies along the
marsh over the vegetating area, which has neither beauty nor freshness. In the
broad wastes open to the wind the sand drifts in hummocks about the stubby
shrubs, and between them the soil shows saline traces. The sculpture of the hills
here is more wind than water work, though the quick storms do sometimes scar
them past many a year's redeeming. In all the Western desert edges there are
essays in miniature at the famed, terrible Grand Cañon, to which, if you keep on
long enough in this country, you will come at last.

Since this is a hill country one expects to find springs, but not to depend
upon them; for when found they are often brackish and unwholesome, or

maddening, slow dribbles in a thirsty soil. Here you find the hot sink of Death Valley, or high rolling districts where the air has always a tang of frost. Here are the long heavy winds and breathless calms on the tilted mesas where dust devils dance, whirling up into a wide, pale sky. Here you have no rain when all the earth cries for it, or quick downpours called cloud-bursts for violence. A land of lost rivers, with little in it to love; yet a land that once visited must be come back to inevitably. If it were not so there would be little told of it.

This is the country of three seasons. From June on to November it lies hot, still, and unbearable, sick with violent unrelieving storms; then on until April, chill, quiescent, drinking its scant rain and scanter snows; from April to the hot season again, blossoming, radiant, and seductive. These months are only approximate; later or earlier the rain-laden wind may drift up the water gate of the Colorado from the Gulf, and the land sets its seasons by the rain.

The desert floras shame us with their cheerful adaptations to the seasonal limitations. Their whole duty is to flower and fruit, and they do it hardly, or with tropical luxuriance, as the rain admits. It is recorded in the report of the Death Valley expedition that after a year of abundant rains, on the Colorado desert was found a specimen of Amaranthus ten feet high. A year later the same species in the same place matured in the drought at four inches. One hopes the land may breed like qualities in her human offspring, not tritely to "try," but to do. Seldom does the desert herb attain the full stature of the type. Extreme aridity and extreme altitude have the same dwarfing effect, so that we find in the high Sierras and in Death Valley related species in miniature that reach a comely growth in mean temperatures. Very fertile are the desert plants in expedients to prevent evaporation, turning their foliage edgewise toward the sun, growing silky hairs, exuding viscid gum. The wind, which has a long sweep, harries and helps them. It rolls up dunes about the stocky stems, encompassing and protective, and above the dunes, which may be, as with the mesquite, three times as high as a man, the blossoming twigs flourish and bear fruit.

There are many areas in the desert where drinkable water lies within a few feet of the surface, indicated by the mesquite and the bunch grass (*Sporobolus airoides*). It is this nearness of unimagined help that makes the tragedy of desert deaths. It is related that the final breakdown of that hapless party that gave Death Valley its forbidding name occurred in a locality where shallow wells would have saved them. But how were they to know that? Properly equipped it is possible to go safely across that ghastly sink, yet every year it takes its toll of death, and yet men find there sun-dried mummies, of whom no trace or recollection is preserved. To underestimate one's thirst, to pass a given landmark to the right or left, to find a dry spring where one looked for running water—there is no help for any of these things.

Along springs and sunken watercourses one is surprised to find such water-loving plants as grow widely in moist ground, but the true desert breeds its own kind, each in its particular habitat. The angle of the slope, the frontage of a hill, the structure of the soil determines the plant. South-looking hills are nearly bare, and the lower tree-line higher here by a thousand feet. Cañons running east and west will have one wall naked and one clothed. Around dry lakes and marshes the herbage preserves a set and orderly arrangement. Most species have well-defined areas of growth, the best index the voiceless land can give the traveler of his whereabouts.

If you have any doubt about it, know that the desert begins with the creosote. This immortal shrub spreads down into Death Valley and up to the lower timber-line, odorous and medicinal as you might guess from the name, wandlike, with shining fretted foliage. Its vivid green is grateful to the eye in a wilderness of gray and greenish white shrubs. In the spring it exudes a resinous gum which the Indians of those parts know how to use with pulverized rock for cementing arrow points to shafts. Trust Indians not to miss any virtues of the plant world!

Nothing the desert produces expresses it better than the unhappy growth of the tree yuccas. Tormented, thin forests of it stalk drearily in the high mesas, particularly in that triangular slip that fans out eastward from the meeting of the Sierras and coastwise hills where the first swings across the southern end of the San Joaquin Valley. The yucca bristles with bayonet-pointed leaves, dull green, growing shaggy with age, tipped with panicles of fetid, greenish bloom. After death, which is slow, the ghostly hollow network of its woody skeleton, with hardly power to rot, makes the moonlight fearful. Before the yucca has come to flower, while yet its bloom is a creamy cone-shaped bud of the size of a small cabbage, full of sugary sap, the Indians twist it deftly out of its fence of daggers and roast it for their own delectation. So it is that in those parts where man inhabits one sees young plants of *Yucca arborensis* infrequently. Other yuccas, cacti, low herbs, a thousand sorts, one finds journeying east from the coastwise hills. There is neither poverty of soil nor species to account for the sparseness of desert growth, but simply that each plant requires more room. So much earth must be preëmpted to extract so much moisture. The real struggle for existence, the real brain of the plant, is underground; above there is room for a rounded perfect growth. In Death Valley, reputed the very core of desolation, are nearly two hundred identified species.

Above the lower tree-line, which is also the snow-line, mapped out abruptly by the sun, one finds spreading growth of piñon, juniper, branched nearly to the ground, lilac and sage, and scattering white pines.

There is no special preponderance of self-fertilized or wind-fertilized plants, but everywhere the demand for and evidence of insect life. Now where there are seeds and insects there will be birds and small mammals and where these are, will come the slinking, sharp-toothed kind that prey on them. Go as far as you dare in the heart of a lonely land, you cannot go so far that life and death are not before you. Painted lizards slip in and out of rock crevices, and pant on the white hot sands. Birds, hummingbirds even, nest in the cactus scrub; woodpeckers befriend the demoniac yuccas; out of the stark, treeless waste rings the music of the night-singing mockingbird. If it be summer and the sun well down, there will be a burrowing owl to call. Strange, furry, tricksy things dart across the open places, or sit motionless in the conning towers of the creosote. The poet may have "named all the birds without a gun,"[1] but not the fairy-footed, ground-inhabiting, furtive, small folk of the rainless regions. They are too many and too swift; how many you would not believe without seeing the footprint tracings in the sand. They are nearly all night workers, finding the days too hot and white. In mid-desert where there are no cattle, there are no birds of carrion, but if you go far in that direction the chances are that you will find yourself shadowed by their tilted wings. Nothing so large as a man can move unspied upon in that country, and they know well how the land deals with strangers. There are hints to be had here of the way in which a land forces new habits on its dwellers. The quick increase of suns at the end of spring sometimes overtakes birds in their nesting and effects a reversal of the ordinary manner of incubation. It becomes necessary to keep eggs cool rather than warm. One hot, stifling spring in the Little Antelope I had occasion to pass and repass frequently the nest of a pair of meadowlarks, located unhappily in the shelter of a very slender weed. I never caught them sitting except near night, but at midday they stood, or drooped above it, half fainting with pitifully parted bills, between their treasure and the sun. Sometimes both of them together with wings spread and half lifted continued a spot of shade in a temperature that constrained me at last in a fellow feeling to spare them a bit of canvas for permanent shelter. There was a fence in that country shutting in a cattle range, and along its fifteen miles of posts one could be sure of finding a bird or two in every strip of shadow; sometime the sparrow and the hawk, with wings railed and beaks parted, drooping in the white truce of noon.

If one is inclined to wonder at first how so many dwellers came to be in the loneliest land that ever came out of God's hands, what they do there and why stay, one does not wonder so much after having lived there. None other than this long brown land lays such a hold on the affections. The rainbow hills, the tender bluish mists, the luminous radiance of the spring, have the lotus charm. They trick the sense of time, so that once inhabiting there you always mean to go away without quite realizing that you have not done it. Men who

---

[1] Austin is referring to the first line of Emerson's 1842 poem, "Forbearance."

have lived there, miners and cattle-men, will tell you this, not so fluently, but emphatically, cursing the land and going back to it. For one thing there is the divinest, cleanest air to be breathed anywhere in God's world. Some day the world will understand that, and the little oases on the windy tops of hills will harbor for healing its ailing, house-weary broods. There is promise there of great wealth in ores and earths, which is no wealth by reason of being so far removed from water and workable conditions, but men are bewitched by it and tempted to try the impossible.

You should hear Salty Williams tell how he used to drive eighteen and twenty-mule teams from the borax marsh to Mojave, ninety miles, with the trail wagon full of water barrels. Hot days the mules would go so mad for drink that the clank of the water bucket set them into an uproar of hideous, maimed noises, and a tangle of harness chains, while Salty would sit on the high seat with the sun glare heavy in his eyes, dealing out curses of pacification in a level, uninterested voice until the clamor fell off from sheer exhaustion. There was a line of shallow graves along that road; they used to count on dropping a man or two of every new gang of coolies brought out in the hot season. But when he lost his swamper, smitten without warning at the noon halt, Salty quit his job; he said it was "too durn hot." The swamper he buried by the way with stones upon him to keep the coyotes from digging him up, and seven years later I read the penciled lines on the pine headboard, still bright and unweathered.

But before that, driving up on the Mojave stage, I met Salty again crossing Indian Wells, his face from the high seat, tanned and ruddy as a harvest moon, looming through the golden dust above his eighteen mules. The land had called him.

The palpable sense of mystery in the desert air breeds fables, chiefly of lost treasure. Somewhere within its stark borders, if one believes report, is a hill strewn with nuggets; one seamed with virgin silver; an old clayey water-bed where Indians scooped up earth to make cooking pots and shaped them reeking with grains of pure gold. Old miners drifting about the desert edges, weathered into the semblance of the tawny hills, will tell you tales like these convincingly. After a little sojourn in that land you will believe them on their own account. It is a question whether it is not better to be bitten by the little horned snake of the desert that goes sidewise and strikes without coiling, than by the tradition of a lost mine.

And yet—and yet—is it not perhaps to satisfy expectation that one falls into the tragic key in writing of desertness? The more you wish of it the more you get, and in the mean time lose much of pleasantness. In that country which begins at the foot of the east slope of the Sierras and spreads out by less and less lofty hill ranges toward the Great Basin, it is possible to live with great zest, to have red blood and delicate joys, to pass and repass about one's daily performance an area that would make an Atlantic seaboard State, and that with no peril, and, according to our way of thought, no particular difficulty. At any rate, it was not people who went into the desert merely to write it up who invented the fabled Hassaympa, of whose waters, if any drink, they can no more see fact as naked fact, but all radiant with the color of romance. I, who must have drunk of it in my twice seven years' wanderings, am assured that it is worth while.

For all the toll the desert takes of a man it gives compensations, deep breaths, deep sleep, and the communion of the stars. It comes upon one with new force in the pauses of the night that the Chaldeans were a desert-bred people. It is hard to escape the sense of mastery as the stars move in the wide clear heavens to risings and settings unobscured. They look large and near and palpitant; as if they moved on some stately service not needful to declare. Wheeling to their stations in the sky, they make the poor world-fret of no account. Of no account you who lie out there watching, nor the lean coyote that stands off in the scrub from you and howls and howls.

WILLIAM JUSTEMA

William Justema was born in Chicago, Illinois, in 1905. After spending several years as a monk in an Oregon monastery, he moved to San Francisco, where he designed wallpapers and fabrics for the home furnishing industry for forty years. He also wrote poetry and in 1944 published a collection of poems inspired by his time spent in military service during World War II, Private Papers: Poems of an Army Year. *He died in 1987.*

"No Place for a Puritan," the title of this anthology, comes from the title of a poem by Justema from Private Papers.

No Place for a Puritan

is the desert. No place to draw curtains.
And what is a hymn in the open
but a pity, or a sin…
Where God does not exist—invent Him.
Regard then the Great American Desert God
morbid and elegant, His kingdom almost emptied.
The sea was here and left its shapes
to wither. Only my "aigrette" (or Smoke) tree;
my rare "flute bush" (?) show vividly
the rush of water. Filigree weeds
reach for they know not what—
finally breaking in the hush-hush wind,
their odd jewelry strewn on the ancient beach.
We have forgotten the islands;
see the mountains (my "Potpourris") each
evening—as an iris, a violet, and a rose
gathering when the sun drops out of reach
and the moon takes over: a lunar
landscape too barren and too bold
for homely virtues…
Our God is a sensuous God—Who
would not be, from a couch of sand?
The soldier on the desert

rears the most worldy altar that he can
and serves in serpent skin, with curses.
It is the right degree of the wrong ecstasy:
do you blame the man? He never sees
the fountains he continually hears,
nor finds the bare feet dancing his mosaic floors.
While far above him clouds are whirled
carnation colored and carnation curled—
hardly Christian.

EDWIN CORLE

Edwin Corle was born in Wildwood, New Jersey, in 1906. A prolific author and playwright, he wrote several books and many short stories and essays about the California desert landscape and its inhabitants. His books include Mojave: A Book of Stories *(1934),* Fig Tree John *(1935), and* Desert Country *(1941). Corle died in 1956 in Santa Barbara, California.*

The following excerpt from Desert Country *is a gripping account of one man's encounter with a rattlesnake.*

from *Desert Country*

If You Don't Get Excited—

The forked stick dropped deftly over the snake's body just back of the head. The reptile was effectively pinned to the sandy bottom of the dry wash. Banning, holding the stick, looked down in concentrated admiration at the prize. The snake, a red diamond rattler, indigenous to the Colorado Desert of southeastern California, about five feet in length or perhaps a few inches short of that mark, seemed stunned, unable to comprehend a new experience. It made no violent movements at all and showed no special fear, and only a slight waving of the tail and a rapid shooting in and out of its blue-black tongue evinced its dull but mounting anger. Banning congratulated himself on his skill. The snake itself couldn't have struck with more accuracy than he had done in capturing it, and moreover, the stick apparently had in no way injured it.

"Hello, baby," Banning said aloud, maintaining his grip on the stick to keep the reptile down firmly, but taking care not to exert too great pressure on the back of its neck. He bent forward.

It was truly a beautiful specimen—a rusty red, with the diamond markings in deeper red, thinly outlined in flat white. The head was darker red than the body, looking triangular from above with the poison pouches plainly visible on each side. The brown eyes stared unknowingly, uncomprehendingly, into space. The tail, about six inches in length, was white, with four black rings ending in nine rattles.

"You're a sweet baby," Banning said, holding the stick with his left hand and opening the sack with his right. Watching the head of the snake and shaping the mouth of the sack into a wide crater, which, from long experience, he could have done in the dark or if he were blind, he let it remain in readiness on the sandy wash within a foot of the snake. His last groping precaution was to feel for the

11

drawstring which he would yank, once the snake's body was within the crater of the waiting sack. The string was there.

Still keeping a firm grip, he let his right hand descend to a point on the red body just back of the forks of the stick. This was the moment. Excitement now would bungle the whole job. He even went over in his mind the details of the operation so that it could be executed quickly, harmlessly, and without a wasted motion: his feet well braced, with the left forward and the right to the rear (for if you are off balance you are only handicapping yourself), thumb and forefinger on each side of the reptile's neck up close to its head, the sack only inches away on his right side. All was perfect.

With a rhythmic gesture he raised the forked stick and tossed it aside, at the same time increasing the pressure of his thumb and forefinger; and lifting the reptile head first into the air, he swung it to the right until the tail was over the crater. This is the instant when the amateur is impelled to drop the prize into the sack, and that's the usual mistake. Banning never relaxed his grip on the neck, but lowered his hand straight down so that the five feet of red body went tail first into the crater and immediately began to coil as it felt the solid foundation beneath. When the head was but a few inches above the rest of the body, and with his own body directly in back of the direction in which the head was pointed, he released the thumb and forefinger grip and with his left hand grasped the drawstring. Giving the string a quick jerk upward brought the sack up vertically around the snake and at the same time pulled the crater together. In an instant the snake had disappeared within.

"Couldn't be better, baby," Banning said aloud.

This was too good to keep to himself. He wouldn't hunt any farther. It was at least a mile and a half, perhaps two, back to the car, and he and Mr. Diamondback would start at once. Parker and Knapp might not be back yet, but he could get Mr. Snake safely ensconced in the station wagon and wait with pleasant anticipation until they returned—empty-handed, too, he half hoped. But that was unfair. He should really wish them all the luck that he had just had. Fondly he looked at the sack. There were slight undulations in its sides as the reptile moved within. This was really great. Probably the best Parker and Knapp would have would be some nasty little sidewinder—poisonous little devil, lacking the grandeur of this creature. He had never seen a red diamond rattler quite as long as this one. Four feet was a good length, but this must be very close to five.

"Now we take a walk," he said aloud. He picked up the sack by the end of the drawstring with his right hand and started down the wash. Then, just in time, he snapped the fingers of his left hand.

"Stupid!"

He turned back. He had made his first mistake. It was a bonehead trick typical of the amateur. He had been collecting snakes for five years, and never before had he started off and left his stick where it had been tossed—like the motorist who repairs a flat tire and then drives off and leaves his tool by the roadside. There was the stick. Wouldn't that have been a dumb trick, to go all the way back to the car and forget the stick? They'd have kidded him for that. But then, would either of them have a red diamond?

"All right, baby, now we're off."

It was about a mile down the wash, easy walking but a little sandy, through the dwarfed smoke trees that looked purplish until you got close to them, to the point where the wash curved to the left. He couldn't miss it, for not only was the curve a landmark, but there were two naked palo verde trees on the bank at that spot. Also there were his own tracks, which he could retrace. He wondered why there wasn't an English name for palo verde. In translation it simply meant "green tree." Well, they were green, all right, even to the bark.

Then, when he reached the palo verde trees, he would have another mile across the alluvial fan from the mountains to the next wash, where they had left the station wagon. They had driven up that wash from the highway four miles below, until the sand made it hazardous, and they had left the car and gone out alone on foot from there. At least, Banning had gone alone, but Parker and Knapp were probably hunting together. Parker, after all, had done this only twice before. Amateur.

And you can't tell; Banning himself might run across something even on this trip back. Luck runs like that.

He plodded on, his heavy laced boots scuffing the sand and the sack swinging at his side. "Say, baby, you're no lightweight"—he shifted the sack to his left hand—"are you, sweetheart?" He wondered about the sex—well, plenty of time to find that out later. He certainly didn't intend to open that sack now.

The bottom of the wash was about six feet lower than the sloping fan through which the wash itself had cut. Back of him lay the Santa Rosa Range, from a canyon of which the dry wash began. Ahead lay the Coachella Valley, the northwestern arm of the Colorado Desert, and across the valley were the painted Mecca Hills. From his present position on the floor of the wash it was impossible to see much of anything over the water-eroded walls except the tops of distant blue mountains. He had been away from the car about an hour when he came upon the rattler gliding across the sand. It was late afternoon, or he would never have had such luck. As a general rule the snakes of the desert country are quiescent during the heat of the day, for the hot sand burns their bellies. Banning and Parker and Knapp had agreed on a three- to four-hour trip, from four in the afternoon until seven or eight in the evening, since that is the time of day when the reptiles begin to forage. Banning reasoned that Parker might tire first; he and

Knapp should be back at the car by six o'clock or thereabouts. It was now five o'clock, and he still had a mile and a half of desert to cover. He would get to the car ahead of them by a matter of minutes, no doubt. He trudged on, wondering what the technical name of his prize might be. *Crotalus atrox* would include it, but then that included most Western diamondbacks.

"Do you know your name, baby?" he asked, grinning. Then a flash of movement caught his eye on the sand ahead. A lizard had darted out of sight under a small rock.

Banning stopped.

He watched the rock, which was about the size of a watermelon and similarly shaped. No scooting shadow emerged. The lizard had sought safety somewhere underneath. He put down the sack and advanced to the rock. He might as well have this, too. In the split second of visible movement he had not been able to discern what species it was. It might have been a speckled gecko. But whatever it was, it was definitely not a Gila monster, which is the only poisonous lizard in the American deserts, and by a quick overturn of the rock it was quite possible that he could catch it in his hands. When the rock was disturbed the creature would flee for the next nearest shelter. Looking around, Banning saw another rock a dozen feet away. In panic the lizard would dart for that objective, and knowing its probable direction, Banning could almost count on its capture. He would tell Parker and Knapp that the lizard was his only prize—and when they hooted at it he would show them the diamondback. He prepared to overturn the rock by a push downhill with his left hand, and he went down on one knee and braced himself with the palm of his right hand on the warm sand.

Then he heard the rattle.

It was a faint, dry, rapid chick-chick-chick-chick-chick, and for an instant he hesitated with his left hand about to push the rock. His first thought was that the rattle came from the red diamondback inside the sack. He turned to look at the sack and immediately felt the impact of lancelike fangs on his right wrist.

He started back, jumping to his feet. A sidewinder was quickly recoiling from its attack. Banning, his concentration on the rock, had put his right hand down on sand under which a sidewinder had burrowed from the heat of the sun.

Both man and reptile were ready for battle. Both were enraged, Banning, if possible, more so than the snake. The little horned rattler, only about a foot in length, was coiling into a loop.

Banning was outraged. His instinct was to kill it. He kicked a heavy boot at its head, missing it as the reptile looped away down the wash with the peculiar side-thrusting S-like motion from which it derives its popular name.

"God damn you!" he said through gritted teeth, plunging after it, stooping for a small rock to hurl at its head. He threw the rock and missed. "Why, you God-damned little—" he began, and then he remembered the wound. He looked at his wrist. Two drops of blood were welling larger and running down into the palm of his hand.

Banning stopped and wiped away the blood. There was no great feeling of pain. It was a scratch, he told himself—nothing more. And at once he knew better. The fangs had punctured a vein. There was no use in trying to kid himself. The little jets of poison had been shot straight into his bloodstream. And he, like a fool, like a rank amateur in herpetology, had got furious with the snake.

He looked around.

The sand-colored sidewinder was gone. The sack with the red diamond rattler lay where he had left it. All was safe there. And the rock with the lizard under it was still there. The lizard may have scooted in the excitement—let him go— Banning had no further interest in the lizard, the unwitting cause of this accident. He looked at his wrist. Idiot that he had been, wasting precious seconds trying to kill the sidewinder while its poison coursed through his system. And he had even expended energy in pursuing the snake and hurling rocks at it, increasing his heart action so that the poison virus was forced more rapidly through his veins than it would normally have flowed. Fool—fool.

Now the thing to do was to calm down. Take it easy. Men have been bitten before—by devilish little sidewinders, too. He had a first-aid kit for just such an emergency. Now use it—don't hurry or get excited—but simply make practical use of it.

He unstrapped the kit from his belt, and he saw that his hands were trembling. Five years and never an accident—well, it's got to happen sometime—or so they say. "Why in the hell didn't I think of sidewinders' hiding in the cooler sand just beneath the surface of dry washes? I know that as well as I know my name."

Banning opened the first-aid kit, instinctively using his left hand and saving his right. Perhaps he imagined it, but it seemed that his right hand felt numb. The first article of equipment that he selected was a ligature for the forearm. This he applied, forcing the tourniquet as tight as possible. He cursed himself for not having done this instantly. It might have prevented most of the poison from being carried to the heart by the blood. But at least he had, perhaps, localized some of it.

Then he examined the two fang punctures. With a small scalpel he made crosscuts over each; then he inserted the tip of the blade into the wounds, slightly deeper than the fangs had penetrated. Having thus enlarged the punctures by scarification, he sucked hard on the wound, spitting out the blood and saliva onto the sandy wash. He did this intensively for several minutes. Next came an application of a solution of permanganate of potash. With the wound thoroughly saturated, an antiseptic bandage finished the treatment. He felt he had done an efficient job—once he had started it.

Now to get back to the car before there was any more trouble. He picked up the sack this time without speaking aloud to the rattler within, found the forked stick, and marched on. The thing to do now was to keep calm. No excitement, no running, just a steady, determined pace that he could maintain for hours if necessary. And he was now only a little more than a mile from the car and from Parker and Knapp—if they were back yet.

Banning trudged on down the wash, carrying the sacked red diamond rattler with his left hand and the forked stick with his right. The ligature was effective, no doubt, for his hand, wrist, and forearm were very numb. The numbness was due to the ligature. Mustn't forget that.

It seemed that he must surely be at the point where the wash curved to the left and where the two palo verde trees stood. But no—not yet—and there was no mistake about it, because ahead were the tracks he had made as he ascended the wash only half an hour or so ago.

He dropped the forked stick.

He hadn't felt it slip from his grasp. But suddenly it was on the ground. All feeling was gone from his right hand. When he put down the sack and retrieved the stick with his left hand, the whole right arm felt heavy. It wanted to hang lifeless from the shoulder down. He couldn't make his fingers grasp the forked stick. The fingers were swelling slightly. And as for the wrist, that had swollen considerably. He wiped the sweat from his forehead and swung his canteen from his left shoulder. A draft of cooling water would help. He took several long swallows. It wasn't very cool, but it was welcome. He had three more, then capped it, and swung it into position over his shoulder again. His left arm was tired from the weight of the diamondback, but there was no relief for it, as his right arm was now useless. It was the ligature doing that, of course. Nothing to be alarmed about....Breathing deep, he tucked the forked stick under his left elbow and picked up the sack with his left hand and marched on.

The sand made speed impossible. But then he must remember that there was no reason for speed. He wasn't trying to hurry; he must always remember that—don't get excited and you'll be all right. But where the devil was that bend and the two palo verde trees?

And why might there not be more sidewinders in this wash? No doubt there were. It didn't matter. No sidewinder or no anything else could strike through a heavy pair of boots such as he wore. He was amply protected up to the knees. But if he saw another sidewinder he'd kick its damned head off and—no—no, that wasn't it. Never get mad at the snake. He had simply been a fool and begged for trouble by kneeling in a sandy wash and not watching where he put his hand. Damn that lizard, anyway—no, never get mad at the snake. But lizards aren't snakes. You can get mad at lizards without breaking the rule. So damn the lizard that had caused all this. Was he being silly? Well—be silly, then.

Whew—the red diamondback weighed fifty pounds—or so it felt when you carried it down a desert wash in a hot sun. Now where were his tracks? Must have rambled to one side somewhere—let's see—there—there they are. We'll go on to those two palo verde trees and then have another drink.

And there they were. God bless 'em—there they were. And there's where the wash curved to the left. Now we climb the bank up onto the harder ground and cut straight across to the next wash. One mile more, and there we are.

Climbing the six feet of shaly bank was not easy. His whole right arm throbbed from the shoulder to the tips of his fingers. Twice he slipped and almost tumbled. And a tumble might injure the red diamondback. Contrary to popular belief, snakes are easily injured. The very nature of their bodies leaves them highly vulnerable. Should he drop the sack and inadvertently give Mr. Rattler a bump on the head, it might even kill him. And that would be a shame after all this effort. Nothing was going to hurt that snake. That was final....Now let's see how we are going to do this.

Banning stood still and inhaled deeply three times to break the rhythm of his panting, almost gasping, breath. He moved to the bank and took two unsteady steps up and swung the sack above like a man swinging a satchel into an overhead baggage rack in a day coach. It rested on the brink. So Mr. Diamondback was safe. He may not like it in there, but he's safe. Now all Banning had to do was scramble up himself. Then he realized that the forked stick was gone again. It must have dropped out from under his left arm somewhere back up the wash. Too bad. He wasn't going back for it now. Quarter of six, with the sun still an hour above the horizon. He *could* go back and find that stick, after all. No—not worth it, but you just hate like hell to let it get away from you. He knew approximately where it must be; he could go back for it tomorrow, or any day.

He wiped the sweat from his forehead and again tried to scale the bank. There was nothing to stop him from scrambling up its unsure footing except his aching right arm, which was rapidly becoming an agony to him. It's that damned ligature that's doing it. Not the wound. Not the wound? Certainly not!

He was making it now, the shale sliding out from under him as he clawed for a root or a rock with his left hand, missed, and helplessly slid down to the sandy bottom again, scratching his left arm and tearing a button off his shirt. Now who would think that a little climb of six feet could be so difficult? He looked at the rock critically. "I know what that is," he thought. "Why can't I get over it?" What he needed was more water. He had planned to drink when he reached the top of the bank, where Mr. Rattler was waiting peacefully in his sack. But if he rested for only a few seconds and had a drink, he could doubtless take the bank with a leap or two. He unscrewed the cap of the canteen and said to the bank, "You are a fissile argillaceous rock, somewhat resembling poor slate, and you're not worth a God damn."

He drank.

"Geologically, you stink," he said to the bank. And he drank more water, tipping the canteen high above his head. "About three quarters gone," he reasoned. "Save the rest until halfway across this desert fan to the car in the next wash. That's what."

Back on the canteen went the cap. Instead of clawing helplessly at the bank in order to ascend at this spot, he backed away from it, looking up and down the wash for any small side wash that might afford an easier, gradual ascent. There seemed to be none at this point; the bank looked much the same in either direction. "Might as well fight it out on this line if it takes all summer," he thought. Now who said that? Somebody. Civil War, maybe. And whenever he thought of that quotation he always thought of another. "I would rather be right than be president." Henry Clay said that one. Well, here goes.

He walked to the bank and looked for the most likely footing and dug his boots into the shale and scrambled up and hung on with his good left arm, sweating, straining, pulling, gasping, until he got his waist over the edge and then one knee and at last crawled a couple of feet from the edge and lay panting for breath. It wasn't the lack of breath that was so bad; it was the pain in his right arm that was almost too much to bear. The final exertion of that six feet of clawing and climbing was concentrated in the arm, which hadn't done any of the pulling, but only had been bumped a few times.

He lay face down on the desert soil and felt a surge of nausea and a throbbing at the back of his head with each pulse. Suddenly, unexpectedly, he vomited the water he had recently consumed down in the dry wash.

Slowly he rolled over on his left side and sat up, avoiding cholla cactus and bull's-tongue and rolling toward the scant shade of the nearer of the two bony palo verdes. He exhaled mightily, really a sigh, and then listened. There wasn't another sound in the desert except his own breathing. Beside him were several brown, angular stalks of bottleweed, and beyond them some bladder sage, and farther still some scabiosa sage. He looked at these desert plants blankly, and said finally, "Whew—Jesus H. God."

Even that exertion seemed to run like an electric shock from shoulder to finger tips, and for the first time in many minutes he examined his arm again. There was no wrist. There was a dark, ugly, swollen, discolored hand swelling to a larger, unrecognizable mass that had been forearm. The pain was so great that there was no enduring it any longer. It was that God-damned ligature that had done its work and done it too well. He could never suffer this for another mile across the desert to the next wash and the relief of seeing that car—and Parker and Knapp. The ligature must certainly have served its purpose by now if it was going to at all—either it had prevented the spread of the poison or it had not. In either case, off with it.

Then he found that he couldn't get it off. It had been perfectly secured a half a mile or so back up that damned wash, and it just didn't seem possible to—well, hell, *cut* it off! Get it off somehow. And get it off quick, for God's—no, no—don't get excited. Just don't get excited, but use your head, and you'll be all right.

He fumbled for the first-aid kit—found it—opened it—and searched for the scalpel. Now where was *that?* Hadn't left *that* somewhere back in that wash, had he? Great God, if he were as batty as that over a little thing like this, why, he must be—No, here it is.

He placed the sharp instrument on the bandage and began to cut into it before he realized that it wasn't the bandage he wanted to cut, but the ligature. On the throbbing, beating arm they were both alike; he had no sensation of either. It was the whole arm that was on fire. Four careful slices with the scalpel, and the ligature was released. It slipped to the ground, and Banning lay back under the palo verde and gritted his teeth while the clotted blood broke free and flowed through the arm.

As he looked up he could see one or two green palo verde branches and pale blue sky. He decided to rest like this for a few minutes and give the blood a chance to circulate.

A man's chances of being bitten by a snake are not more than 10 per cent, he reviewed in his mind. One chance in ten, call it; and then of all the snakes that may bite, only about one in five of those in the desert are poisonous. So you might carelessly pick up fifty different snakes without even looking at them and your margin of risk was only one in fifty, or 2 per cent. So it's not such a dangerous affair after all. Five years in it and never a scratch—up until today, when he had become careless over his glorious red diamondback and invited trouble from the meanest little snake in America—*Crotalus cerastes* was his name—he had an armful of him right now.

Why couldn't he have put his hand down where a red racer lay hidden—or a nice friendly gopher snake? Why had he had to hit the 2 per cent margin at last and the fangs of the most vicious little varmint that bred in the desert? Sidewinders were the Apaches of the snake family. He had had a red racer sleep on his desk all afternoon—many afternoons, and it was one of the nicest, cleanest, friendliest pets you'd ever want to see. He had a gopher snake named Ludwig which liked to be read to, and a beautiful milk snake which enjoyed Beethoven. People laughed at that, but it was absolutely true. But then few people (considerably less than one in a hundred, in fact) understood the first thing about snakes. Most people had an instinctive horror of them—a fear born of ignorance. To them all snakes were evil and should be killed on sight. Damn fools! And most of all he hated silly women who screamed and ran whenever they saw a snake. That was the nadir of something or other. Why, only touch a snake—any snake—once—just one touch, and instantly you lose all fear and horror. They're not slimy, they're not repellent. They're cleaner than dogs or cats, they never have parasites, they exterminate rodents, and if you will only let your mind admit it, nearly all of them are beautiful, and many of them gorgeous. And what's more, they don't crawl—they take steps; only instead of having two or four feet they have scores of plates which do exactly the same thing that a foot does. People are such damned fools.

A snake's progression over ground is a beautiful synthesis of correlated rhythms. If only Banning himself had a few thousand plates on his belly he could move on to the car without each step being a throbbing agony.

Now how was that arm? Bad. Just plain bad. And it was time to start moving— he could make it after that rest, surely.

Banning got to his feet. For a moment he was dizzy, and he couldn't see anything but revolving black spots, and there was a ringing in his ears. Then he saw the sandy wash six feet below. A few steps away was the sack containing his prize; he walked uncertainly to it, picked it up, and started across the desert. His route lay around and through a miniature forest of cholla cactus—thousands of sharp yellow needles on stalks growing to a height of two to four feet—and underfoot were the brown and black detrital rocks of an alluvial fan from the Santa Rosa Mountains.

The next hundred yards were an agony, and as for the distance yet to go, he tried not to think of it. He clenched his teeth and vowed he'd make it. Slowly—slowly—if you don't get excited you'll be all right. He stumbled on, counting his steps in series up to twenty, and then starting over—it gave his mind something to do. When he had counted to twenty four or five times, red racers were fighting sidewinders, and a great lazy boa encircled half a dozen sidewinders and constricted them all at once. A little garter snake, like laughter, was holding both his sides.

Banning caught his breath and shook his head. "I don't know," he thought. "Maybe not." It wasn't the pain any more; it was drowsiness now. It was all a confusion that somehow wasn't worth the exertion. And yet, somehow, he was still walking, and the sun was sinking, and he could tell by the shadows of the cholla cactus that he was still stumbling on in the right direction.

He felt for the canteen, but it had slipped from his shoulder. But that was all right. It wasn't lack of water that was dangerous; it was lack of mental control.

He was attacked by more vicious enemies than thirst. He was waging a single-handed war against sand, rocks, cactus, heat, fear, and time.

It was all those allies of the venom and the wound that were making a real fight of it. His mind was now his last defense....If that goes, the others have got me. But you must remember not to be afraid. Fear of fear is the worst weakness of all. Don't be afraid you won't make it. Just count the steps again and keep on going, and surely Parker and Knapp will be back now. It must be all of six-thirty by this time. That sun is going down.

Stumbling on, trying to keep the sack from being jostled too severely, Banning made the mile between the two dry washes. The one in which the car had been left had more gradual banks. There would be no troublesome descent. He reached the point where the cholla cactus gave way to sand verbena and desert mallow, and there was the wash where they had parked the station wagon earlier in the afternoon. Only one thing was wrong. The car was gone.

Banning stumbled down into the wash. He put the sack on the ground and look around. Had he misjudged the wash? Was he too far north or south? No—those marks were tire tracks. And a dozen yards away a few blobs of black grease on the sand showed plainly that the car had stood for some time at that spot. But it had been backed around and driven away down the wash. The tracks showed that, too.

Why?

Banning went over to the sack and sat down beside it. This was the last ally of the venom. This was the last unexpected little joker that meant the game. He had done everything he could be expected to do. If the car didn't come back—well, that was that. He lay back slowly, his left hand near the sack. He wouldn't think any more.

His hand touched the sack, and within the reptile squirmed. Banning moved. There was one thing more he wanted to do—just one final thing. He fumbled with the drawstring of the sack. He couldn't do anything with it. There was that scalpel. He fumbled in the first-aid kit again and found it. Then he cut the drawstring and pulled open the mouth of the sack.

"Come on," he whispered. "It's your game. You won."

He lay quietly beside the sack, his head but a few feet away as the red diamondback rattlesnake emerged. It was utterly confused and very frightened. It didn't rattle; it didn't even coil. It crawled hastily out of the sack on the opposite side from Banning.

"Don't mind, baby," Banning whispered. "Don't be scared. You're free. Go on."

Quickly the rattlesnake crossed the wash to the far side and in the half-light disappeared into the desert.

The twilight was deepening when a pair of headlights bounced crazily up the wash. The car, laboring in second gear, came on until the lights picked up Banning and the empty sack beside him. Parker and Knapp were out of the car in a second.

Knapp read the story almost at once—only the empty sack mystified him. He could feel Banning's pulse, but he could not rouse him. He ordered Parker to turn the car around and get set for a dash to the highway. Wait—there was something in the car to be tried as an emergency treatment. He had never had occasion to use it, but this, surely, was the time. It consisted of a hypodermic injection of subbichlorate of potassium.

"It's been said," he told Parker, "on no authority at all, that this is a sure antidote for any snake venom."

He made the injection.

"But is it?" asked Parker. They were lifting Banning into the station wagon.

"We'll damned soon know," said Knapp. "You drive down to the highway and then let 'er rip for the emergency hospital at Indio. You ought to do it under forty minutes. I'd say he's got a little better than a fifty-fifty chance. It couldn't have been over two hours ago. But why did he put that empty sack there? Must have gone nuts. Step on it. What do we care if we break a spring or two?"

The station wagon careened wildly down the wash, its lights cutting into the spreading darkness.

SYLVIA PLATH

Sylvia Plath was born in 1932 in Jamaica Plain, Massachusetts. An internationally established poet, she authored the critically acclaimed books of poetry Ariel *(1965) and* Crossing the Water *(1971), and the novel* The Bell Jar *(1963). Plath died in 1963.*

The following poem was first published in 1960 and was inspired by a night Plath spent camping in the California desert while on a cross-country vacation with her husband, the poet Ted Hughes.

Sleep in the Mojave Desert

Out here there are no hearthstones,
Hot grains, simply. It is dry, dry.
And the air dangerous. Noonday acts queerly
On the mind's eye erecting a line
Of poplars in the middle distance, the only
Object beside the mad, straight road
One can remember men and houses by.
A cool wind should inhabit these leaves
And a dew collect on them, dearer than money,
In the blue hour before sunup.
Yet they recede, untouchable as tomorrow,
Or those glittery fictions of spilt water
That glide ahead of the very thirsty.

I think of the lizards airing their tongues
In the crevice of an extremely small shadow
And the toad guarding his heart's droplet.
The desert is white as a blind man's eye,
Comfortless as salt. Snake and bird
Doze behind the old masks of fury.
We swelter like firedogs in the wind.
The sun puts its cinder out. Where we lie
The heat-cracked crickets congregate
In their black armorplate and cry.
The day-moon lights up like a sorry mother,
And the crickets come creeping into our hair
To fiddle the short night away.

EDWARD ABBEY

Edward Abbey was born in 1927 in Indiana, Pennsylvania. As a young man he moved to the southwestern United States, where he lived for the rest of his life. An author and essayist, he was passionate about environmental preservation and called for public land policies to protect the American Southwest. Abbey's love for the land is reflected in his most famous work, Desert Solitaire, *a 1968 memoir of a season he spent working as a ranger in Arches National Park. Writer Larry McMurtry called Abbey the "Thoreau of the American West," and as a result of his often inflammatory writing, Abbey has also been deemed a "desert anarchist." He died in 1989 at his home in Oracle, Arizona.*

In the excerpt that follows, taken from the essay collection The Journey Home: Some Words in Defense of the American West, *Abbey writes about Death Valley National Park, where he lived in 1961 while working as a school bus driver.*

from "Death Valley"

Badwater, January 19. Standing among the salt pinnacles of what is called the Devil's Golf Course, I heard a constant tinkling and crackling noise—the salt crust expanding in the morning sun. No sign of life out there. Experimentally I ventured to walk upon, over, among the pinnacles. Difficult, but not impossible. The formations are knee-high, white within but stained yellow by the dusty winds, studded on top with sharp teeth. Like walking on a jumble of broken and refrozen slabs of ice: At every other step part of the salt collapses under foot and you drop into a hole. The jagged edges cut like knives into the leather of my boots. After a few minutes of this I was glad to return to the security of the road. Even in January the sun felt uncomfortably hot, and I was sweating a little.

Where the salt flats come closest to the base of the eastern mountains, at 278 feet below sea level, lies the clear and sparkling pool known as Badwater. A shallow body of water, surrounded by beds of snow-white alkali. According to Death Valley legend the water is poisonous, containing traces of arsenic. I scooped up a handful and sampled it in my mouth, since the testing of desert waterholes has always been one of my chores. I found Badwater lukewarm, salty on the tongue, sickening. I spat it out and rinsed my mouth with fresh water from my canteen.

From here, the lowest point in all the Americas, I gaze across the pale lenses of the valley floor to the brown out-wash fan of Hanaupah Canyon opposite, ten miles away, and from the canyon's mouth up and up and up to the crest of Telescope Peak with its cornices of frozen snow 11,049 feet above sea level. One would like to climb or descend that interval someday, the better to comprehend what it means. Whatever it means.

I have been part of the way already, hiking far into Hanaupah Canyon to Shorty Borden's abandoned camp, up to that loveliest of desert graces, a spring-fed stream. Lively, bubbling, with pools big enough and cold enough, it seemed then, for trout. But there are none. Along the stream grow tangles of wild grapevine and willow; the spring is choked with watercress. The stream runs for less than a mile before disappearing into the sand and gravel of the wash. Beyond the spring, up-canyon, all is dry as death again until you reach the place where the canyon forks. Explore either fork and you find water once more—on the right a little waterfall, on the left in a grottolike glen cascades sliding down through chutes in the dark blue andesite. Moss, ferns, and flowers cling to the damp walls—the only life in this arid wilderness. Almost no one ever goes there. It is necessary to walk for many miles.

Devil's Hole, February 10. A natural opening in the desert floor; a queer deep rocky sinkhole with a pond of dark green water at the bottom. That pond, however, is of the kind called bottomless; it leads down and down through greener darker depths into underwater caverns whose dimensions and limits are not known. It might be an entrance to the subterranean lakes that supposedly lie beneath the Funeral Mountains and the Amargosa Valley.

The Park Service has erected a high steel fence with locked gate around the hole. Not to keep out tourists, who only want to look, but to keep out the aqualung adventurers who wish to dive in and go all the way down. Within the past year several parties of scuba divers have climbed over and under the fence anyway and gone exploring down in that sunless sea. One party returned to the surface one man short. His body has not been found yet, though many have searched. If supposition is correct, the missing man may be found someday wedged in one of the outlets of Furnace Creek springs.

Death Valley has taken five lives this year—one by water, two by ice, and two by fire. A hiker slipped on the glazed snow of the trail to Telescope Peak and tumbled 1,000 feet down a steep pitch of ice and rock. His companion went for help; a member of a professional mountaineering team, climbing down to recover the victim, also fell and was also killed.

Last summer two young soldiers from the Army's nearby Camp Irwin went exploring in the desert off the southwest corner of Death Valley. Their jeep ran out of gas, they tried to walk home to the base. One was found beside the seldom-traveled desert road, dead from exhaustion and dehydration. The body of the other could not be found, though 2,000 soldiers hunted him for a week. No doubt he wandered off the trail into the hills seeking water. Absent without leave. He could possibly be still alive. Maybe in a forgotten cabin up in the Panamints eating lizards, waiting for some war to end.

Ah to be a buzzard now that spring is here.

CRAIG CHILDS

Craig Childs was born in Tempe, Arizona, in 1967. He is the author of The Animal Dialogues, House of Rain, *and* The Secret Knowledge of Water, *books that explore unique and often ironic occurrences in the natural world. Childs is a commentator for National Public Radio and his articles have appeared in the* New York Times *and many other newspapers and journals. He lives in Colorado.*

The following is the surprising true story of the immense damage that a flash flood in August 1997 caused to a railroad bridge and a train in the desert. It is taken from his book The Desert Cries: A Season of Flash Floods in a Dry Land.

from *The Desert Cries*

North of the Old Woman Mountains in the desert of eastern California, the first rain comes as a wall. The train's skin shudders as if passing into a solid object. Inside the cars, a steely hiss rises over the sound of the tracks. Some people wake to the change. They feel around in the dark, sit up, and then realize that it is only rain. They go back to sleep.

This night, wind has been up around 80 miles per hour. A flash flood earlier hit the outskirts of the Arizona town of Kingman, where helicopters rescued two people off the roofs of their stranded vehicles.

This wall of rain has been on the move all night and into the morning. People who have been manning the satellite computers are ready to give way to the incoming day shift. These exhausted weather watchers have been tracking thunderheads back and forth across the borders of Arizona, Nevada, and California throughout the night, then making calls and typing out messages before returning to check the monitor screens again. They have been sending out flash flood warnings much of the night, leap-frogging their messages across counties.

Rain, when it comes to this desert, falls out of the sky like bricks. Storms hit the ground and are gone. Even satellites can lose the trail as thunderheads bloom out of nothing and disappear.

In the coming morning, a truck will be found cocked on its side—the driver's door hanging open and draped with flood debris—in a drying wash near Kingman. A flood pushed the abandoned truck off a nearby road that crosses the wash. The driver, a 41-year-old woman, will be found the next day a few miles downstream, her lifeless body half-buried in sand and mud.

All night, storms had jumped from Kingman to Las Vegas, Nevada, and back. They filled Las Vegas streets with water, washed cars away, and toppled power lines. Then, wiping their hands clean of violence, the storms lifted

26

and vanished. Their faint blue flashes of lightning trailed into the distance. But the storms never went far before touching ground again. Shortly after the Las Vegas floods, storms erupted to the south at Lake Havasu City, Arizona, and again the cars turned into boats. Rescuers worked into the night pulling people out. Then fresh thunderheads built once more over Las Vegas, sending water back down the drying streets.

The night went on like this, thunderheads building, dumping, and fading in an hour's time, or even in 15 minutes. Each time a storm would leave, silence settled in, a pure, unadulterated quiet that made the previous storm seem impossible. Abandoned cars rested sideways along streets that now carried only trickles of leftover rainwater. With the power out in parts of Las Vegas, stars reflected brilliantly against wet asphalt.

At 3 a.m., the Southwest Chief stops in the desert town of Needles, California.

Electricity still serves the town because crews have been out repairing a transformer struck by lightning. The sky is strangely clear in Needles. There is no sign of rain, except when the train doors open and the strong smell of a desert thunderstorm enters the cars. It is a newborn scent, sweet with creosote bushes and the steaminess of cold water against hot earth.

Needles still shimmers from a recent rain, a gully washer the locals call it. The man working the station complains to the Southwest Chief's conductor, saying it's been raining off and on like all hell's broken loose. And now nothing. The last storm headed northeast. He points that way, as if relieved to be free of it.

The train starts up again. First there is a forward jolt that brings most people awake for just a second. Then a couple of lesser surges, and the train is moving freely, heading northeast.

An hour or so before the Southwest Chief reaches western Arizona's Hualapai Valley, the Burlington Northern Santa Fe tracks are deluged in rain. This particular thunderhead gathers around the nearby Peacock Mountains as if taking up residence, hurling pulsing rain down at the desert.

A light appears on the tracks below the mountains. The beam, slivered by rain, is accompanied by the sliding rattle of steel wheels. A spotlight sweeps the ground, dipping to one side, then the other.

The light and sound come from a railroad vehicle sent ahead to check the tracks because flood warnings had dictated that the line be surveyed in person. The vehicle, a truck hoisted onto drop-down track wheels, slows as it crosses each trestle. A track inspector shines the light into the bare washes below. Everything looks good. No sign of flooding or damage. The vehicle keeps on, its red tail lights becoming smaller, more faint, and the noise of its wheels fading into the pulse of the storm.

The thunderhead rises with hot updrafts off the back of the Peacock Mountains, scattering lightning inside of its own folds. The rain up there comes too fast to soak back into any of the small springs. It is like dumping a bucket. The water grows.

Small floods seek every possible wash, crack, arroyo, rill, and hole off the flanks of the Peacocks. They flow around tight bundles of drought grass until the grass uproots. Objects such as bones or branches either impede the water flow or gather into the momentum and are carried away. Any grass available to hold the water back is bunched and isolated. Rainwater runs across bare rock and cascades into larger and larger accumulations.

Thus orchestrated, larger floods emerge from the Peacock Mountains, running through a washboard of drainages. By the time they pour to a lower elevation, they have settled into a matrix of atavistic flood channels. These courses form a web over the entire desert slope beneath the mountains, sorting water with ancient familiarity.

The front edges of these floods snap and growl with debris—pieces of manzanita, fallen juniper branches, coarse gravel. They are foamy, gurgling shelves that slow down, stagnate, and abruptly quicken. They gather into a broad wave that sweeps the desert, caroming into the Burlington Northern Santa Fe tracks, which lie at a 90-degree angle to the water.

The flood courses northeast along the tracks until it can fall into an arroyo and pass under its only exit, a channel under a small bridge called 504.1S in railroad lingo. The entire surrounding desert feeds its water into this narrow passage, a funnel too small for the volume. The channel gulps and gasps, trying to swallow the flood and pass it through. Fast, rumbling water surrounds the bridge's support timbers. Rising 8 feet to the underside of the tracks, the water shakes the timbers. Critical parts of this 75-year-old wooden bridge weaken. Bank walls cave in, sucked under in dark roils. Buried timber foundations are swept open and carried away. It is 4:45 a.m.

At just about 5 o'clock, the Southwest Chief settles into top speed after the latest stop in Kingman, Arizona. There was no time wasted, but the train is now an hour and a half behind schedule. The dining compartment, decorated with fresh flowers and linen, is closed until breakfast. The car attendants are just getting about their early tasks, arranging newspapers that they will slip under the cabin doors. Passengers who are awake on the Southwest Chief this morning roll through the desert, looking out into predawn darkness as the rain tapers off. They can see lightning outlining the Peacock Mountains in the distance.

Just 150 feet from the bridge, the engineer glimpses a sag in the line. The tracks are warping downward, and they begin to buckle before his eyes. Below the misshapen tracks, he sees a flood, and he hits the emergency brake. The train is too close. Crossing the damaged bridge at 90 miles per hour, three locomotives uncouple. They break free from the remainder of the train, actually leaving the tracks only to fall back into place.

As the following cars pass, flood-weakened timbers collapse. The bridge explodes beneath the weight, hurling wood into the flood. Sixteen cars jump the track as the bulk of the Southwest Chief goes from 90 miles per hour to a complete stop in six seconds.

Train accidents are nothing like car accidents. There is so much momentum that steel disintegrates on impact, weight from behind still racing forward at full speed. Sturdy lengths of iron track twist as if made of hot plastic. Blocks of solid oak railroad ties rupture into splinters, chewed into the machinery. In a blizzard of sparks, a half-ton axle sails away to plant into the wet earth like a javelin. People in their beds fly into the ceilings like rag dolls without even a moment to wake. In coach class, people are ripped from their seats, thrown row after row toward the front.

The two passenger cars ahead of the bridge jackknife into each other, hinging 20 feet off the track. Two locomotives plow to one side, leaving the third to roll ahead, uncoupled and alone. Portions of track peel back and vivisect the baggage car. Arriving cars slam into one another in rhythm.

At last, the train does not move, compressed and cantilevered as it is amid the jumble of rails and ties. Now there is no rumble of wheels on tracks, only the sounds of injured people shouting, of panicked passengers struggling toward exits in the dark, and of the surging hiss of flood waters. Lights are out in most of the cars. Some people find flashlights. They kick at the metal doors and windows to open them. In one car, a group finds an emergency sledgehammer and breaks down the door, only to find their car suspended 12 feet above the ground. Part of the bridge hangs 45 degrees into the flood, forming a steep rapids, while a sleeping car spans the arroyo where the bridge had been. In the first faint light, passengers emerge from the sleeping car with desperate hesitancy, balancing over the flood on what remains of the opposite, westbound track. They grab the side of the train for support, taking one step at a time to get out, and fix their eyes on the dark water and foam rushing beneath their feet.

Amid the wreckage, 116 people are injured. People from all over the continent, but most have never been to a desert. Those still able to walk get out into the morning heat, into the smell of wetness. Tossed into an ironic, vacuous land, they mill aimlessly about the wreckage. This train is famous for the scenery it passes through, being promoted like a traveling postcard. These people walk around in the irony, face to face with the desert. Sirens sound from the distance.

Without pause, the water passes from below the Peacock Mountains and looks for still farther channels. The water is beyond tearing away the last of the bridge and the sleeping car above it. As it is now, the water will travel to the sand dunes, fanning window glass and sheared steel bolts into the desert. Out there, as the flood spreads into the sand, the desert steals the precise definitions of water courses. There will be no destination for this water. It will not reach the Colorado River by this path. It will not sustain cattle or be pumped into bottles to be sold as drinking water. No corn or cottonseed or soy will grow from it. There will be no legal battles between states as to who owns it. The water ebbs and vanishes—evaporates or sinks—before the last passenger is taken from the Southwest Chief in a stretcher. No one has died. Some say that it is a miracle.

The decision-makers come a few hours after the rescuers. They stand in the damp wash and look up at the damage. An FBI official is flown in. He walks up to the fire chief from Kingman and asks his opinion on what happened.

The fire chief, Chuck Osterman, who has been awake since midnight on flood rescues, regards the FBI man. Osterman is exhausted and not in the mood for someone seeking complicated answers to a simple question.

"Looks like it rained hard, flooded, and the bridge failed. Then the train came," he tells the federal agent.

The FBI man nods stoically, thinking that he will have to find his answers elsewhere.

For those officials and engineers brought from other states, other climates, the desert must seem peculiar. The flood must seem a lie. Pieces of track and splintered timbers are now scattered across the desert plain miles beyond, unaccountably abandoned in curious, desolate places that are already almost dry.

RUTH NOLAN

Ruth Nolan was born in 1962 in San Bernardino, California, and was raised in a remote area of the Mojave Desert. She is a writer and poet whose poetry collections include Wild Wash Road *(1996) and* Dry Waterfall *(2008). She is the recipient of a Joshua Tree National Park Affiliate Writers Residency, 2007-2008, and is the founder-editor of the California desert literary magazine* Phantom Seed. *She is an associate professor of English at College of the Desert, where she teaches poetry, creative writing, desert literature, and Native American literature. She lives in Palm Desert.*

The following poem is taken from Wild Wash Road.

Mirage

near Kelso Sand Dunes, East Mojave

My ten-year-old daughter is feeling brave.
We go rock hunting today, exploring far
beyond the last dirt road, just she and I.

We see the sand dunes from miles away,
some hallucinogenic scene from the Sahara,
camel humps rising from the flat desert floor.

My daughter wants to climb them, but
there's no way to guess how far away
they are, no sure measure to tell how tall.

I tell her it's not safe to hike mountains
that unstable, hills that shift in the wind.
Our boots would fill with sand, and we'd

sink like thirsty prospectors, come to find
buried treasure, lured by promises of silver
and the hollow chant of a rattlesnake's charms.

HANNAH NYALA

Hannah Nyala, a writer and tracker, was born in rural Mississippi and has a background in ethnography and maritime environmental history. She is the author of Leave No Trace *(2002),* Cry Last Heard *(2004), and the forthcoming* Tracking at the Future. *As a young woman, she escaped an abusive ex-husband, taught herself the skills of tracking in the Mojave Desert, and became a search-and-rescue worker, dedicating herself to saving the lives of the lost.*

The following excerpt, taken from her memoir, Point Last Seen: A Woman Tracker's *Story of Domestic Violence and Personal Triumph (1997), is an account of the search for a child lost in Joshua Tree National Park, where Nyala worked in the 1980s.*

from *Point Last Seen*

0955

Here she is again, finally. Mandy's footprint, still headed away from the campground. I am scared to death right now. It's one thing to practice tracking by yourself—which I've been doing for three months here—and an entirely different thing to be tracking by yourself when a child's life is involved and it's your first real search. Jason was right when he said none of us were good enough to be doing this. The choppers are in the air. And I think I feel the scent of death in this place.

"Tracker One to Base."

"Go ahead, One."

"Tracks have changed direction and are heading northeast. It'd be good if we could get somebody to cut sign on the old mining road out of the north end of the campground, One out."

"We copy, Tracker One. We'll put the choppers over that area now and get people on it as soon as we have some to spare. Base over and out."

"Headquarters to Search Base."

"Go ahead, HQ."

"Catering truck is on the way with 100 lunches, ETA 0100. Reporter from the local paper should arrive at your location in driving time. HQ over and out."

"Base copy and out." This search is expanding fast.

"Chopper One, Search Base."

"Go ahead, Base."

"Tracker One reports change in track direction. There's an old mining road that leads north out of the top loop of the campground—we'll put somebody out to cut a perimeter on the ground but can you take a look over there for us now? Base out."

"Sure, no problem—Chopper One copies and is outta here."

All right. Dale is in the air. He's the best helicopter pilot around. If anybody can see Mandy from the sky, it'll be him.

1030

Mandy must be following a bunny or a series of bunnies and lizards or lord knows what else that moves: She is wandering all over the place. We've been west and northeast, and now we're heading southwest again. Both choppers have been working back and forth over this area, but no one has seen her yet. The fixed wing hasn't either. I'm still tracking by myself, still scared, but I'm even more afraid of what could happen to Mandy if I bungle my assignment.

The car searches at the park entrances have yielded nothing, which means nothing. If Mandy went to the restroom at 5 a.m., she could've been outside this park by 5:15. Given that her tracks are *here*—and no one else's are as yet—we could stretch that by an hour or so. I'm just hoping and praying, little girl, that you don't wander anywhere near a road: Out here, nobody else is likely to have seen (or threatened) you. A few snakes, a lot of cacti, uneven ground where you might have fallen and sprained an ankle...that's it. On the road, we're dealing with an entirely different set of dangers: I'm an adult and quite capable of taking care of myself, yet I wouldn't choose to walk down a road alone in a National Park for anything in this world. You are nine, and two-legged animals are a lot more likely to hurt you than any others, Mandy. So *please* stay out here 'til we can catch up with you, kid! Don't let the scent of death on my shoulder become real again.

Thanks for staying in this wash for a while too—that's a kind thing to do for a green tracker, lets me move a little faster than I could on those rocks.

1105

"Search Base to Tracker One."

"Go ahead, Base."

"Riverside's here—we're briefing them now. As soon as we can get a chopper back here we'll be bringing out two trackers to join you and we'll put the others on perimeter cuts. ID your location for the bird. Base over and out."

"Tracker One copy and out."

1115

"Hi there—boy, am I ever glad to see you two!" I say to the two strangers who trot away from the helicopter toward me. Dr. Ed Smith and Jim Calloway, trackers. (They even look like trackers.) I circle one of the better footprints and wait for them.

"What've you got?" Ed asks, pulling out his notebook.

I point to where Mandy's tracks have come from and where they've been heading recently and Jim says, "All right, let's hit it." He motions for me to take point, the lead position.

"Look guys, I think it's time for me to say up front that this is my first search—and I'm sure you two are much better and much faster at this than I am, so if you want me on flank, that's fine, but I'd just as soon one of you take point. We need to find this kid."

Ed looks at me for a second, then says, "Well, from what we've seen so far—from the briefing at the base and here—you're as much a tracker as anybody and that includes us. You've already done half of the work, and we need you on this team, so if you want to flank awhile because you're tired, that's okay, but otherwise stay on point—it'll give us a chance to get to know this kid's tracks."

So that's that. They have more confidence in me than I do in myself. "Tracker One to Search Base—Team One is on the tracks again. One over and out."

"We copy, Team One. Base out." Ed and Jim flank out behind me on either side, their eyes scanning the ground for any signs that Mandy has changed direction, and I move ahead on her tracks. One step at a time.

1130

Wow! What a difference it makes to have three people working together as a team: We've been racing over the ground like hounds. But just up ahead there's a nasty stretch of rocks, and we're about to lose Mandy for the first time since Ed and Jim got here.

"We need to cover some ground—maybe we ought to start jumping," Ed says.

"Jumping?" I ask, since I'm fairly certain he's not talking literally.

"Yeah," Jim says. "One flanker will fan out left, while the other fans out right—way out, both of them—and then they'll cut for sign while moving forward and back toward each other [and with luck, toward an intersection with the tracks]. Point stays put, step by step, until we get a confirmed track ahead—then we all jump to it—kind of a modified leap frog. It's fast when you have three really good trackers. Shall we?" Although I've never tried it before, the concept makes sense. The flankers will basically cut for sign in two concentric circles forward from the last track, while the point person moves forward from that track as usual. This should speed things up considerably.

"Okay by me, but I want to stick with the step-by-step since that's the only formal training I've had—y'all take off!" I say, grinning but feeling nervous again. "I'll just bring up the rear!"

Ed and Jim disappear to the right and left, and I restart Mandy's track just beyond the rocky section.

1140

Ed calls out, "Jim, come take a look!" Then they both call me, "Hey Hannah, we're somewhere near your ten o'clock!" I circle Mandy's last track and trot toward their voices, looking back occasionally so I can retrace my steps more easily.

"Wow," I say, standing between them looking at Mandy's print. "That's her!"

"All right, let's hit it again, team!" Jim says, and he and Ed disappear once more. For the first time today, I'm beginning to have faith in tracking as a technique. It is working. I start calling Mandy's name every few minutes, as do Jim and Ed. We are going to find this little girl.

1220

We're moving fast. This jump track method is amazing—and fun. But the urgency in our voices as we call for Mandy suggests that we're not enjoying the fun very much. Mandy's tracks are wandering almost erratically now, and the sun's high. Base probably has ten or twelve other trackers out, people on practically every road in this vicinity, but we're the only team to make contact with Mandy's tracks yet, which means we could be very close to finding her. Or very far away.

"Search Base to Tracking Team One."

"Go ahead, Base."

"We've got another tracking team to dispatch. Report current direction of tracks."

"North-northeast. We still need a perimeter cut on that old mining road out of the campground. Team One out."

"Search Base to Chopper Two."

"Go ahead, Base."

"Have a transport from the LZ to the far end of that old mining road, what's your ETA to base?"

"Five minutes, tops. Chopper Two over and out."

"Hey Hannah, we're at your nine o'clock!" I circle Mandy's print and start trotting toward Ed's voice.

"What do you think?" Jim asks me as I crouch down to look at the indistinct print on the ground.

"That's not her," I say slowly, a little uncomfortable to be disagreeing.

"It's not got the detail, maybe, but it's the same size—and we haven't been getting detail for a long time now," Ed says.

"Besides, nobody else would be wandering around out here—we've seen what, maybe ten other people altogether?" Jim adds.

"It's not her, guys. Look how her foot hits the ground here on the outside—Mandy's doesn't do that. This is somebody else." I say, getting surer by the second, looking ahead at the next few indistinct tracks to check my conclusion.

"Are you sure?" Ed asks.

"Positive. Look—even the stride is too long. And it's too old, may have been made several days ago." (To someone who hasn't spent a lot of time in this desert, that set of tracks probably looks like it could've been made this morning, but footprints can be deceptive here.)

"All right then, we'd better cut this one again."

"Well, I'm going back to point," I say.

1245

"Hannah, let's try this again—we're at your two o'clock."

"On my way."

Ed and Jim are crouched down over the track, but they move aside to let me join them.

"That's her again—good job, you guys!" I say, genuinely happy now. "We're going to find this kid yet! *Mandy!* Can you hear us?"

We hit it again, jump-tracking for all we're worth.

1345

"Tracking Team Eight to Search Base."

"Go ahead, Eight."

"We've done a sign cut of the old mining road all the way to the campground—she hasn't crossed this yet. Do you copy?"

"Base copy, Tracking Team One do you copy?"

"That's affirmative. Our tracks are heading straight for that road right now and we're within an eighth of a mile of being on it ourselves. We'll keep you advised. Team One out." Then to Ed and Jim, I yell, "Hey, did you hear *that*?!" If Mandy hasn't crossed that road yet—and hasn't doubled back on us—we may be very close to finding her.

Ed and Jim both yell yes, and then we all call out, nearly in unison, "Mandy, can you hear us?"

I'm on point in the middle of a sandy wash which will intersect with that mining road. Come on, Mandy—answer us! *Be here*, little girl!

1400

"Hey Hannah—we're at your nine o'clock!"

Oh no, Mandy, please don't go off in that direction—it's miles to anywhere! For the first time all afternoon I don't trot toward Jim and Ed. I stand still, eyes shut but turned toward the sky. "Mandy, can you hear me?" I call out hoarsely and then turn to walk toward the guys.

Then all of a sudden, this waif of a girl appears, running down the center of the wash toward me:

"I'm Mandy and my parents have gotten lost!" she calls out.

"Oh really?" I say, dropping to my knees and hugging her close. "Whew, am I ever glad to see you, young lady!"

Mandy insists, "My parents got lost and I can't find them anywhere!"

"They're over at my house, not too far from the campground. We'd better give them a call."

"Tracking Team One to Search Base."

"Go ahead, Team One."

"We've just met a young lady out here named Mandy Evans who's looking for her parents—who've gotten lost—end quote!"

There's a general hurray from a lot of locations in this desert, I know, even though I can hear only those within a quarter mile or so. Ed and Jim are here before I even sign off the radio. Mandy agrees that a helicopter ride would be nice, so she and Ed take the first ride out. Jim and I sit on a rock and wait for our turn. "It's been a good day," I say. "We'd never have found her without you and Ed."

"Sure you would've," Jim says, grinning. "But with only a one-man team, it would have been several hours later!" He's quiet for a few seconds, then adds, "Nobody can jump alone."

MARY SOJOURNER

Mary Sojourner was born in 1940 in Rochester, New York, and moved to Flagstaff, Arizona, in 1985 with two intentions, she said: "to write and fight for the earth." An active desert environmentalist, she is a fiction writer and poet, author of Sisters of the Dream *(1989), a novel of twelfth-century northern Arizona;* Bonelight: Ruin and Grace in the New Southwest *(2002), a collection of personal essays about the Southwest; and* Solace: Rituals of Loss and Desire *(2004), a memoir. Also a National Public Radio commentator and a writing instructor for the Hassayampa Institute for Creative Writing, in Prescott, Arizona, Sojourner is currently at work on a new novel. She lives in the desert near Joshua Tree National Park.*

The following essay, "bitch bitch bitch," was first published in Sojourner's collection Bonelight.

bitch bitch bitch

I jump in Ev's ancient pickup. "We're outta here." I shake my genuine cowhide Jackpot bag under his nose. "Forty-two dollars and seventy-five cents in nickels. That's an investment. I *gotta* hit."

Ev pulls onto the dirt road that leads to the straight-shot you're-gonna-get-lucky white line that's been pulling people west since Old Route 66 got reamed, razed, and reduced to a few two-lane stretches sprouting cheatgrass.

"You don't want to win," he says. "You want to saunter in, hunker down, and play till you get those squiggly coyote eyeballs."

"Hang on," I say. I bet Ev thinks I'm going to remind him that part of our deal is that we never say, "I think your gambling is getting a little out of hand." He shuts up, which you're supposed to do when your best friend calls you on your faint air of superiority. Instead, I jerk my thumb toward the mailbox. "I want to grab my mail."

Ev surfs March mud into the barrow ditch. I tug the box open and pull out the mail.

"Anything?" Ev looks chipper, the way you're supposed to look when your friend is a somewhat-known writer who might actually find a check in her mail.

"Oh, yeah, the NEA has set aside a special fellowship for me—a hundred K and a snake-hipped houseboy."

"The usual," Ev says, and we're off. We're loners who found ourselves neighbors in a cluster of substandard cabins near an Arizona mountain town, from which in thirty minutes in any direction you used to be able to drive to places of such beauty that you had to stop immediately, climb out, and thank the Big Whatever. Persimmon sandstone; black-sand washes whose basalt walls were filled with old scratchings of bighorn sheep, zero-eyed gods, and busty goddesses; a volcanic caldera where kids smoked pot and leaned back on the ruined foundation of an old dairy to watch a bald eagle swoop into her nest—all gone or threatened.

Now, when we head toward the horizon, we drive past Italian takeout clones owned by Pacific Rim corpo-mafiosi, bad espresso in every Stop-'n'-Run, and big gates that read, "Highlands Golf Preserve."

"Why do we need to preserve golf?" Ev always says. "We need to thin the herd."

I tuck a sheaf of unopened bills into my backpack. "I'm so glad we're getting out of here. Ten years ago, I would have never believed I'd ever say that." I open the map. "The Mojave. Cadiz and Amboy. Roads we've never run, pink hotels crumbling into the sand and Joshua trees and 1973 trailers with broken windows and the Old Woman Mountains and—"

"Laughlin."

"All-you-can-eat buffets for $4.99 and free breakfast because you're traveling with a senior citizen and then—"

"We saunter in and hunker down."

"Ev, my friend, it doesn't get better than this."

Laughlin spanks us. We're up, up, up, don't bother to eat, me crouched in front of Winning-Touch, Ev going red-eyed in front of Double-Wild-Cherry. Ev staggers up to our room at 3 A.M., $12.45 down. I drop on the other bed around five.

"Fuck fuck fuck," I say. Ev doesn't ask.

Lou Jean, our River-O'-Gold Breakfast Buffet hostess, brings us coffee. She taps my Lucky Jackpot bag. "That true?"

I raise my cup. "Let me get some of this in my system."

Lou Jean shakes her head. "Never mind, honey." She sets Ev's orange juice in front of him. She's got a ring on every gnarled finger, gold filigree and zircons, silver and chunky turquoise. "At least you got a free meal."

"Not exactly," I say. "More like a hundred fifty bucks."

Lou Jean shakes her head. "I lose that much every week."

An hour later, by some miracle, we are not in front of our machines. We are in Needles, California, looking at the map. Ev traces not the arteries, but the delicate blue veins. "Highway 95 to Parker, Highway 62 to Twenty-Nine Palms, cut northwest on the dotted line to Cadiz." Cloud shadow moves over the map.

"Incoming," I say. We're back in the truck before the hail hits.

I open the window, cup my hand. "Ouch!" I yank my hand in, set hail on the dashboard between the two-pound chunk of obsidian and the pink plastic My Little Horsie that Ev found in the bottom of his Circle-K Big Gulp on our famous Big Rock Candy Mountain trip.

Needles goes silver, City Midas frozen in light. What I love about this friendship is neither of us turns on the tape deck. We don't talk. Hail rattles on the truck roof. Mist rises from the dirt street, the busted sidewalk, Ray's Car Stereo/Auto Parts, and the old cottonwood arching over it all.

Silence falls as sudden as Mojave weather. I slide in the *Spring '97* tape, featuring five reruns of Robert Plant's "29 Palms." "Let's go."

We almost make it to Parker. The tribal casino sign isn't that big, but it's right where we stop to make sandwiches. I feed bread crusts to the humongous raven perched on the sign. It studies us calmly with one glittering eye. "Here, dude," I say, "or dudette. You, yourself, must be a sign." The raven flops down onto the sand and grabs a crust. Ev and I look at each other wisely.

"It's a sign."

We head down a narrow road. Lake Fake glows like a Website beyond the little casino. I poke through the ashtray, find three dollars in quarters, and pry nickels out of the candle wax on the dash. "We *are* cashy."

We pay the gatekeeper a dollar and park.

There are thirty identical slot machines. You can play nickels, quarters, dollars, all on the same machine, which would be fascinating if we had more than three dollars and twenty-five cents, and if the screen didn't most resemble the beginning of a migraine. We split our bankroll and go to it. Four minutes later, Ev is up seventy-five cents and I am yelling, "I just hit twenty bucks! Let's go." We take our pay slips to the sweetly cheerful cashier and go into the glare of the hallway.

Ev stops at the door. "Wait a minute, let's check this out." He heads down a second hallway to a cashier's counter surrounded by blue glass windows. In the delicate gloom, we could be undersea, or in the middle of Manhattan. The air is heavy with cigarette smoke, bathroom disinfectant, and desperation, and I desperately want a straight gin. I have not had a gin and lime for nine years, three months, one week, and four days, nor have I found myself waking next to the kind of sub-animal lured by straight gin. Ev goes ahead of me into the restaurant. He is suddenly a black Ev cutout against an aniline brilliance of tinted windows, lake view, and desert hills crammed with mobile homes on the far shore. He turns. For an instant, I don't know who he is. "Hey," he waves.

I am paralyzed. Ev grabs my elbow. "Let's get out of here." He steers me to the door and through.

I sit on the cinder-block border of a dozen carefully spaced petunias. "Jesus, I want a drink."

"We are leaving *now*." Ev starts to walk away.

I stick my finger in the petunia dirt. It's wet. "Hey," I yell, "but they take such good care of this place!"

We are back on the road. I tuck the twenty under the obsidian.

"What happened?" Ev says quietly.

"You know." I look straight ahead, a stunningly ineffective way to outrun demons.

"You want to talk?"

I make myself turn my head left, take in lilac shadows, turn right toward Ev's profile against a dove-gray tatter of mountains. "Sure. I'm fifty-seven years old and I gave away twenty-five years of my life and I want it back. No big deal."

"Remember," Ev says, "the Mojave taketh, the Mojave giveth back." He shakes his head. "What am I talking about? I'm forty-two, fifteen years gone. I'm gonna end up being one of those old guys on a bicycle with a milk crate duct-taped to the handlebars and a mangy old dog riding in it."

This worries me. "What's so bad about that? I thought that was what you wanted."

"You ever really look at them? They're lonely. All they've got is that dog and the road, no friends, no one to come home to."

"You mean you won't get laid."

His silence is as hard as the land around us. I shut up. We miss the turnoff to Cadiz three times, get it right on the fourth. Long winter light holds, a dry lake shimmers moonstone.

I sing along with Plant.

Ev gives me the same look my kids used to when I would begin to sing.

"Hey," I say, "guess what was in *Modern Maturity*?"

"Huh?"

"The magazine for seniors. I joined. I did it for the discounts. Come on, Ev, ask me."

"What was in, I'm not sure I want to know, *Modern Maturity*?"

"Robert Plant's picture. He just turned fifty. His hair's all permed out and he looks pretty good."

"That, in case you're trying to cheer us up, is the most depressing thing you could possibly have said."

"Ev, how bad is our life? How bad is this?" I wave at the opal-shadowed big empty around us. An old salt plant rises from the glittering sand, rusted-out trucks and trailers gone bronze in the dying sun, broken windows shards of diamond.

"What I love about this Mojave," I say, "is that sometimes you can't tell if a thing is abandoned or still operating."

"Kinda like our luck," Ev says.

He snorts. I realize I'm grinning.

"Look." Ev points west. A great white gash shines across the blue-black mountains. "How can a mine be so beautiful?"

We come to the dugout as the sun begins to drop behind a rolling mystery range.

"Stop!" I yell.

We climb out, and Ev goes into the broken doorway of the dugout. Later, he'll tell me it was cool inside. The air smelled, to his surprise, not of piss, but only of damp earth. He imagined miners living there, taking refuge from the withering sun.

"I wanted to crouch on the dirt floor," he'll say, "rest my arms on the low plank table, rest my head on my arms, and just go to sleep." He'll say he saw my legs moving just beyond the entrance and he realized he was glad, for once, he wasn't alone.

I don't follow him in, not out of any particular respect for his privacy but because I have just stubbed my toe on a battered KLM flight bag.

"Ev," I whisper, "Ev, get out here. It's happened again."

He surfaces from the shadows. "I don't believe it."

I stare up at him. I am crouched over the flight bag, and I hold a fistful of magenta satin. It is a low-cut chemise trimmed with black lace.

"The bag is filled with this stuff," I say. I pull out ivory silk and emerald velvet, black crotchless panties, a white underwire bra laced with purple satin.

Ev starts to touch the chemise and pulls his hand back. "The white Levi's," he says.

We'd camped near a little casino in early winter. Just past dawn, walking, swilling camp coffee, chewing over our mutual disasters, we found a rusted truck chassis shot to lace and, scattered over it, a woman's Levi's and blouse, bra and panties, sneakers and socks—maybe once white, maybe bleached by the implacable Mojave sun. We looked for bones and found nothing.

Now I laugh, a high, choked sound. "What is going on?" I rummage in the bag and pull out one black stocking. Woven in white, the word *bitch* repeats again and again. You can imagine *bitch bitch bitch* spiraling up a long, slender leg, and you can guess how the watcher might feel.

I look past Ev and hold out the stocking. "Where is the other one? Ev, where could it be?" I stretch the stocking out, tie a knot, and pull it tight. Ev flinches.

He goes back into the bunker. I hold the clothes. I consider taking them home, washing them, giving them to someone small and voluptuous. *Bitch bitch bitch*. I know I must not. I think of Los Angeles a few hours away, Vegas, how a girl, a daughter, a sister, can disappear in the space of a bad blow job.

Ev comes up out of the dugout.

"There were a couple candles," he says. "And a bag half-full of corn chips. There was a big sheet of white cardboard, not much else except somebody had spray-painted stuff on the rafters."

"Stuff?"

"Nothing bizarre, just gang stuff."

"Nothing bizarre. Just gang stuff. Yeah."

I realize I have the chemise and stocking gathered close to my heart. We check out a rusted trash can near the doorway. Ev reaches in, pulls out a rubber amputated arm, the kind of creep fetish that kids go for on Halloween.

"Movies?" he asks. "White cardboard for the light meter. Gothic porn?"

"We aren't ever going to know."

I gather up the silk and velvet and satin, and I head west. The sun is nearly gone, the air silvery except for a streak of brilliant orange along the horizon. I walk through skeletal Joshua and backlit cholla and creosote bush scenting the small breeze. I watch the ground. A ravine curves toward the mountains. I walk its rim, studying shadows, boulders. Nothing moves. There is nothing bone-white, no stink of rotting meat.

"Just the desert," I call to Ev. He goes to the truck, sits on the back bumper. I see the flare of a match, the bright coal of a cigarette.

I drape the stocking around my neck and I go to work. I feel my arm move, my hand release. Up. Out. Satin shimmers in last light, silk catches on yucca, lace on creosote, the magenta chemise on a Joshua. When I am done, I begin to walk toward Ev. He is nothing but shadow and a tiny red star that flares and dims. I unwind the stocking from my neck.

For an instant, I think about taking it home. I run it through my fingers. *Bitch bitch bitch.*

"Whoever you are, whoever you were, whatever has happened, I won't forget this." I crouch and bury the stocking in the cool sand.

Two days later, I call the sheriff in Twenty-Nine Palms. She takes the report. A deputy will drive out and check the place. I tell her the wind may have scattered the clothes.

She tells me she will call if they find anything.

CROSSINGS, PAST AND PRESENT

PEDRO FONT

Father Pedro Font, born in 1737, was a Franciscan priest from Spain who served as a chaplain on the 1775-1776 expeditions from Mexico to northern California with Juan Bautista de Anza. Font died in Spain in 1781.

It was on the Anza expedition that Font wrote the following journal entries, which describe the hardships and surprises facing the party of men, women, and children as they crossed an extremely rugged, mostly sandy and waterless section of the southernmost California desert along an old Indian trail. The route taken by Anza's party later became known as the Butterfield Stagecoach Trail and was used by nineteenth-century California gold seekers and settlers. It is now California's Highway 78 and remains one of the state's most remote roads.

from *Font's Complete Diary: A Chronicle of the Founding of San Francisco*

Monday, December 11.—[…]The road would not be very bad if it were not so long, but for this reason we arrived very tired out. One sees along the way many piles of mussels and an infinity of sea snails, very small and spiral shaped, and in places as white as flour. This confirms my opinion that this is a sea beach, and although the sea has reached here sometime, yet no barrancas are found like yesterday. Only in the afternoon, when we came opposite the Cerro del Imposible (so-called since the last expedition, because the soldiers found it impossible to reach it) and which we passed at our left, did we enter a very difficult terrain, where all the country is full of little mounds of hard earth which they call *almondigones,* very apt for tiring the riding animals; after which we crossed a sand dune for about a league. This road from El Carrizal to Santa Rosa was discovered by some soldiers who were sent ahead by the commander for this purpose from the Gila River, on the 16th of November. By it the sand dunes are circumvented, leaving them on one side or the other, for only a small piece now and then is crossed.

It is very cold in these plains at this season, and we had a cloudy sky. On the 9th the clouds were like cobwebs; on the 10th they were somewhat heavier; on the 11th the sky was covered all day with thicker clouds; on the 12th, day dawned with thick and low clouds touching the sierra, and I thought perhaps this was an indication that we were approaching a country where the rains come in the winter.

* * *

Tuesday, December 12.—I said Mass. We set out from the wells of Santa Rosa at a quarter to two in the afternoon, and at a quarter to five we halted at a dry arroyo, having traveled three leagues to the north.[1] —Three leagues.

At Santa Rosa we left the six wells opened and with water for those who were coming behind. This dry arroyo comes from a range not very far distant, which appears to be a spur of the Sierra Madre, and runs through the plains and sand dunes, which we had on the right, the range being on the left. It has no water, but there is some galleta grass, some of which also is encountered on the way; and it has also some firewood with which to warm us, which was lacking at Santa Rosa. The road is fairly good, having only some ups and downs over some hills on leaving Santa Rosa, ridges of sandy and hard earth, with many black, flat stones that are not very large.[2] After noon a west wind blew up very strong and cold, coming from the Sierra Madre de California, where apparently it was raining, for it was all covered with thick clouds, and the wind continued stiffer in the afternoon and almost in the same way all night until daybreak. This strong wind, which perhaps is usual in these plains, is what forms the sand dunes, with their various shapes. They are mountains of fine sand which the wind moves from one side to another, as I observed today; for with the wind they looked in the distance like clouds of very thick dust, low and even touching the earth.

Wednesday, December 13.—In the morning it was cold, as if it were going to snow, and it continued so and got even colder until afternoon, there being a light sharp wind which cut our faces. We set out from the Arroyo Seco at nine in the morning, and having traveled some seven long leagues to the north-northwest, with some inclination to the north,[3] at half past three in the afternoon reached San Sebastián, which is a small village of mountain Cajuenches, or more properly, of Indians of the Jecuiche tribe.—Seven leagues.

The road is level and without sand dunes, but the footing in places is treacherous, for on traveling across it the animals in some places suddenly buried all four feet. This place of San Sebastián is a spring of water that is rather hot or warm when it emerges, deep and permanent, like a marsh, and flowing very little. It has its carrizo and some grass, although it is not very

---

[1] Anza gives the distance as four leagues. Camp was about three miles north of Plaster City, at Coyote Wash. It was near Sackett's Well on the old stage road from Mexicali to Carrizo Station. [All footnotes are from Bolten's *Font's Complete Diary* (Berkeley: UC Press, 1931).]

[2] These ups and downs over the small hills and the black, flat rocks last for two or three miles, past Yuha Drill Hole.

[3] The march was nearly north over the flat desert plain between Superstition Mountain on the east and Fish Creek Mountain and Coyote Hills on the left. The stumbling of the horses was due to rat holes with which the plain here is still honey-combed in places.

good, because the soil is so saline in all this flat that in places the salt whitens it like flour. But the water does not appear to be very bad, although near the spring there is a ditch which is very miry, with the worst kind of water and very injurious. There is also some firewood of scrubby mesquite.[1] [...]

Thursday, December 14.—In the morning the weather was very cold, and the sierra was covered with clouds. There was a very strong wind, and in the middle of the forenoon it snowed. While it was snowing arrived the cattle which set out on the 10th from the Laguna de Santa Olalla directly for this place, and since they had not drunk in all these days they made for the water like a streak of lightning. On the way eleven beeves had been lost. With them arrived the cowboys and soldiers who drove them, half dead with cold and hunger, for by now their provisions had become exhausted. It snowed for about an hour, the wind slackened, and then it rained all day until late at night. The second division of people under the sergeant ought to have reached here today, and seeing that it did not come we surmised that perhaps yesterday the rain had caught them at Santa Rosa, judging from the clouds which we saw in that direction while on the way, and that therefore they had not left that place.

When the cattle arrived I was in the tent of the commander, where I spent most of the day because it was more sheltered and had a fire in it. Seeing the severity of the weather I said to him that since the grass of this place did not appear to be very bad, and the water was abundant, it seemed to me better to wait here for the two divisions of people who were behind and all reassemble here, than to go to wait at the arroyo of Santa Catharina three days farther on, as formerly had been planned; because in case of some necessity or delay it would be easier to aid them from here than from farther on. The commander replied that he had already planned to do this, and so it was decided to wait here until all the people of the expedition who remained behind in the two divisions should join us.

Friday, December 15.—In the morning the weather was good, although there were a few clouds. The Sierra Madre de California showed itself white with snow, the Sierra de San Sebastián which we had in front of us was all snow-covered from top to bottom, and the rough range which we had on our right on the other side of the sand dunes and plains, which above here joins with

---

[1] San Sebastián, where camp was pitched, was at Harper's Well, at the junction of Carrizo and San Felipe creeks, and some four miles west of Kane Spring on the highway that runs from Brawley to Indio. Harper's Well is a modern drilled well. About four hundred feet northeast of it, on the east bank of the creek, there is a natural well (now dry but active as late as 1915), with carrizo round about, and pottery and other signs of Indian occupation near by. To the west in the mesquite flat there are numerous salty springs, as Font states.

the Sierra Madre, was likewise snow-covered, so that we found ourselves in this plain surrounded by snow, and the weather quite cold.

In the morning we found eight beeves and one of the vaqueros' mules frozen to death, for since they came so thirsty, and gorged themselves with water, the bitter cold of the night killed them. At noon the sergeant arrived with the second division of the people of the expedition and the second pack train. They came half dead with cold from the cruel weather which caught them yesterday on the way, several saddle animals remaining behind, used up, and out of commission. After noon the whole horizon was covered with the fog which came from the Sierra Nevada de California, and the day remained dark with threats of a bad night and of a repetition of the snow or rain.

ALFRED L. KROEBER and CLIFTON B. KROEBER

Alfred L. Kroeber, born in 1876 in Hoboken, New Jersey, was an anthropologist and prolific author known for his pioneering work with indigenous people throughout the southwestern United States. Best known for his early-twentieth-century studies of a Native American man named Ishi, Kroeber also conducted extensive fieldwork with many of the California desert tribes. In the 1950s, picking up on fieldwork Kroeber had conducted in 1903 with Mohave elder Jo Nelson, Kroeber's son Clifton completed his father's A Mohave War Reminiscence: 1854–1880, *which was published in 1973.*

The following section, "First Conflict with Americans," tells a story given to Kroeber by Jo Nelson, or Chooksa homar. Chooksa homar was born in the nineteenth century and lived his life on his tribe's ancestral land, called Aha macave, located in the Colorado River Valley near Needles, California, and the site of the present-day Fort Mojave Indian Reservation.

First Conflict with Americans

After this it was about seven years that the whites came from the east with wagons and cattle, to [Hardyville crossing] about three miles above Fort Mohave.[1] I was then a boy about so high [point: ca. twelve years], living opposite Needles. I [went up river and] saw two white women and a boy, then two more with a child, and more coming behind. There was a man with them who had calico and was cutting off lengths and giving each Mohave man a piece, and a ring and two or three little bells: I saw that. Then they gave me about four feet of calico for a breechclout and a ring and a small bell.

By afternoon, all the whites had reached the river and drove the cattle down to where they would have grass at [the foot of where] Fort Mohave [was later]. All the Mohave stood on the mesa looking; [many of them] had not seen whites or horses [sic] or cattle before. They looked at the cattle but would not go near them, fearing to be hooked. In the evening they returned to where they slept, which was where some Mohave had their houses, near by.

Next morning, some of the head men said: "Do not go to the whites today." Aratêve was then at Yuma; but the five brave men who had been given letters were there and said: "I tell you: I want to fight the whites." Other Mohave said: "That paper you got does not tell you to fight. It says to be friendly to the whites: you

---

[1] This first emigrant party to try the 35th-parallel route to California had accumulated eighteen wagons, more than two hundred people, about three hundred seventy-five cattle and about forty horses. The attack occurred Aug. 30, 1858. Probably nine whites were killed outright. The survivors took two wagons, about twenty cattle and about ten horses back toward Albuquerque, and survived because of meeting two other emigrant trains and because of the army's prompt and solicitous efforts on the trail and after the destitute people reached Albuquerque.

51

brought it to our land from Yuma." They answered: "Oh well, I will tear it up. They just wrote on the paper, but that will not stop me, I did not want their paper; I did not ask for it, they gave it to me. I do not want the whites to come and own the land. They will take it and keep it. I want to stop them, to kill them all."

The other Mohave said to them: "Well, if you five want to fight, go fight. But we will not help you. If you think you can fight them [successfully], go ahead."[1]

The five answered: "If we let the whites come and live here, they will take your wives. They will put you to work. They will take your children and carry them away and sell them. They will do that until there are no Mohave here. That is why I want to stop them from coming, want them to stay in their own homes. The eastern Indians, I hear that is what the [whites] did to them there: they took their children and said to them: 'You are not to see your parents.' And they keep birds eggs and coyotes and bears and every kind: maybe they will keep you all [confined] in a place too. As for me, I do not want them to do that to me. The whites will not listen to the Mohave. If you tell them to do something, they say No."[2]

The old men spoke against the five. [Their arguments are omitted because they are a repetition about the papers: the letters do not say to fight, but to be friendly.]

The five answered: "The whites, when you come near them, push you away; they kick you. A woman, if she is kicked, cries. I am a man: I do not cry: I do not want to be kicked."[3]

In the afternoon, the other Mohave said: "I have heard that these whites are everywhere, on all sides. You have heard that too. Nevertheless you want to fight them. Well, we will follow your counsel: we will go to fight."[4]

"That is what we want. We are not like mountains: we do not stay forever. We are not like the sky, always there; not like the sun or the moon: we die. Perhaps in a year, in a month, in two or three days. I want to die fighting."[5]

"Well, how many times do you wish [expect] to die? You die once and do not come alive again. We will fight with you and die too. No one likes to die. If you like it, why not tie your hands and feet and jump into the river? No one

---

[1] The argument shows how little organized control of tribal action did exist. [All footnotes and bracketed text are from Clifton B. Kroeber's *Mohave War Reminiscence* (Berkeley: UC Press, 1973).]

[2] The warriors may have been actuated by pride and need to maintain their repute, but their apprehensions were largely justified. Their pacifist elders probably sensed defeat if it came to a contest; but in their answer in the next paragraph they carefully do not avow this, but keep harping on the Yuma letters as if these were fetishes.

[3] We do not often get this consideration expressed in white-native relations, though it unquestionably was a factor in many cases (A.L.K.).

[4] The peace party begins to yield to nationalistic sentiment, but still with a current of bitterness.

[5] These are stock Mohave sentiments and phrases which we will encounter again.

does that way: that is killing oneself. So you say you want to die soon: well, good, we will go along and help you."[1]

So they all got their clubs and bows and painted and put on feathers. When they became angry like this, they used to tie their hair tightly together at the nape,[2] letting the ends hang down loose. They painted the hair red, the face black;' that is how they liked to die.

The whites were camped about a mile away, at the river among the cottonwoods at the foot of Fort Mohave terrace, west of it. Now the five said: "Are you ready?" and when they were, they all went toward the whites, the five leading.

And they notified the Chemehuevi who were living in Nevada across the river from Fort Mohave. There was a Chemehuevi named Ahwetaraðme" who had said: "When you decide to fight, send word: I want to fight too."

But when they came nearer, some of the Mohave did not want to fight after all[3] and went up on the [Fort Mohave] mesa and stood there: they wanted to see it. Those who wanted to fight, a good many, went on with the five.

When they came within two or three hundred yards, the whites saw them, went into their canvas-covered wagons, and got their guns. They did not shoot but stood there looking: they thought the Mohave might be coming to shake hands. The Chemehuevi were approaching from the west, up the river bank, from behind the white camp. When they came close, the Mohave ran forward, to seize the whites; then these shot, and the Mohave shot arrows, and they fought there. And the Chemehuevi stood and shot arrows.[4]

Savêre,[5] the leader of the whites, had ridden downstream with one companion to look for a good place [to cross]. He did not see the fight start, but heard the shooting, and rode back, when he ran into the Chemehuevi. These recognized him as the leader and all shot at him. He was hit by four or five arrows, and though he reached his camp, he fell and died: the whites put him into a wagon.

---

[1] They are more prudent, but not ready to admit less bravery, so they agree to go along though suggesting it is mere suicidal folly.

[2] Probably to render it less easy for opponents to grasp. The cylinder-headed war club was made for an uppercut thrust into the face while the foe's head was being pulled forward by its long hair.

[3] More disunity. The effect of guns was no doubt known by repute if not in the experience of the majority.

[4] It would seem that although these were nonwarlike emigrants, the guns sufficed to keep the Mohave from coming to hand-to-hand grip as their brave men prided themselves on doing. It is of interest that none of the five professional braves was among the casualties.

[5] Savedra in the records, an experienced guide whom the U.S. Army commander in New Mexico induced the party to take, before he recommended that they proceed—probably José M. Savedra (more likely correctly spelled as Saavedra). The wagon master was Alpha Brown.

"All right, you who say how many times you want to fight: we die once, but you also will die once only"—the Mohave, who had said that to the five brave men got shot in the breastbone and killed. One Mohave was hit in three places: the right ankle, above the right knee, in the left thigh. Another was shot through the right thigh. Still another was hit in the flank: the bullet stayed in.

When Savêre began to be struck by the arrows, the Mohave saw it and wanted to seize him and pull him off his horse.[1] One of them got under the horses and was running between them when Savêre's companion shot at him, but only tore the skin on his back.

Two Mohave got two horses without being seen, during the fighting, and rode them off. One of them got a blue horse, the other a yellow with a stripe on its nose. I saw them do this.

So four Mohave were shot with bullets, one of them dead. They carried him and the wounded back to camp. The Chemehuevi jumped into the river and crossed back to the west side; also some Mohave who had come over with them. As these had been behind the whites, when the two lines of Indians shot with their bows and missed the whites, they could hit each other; and so the Mohave wounded in the thigh was shot by a Mohave.[2] When the dead man—his name was Tšapotire—was brought in to his camp, everybody wailed.

All four of the wounded lived on the west side of the river. So all the Mohave crossed over there, leaving their effects on the Arizona side.

Then one man, Soli was his name, said, "I can cure those shot: I dreamed of powder, bullets, cartridges." So he treated them.[3]

That night they sent two men back to scout the whites. They crossed to the Arizona side and came near the camp. The whites were loading everything into their wagons, and then turned back on their way, back to the east, leaving their cattle.[4]

About daylight the two scouts returned. "Nobody is there, only cattle." Then the five fighters who had the letters said: "Well, we will all return to the other side, take the cattle, kill and eat them." Those of the Mohave who had come up from Needles and elsewhere downstream said: "We will not kill ours

---

[1] To club: the proper termination of combat, as in our old Infantry Manual it used to be by bayonet.

[2] By an arrow. The others were hit by bullets.

[3] Guns were known though not yet possessed; so someone promptly dreamed of them and had his cure ready.

[4] The Rose accounts show that the whites were in fear of their lives, deeply shocked by the unexpected hostilities; and in any case they had lost sight of the cattle and thought these had all been spirited off hours before.

here, we will drive them home to eat." So they crossed that morning to take their cattle: I went with them. Some got two cows, some several: they drove them into the river to cross. If ten or twenty men lived in one place, they got several cattle: by afternoon, everybody had some and ate meat.

I had none for myself: I was too young, and afraid. But every day I went about the settlements and was given meat.

Now they thought that the whites had left for good.[1] The five fighters said: "You see, no more whites are coming. That is why I wanted to fight. If they had stayed in the country, they would never have left and would have taken everything: that is what I heard about them. Now we have fought and beaten them and they have gone back, far away; it is well, we own the land."

.

---

[1] This incident tended to be recalled by whites whenever the Mohave name was mentioned. It was crucial in setting the white man's view of the tribe and it reverberated on in official circles because of a claim entered against the U.S. government by some of the emigrants for losses incurred here (C.B.K.).

WILLIAM LEWIS MANLY

William Lewis Manly (1820–1903) was born in St. Albans, Vermont. His Death
Valley in '49, *first published in 1894, is a gripping story of a group of westward-
bound pioneers' perilous crossing of Death Valley, which has one of the most severe
climates on earth.*

*At one point during the crossing, the group took a wrong turn and the wagon
train was stranded. The following excerpt recounts part of the rescue effort by
Manly and another party member with whom he had set off on foot to get help.
Miraculously, every member of the party was saved.*

from *Death Valley in '49*

The range was before us, and we must get to the other side in some way. We
could see the range for a hundred miles to the north and along the base some
lakes of water that must be salt. To the south it got some lower, but very barren
and ending in black, dry buttes. The horses must have food and water by
night or we must leave them to die, and all things considered it seemed to be
the quickest way to camp to try and get up a rough looking cañon which was
nearly opposite us on the other side. So we loaded the mule and made our way
down the rocky road to the ridge, and then left the Jayhawkers' trail, taking
our course more south so as to get around a salt lake which lay directly before
us. On our way we had to go close to a steep bluff, and cross a piece of ground
that looked like a well-dried mortar bed, hard and smooth as ice, and thus got
around the head of a small stream of clear water, salt as brine. We now went
directly to the mouth of the cañon we had decided to take, and traveled up its
gravelly bed. The horses now had to be urged along constantly to keep them
moving and they held their heads low down as they crept along seemingly so
discouraged that they would much rather lie down and rest forever than take
another step. We knew they would do this soon in spite of all our urging, if we
could not get water for them. The cañon was rough enough where we entered
it, and a heavy up grade too, and this grew more and more difficult as we
advanced, and the rough yellowish, rocky walls closed in nearer and nearer
together as we ascended.

A perpendicular wall, or rather rise, in the rocks was approached, and there
was a great difficulty to persuade the horses to take exertion to get up and over
the small obstruction, but the little mule skipped over as nimbly as a well-fed
goat, and rather seemed to enjoy a little variety in the proceedings. After some
coaxing and urging the horses took courage to try the extra step and succeeded

all right, when we all moved on again, over a path that grew more and more narrow, more and more rocky under foot at every moment. We wound around among and between the great rocks, and had not advanced very far before another obstruction, that would have been a fall of about three feet had water been flowing in the cañon, opposed our way. A small pile of lone rocks enabled the mule to go over all right, and she went on looking for every spear of grass, and smelling eagerly for water, but all our efforts were not enough to get the horses along another foot. It was getting nearly night and every minute without water seemed an age. We had to leave the horses and go on. We had deemed them indispensable to us, or rather to the extrication of the women and children, and yet the hope came to us that the oxen might help some of them out as a last resort. We were sure the wagons must be abandoned, and such a thing as women riding on the backs of oxen we had never seen, still it occurred to us as not impossible and although leaving the horses here was like deciding to abandon all for the feeble ones, we saw we must do it, and the new hope arose to sustain us for farther effort. We removed the saddles and placed them on a rock, and after a few moments' hesitation, moments in which were crowded torrents of wild ideas, and desperate thoughts that were enough to drive reason from its throne, we left the poor animals to their fate and moved along. Just as we were passing out of sight the poor creatures neighed pitifully after us, and one who has never heard the last despairing, pleading neigh of a horse left to die can form no idea of its almost human appeal. We both burst into tears, but it was no use, to try to save them we must run the danger of sacrificing ourselves, and the little party we were trying so hard to save.

We found the little mule stopped by a still higher precipice or perpendicular rise of fully ten feet. Our hearts sank within us and we said that we should return to our friends as we went away, with our knapsacks on our backs, and the hope grew very small. The little mule was nipping some stray blades of grass and as we came in sight she looked around to us and then up the steep rocks before her with such a knowing, intelligent look of confidence, that it gave us new courage. It was a strange wild place. The north wall of the cañon leaned far over the channel, overhanging considerably, while the south wall sloped back about the same, making the wall nearly parallel, and like a huge crevice descending into the mountain from above in a sloping direction.

We decided to try to get the confident little mule over this obstruction. Gathering all the loose rocks we could we piled them up against the south wall, beginning some distance below, putting up all those in the bed of the stream and throwing down others from narrow shelves above we built a sort of inclined plane along the walls gradually rising till we were nearly as high as the crest of the fall. Here was a narrow shelf scarcely four inches wide and a space of from twelve to fifteen feet to cross to reach the level of the crest. It was all I could do to cross this space, and there was no foundation to enable us to widen it so as to make a path for an animal. It was forlorn hope but we made the most of it. We unpacked the mule and getting all our ropes together, made a leading line of it. Then we loosened and threw down all the projecting points of rocks we could above the narrow shelf, and every piece that was likely to come loose in the shelf itself. We fastened the leading line to her and with one above and one below we thought we could help her to keep her balance, and if she did not make a misstep on that narrow way she might get over safely. Without a moment's hesitation the brave animal tried the pass. Carefully and steadily she went along, selecting a place before putting down a foot, and when she came to the narrow ledge leaned gently on the rope, never making a sudden start or jump, but cautiously as a cat moved slowly along. There was now no turning back for her. She must cross this narrow place over which I had to creep on hands and knees, or be dashed down fifty feet to a certain death. When the worst place was reached she stopped and hesitated, looking back as well as she could. I was ahead with the rope, and I called encouragingly to her and talked to her a little. Rogers wanted to get all ready and he said, "holler" at her as loud as he could and frighten her across, but I thought the best way to talk to her gently and let her move steadily.

I tell you, friends, it was a trying moment. It seemed to be weighed down with all the trials and hardships of many months. It seemed to be the time when helpless women and innocent children hung on trembling balance between life and death. Our own lives we could save by going back, and sometimes it seemed as if we would perhaps save ourselves the additional sorrow of finding them all dead to do so at once. I was so nearly in despair that I could not help bursting in tears, and I was not ashamed of the weakness. Finally Rogers said, "Come Lewis" and I gently pulled the rope, calling the little animal, to make a trial. She smelled all around and looked over every inch of the strong ledge, then took one careful step after another over the dangerous place. Looking back I saw Rogers with a very large stone in his hand, ready to "holler" and perhaps kill the poor beast if she stopped. But she crept along trusting to the rope to balance, till she was halfway across, then another step or two, when calculating the distance closely she made a spring and landed on a smooth bit of sloping rock below, that led up to the highest crest of the precipice, and safely climbed to the top, safe and sound above the falls. The mule had no shoes and it was wonderful how her little hoofs clung to the smooth rock. We felt relieved. We would push on and carry food to the people: we would get them through some way; there could be no more hopeless moment than the one just past, and we would save them all.

It was the work of a little while to transfer the load up the precipice, and pack the mule again, when we proceeded. Around behind some rocks only a little distance beyond this place we found a small willow bush and enough good water for a camp. This was a strange cañon. The sun never shown down to the bottom in the fearful place where the little mule climbed up, and the rocks had a peculiar yellow color. In getting our provisions up the precipice, Rogers went below and fastened the rope while I pulled them up. Rogers wished many times we had the horses up safely where the mule was, but a dog could hardly cross the narrow path and there was no hope. Poor brutes, they has been faithful servants, and we felt sorrowful enough at their terrible fate.

We had walked two days without water, and we were wonderfully refreshed as we found it here. The way up this cañon was very rough and the bed full of sharp broken rocks in loose pieces which cut through the bottoms of our moccasins and left us with bare feet upon the acute points and edges. I took off one of my buckskin leggins, and gave it to Rogers, and with the other one for myself we fixed the mocassins with them as well as we could, which enabled us to go ahead, but I think if our feet had been shod with steel those sharp rocks would have cut through.

Starting early we made the summit about noon, and from here we could see the place where we found a water hole and camped the first night after we left the wagons. Down the steep cañon we turned, the same one in which we had turned back with the wagons, and over the sharp broken pieces of volcanic rock that formed our only footing we hobbled along with sore and tender feet. We had to watch for the smoothest place for every step, and then moved only with the greatest difficulty. The Indians could have caught us easily if they had been around for we must keep our eyes on the ground constantly and stop if we looked up and around. But we at last got down and camped on some spot where we had set out twenty-five days before to seek the settlements. Here was the same little water hole in the sand plain, and the same strong sulphur water which we had to drink the day we left. The mule was turned loose dragging the same piece of rawhide she had attached to her when we purchased her, and she ranged and searched faithfully for food finding little except the very scattering bunches of sagebrush. She was industrious and walked around rapidly picking here and there, but at dark came into camp and lay down close to us to sleep.

There was no sign that anyone had been here during our absence, and if the people had gone to hunt a way out, they must either have followed the Jayhawker's trail or some other one. We were much afraid that they might have fallen victims to the Indians. Remaining in camp so long it was quite likely they had been discovered by them and it was quite likely they had been murdered for the sake of the oxen and camp equipage. It might be that we should find the hostiles waiting for us when we reached the appointed camping place, and it was small show for two against a party. Our mule and her load would be a great capture for them. We talked a great deal and said a great many things at that campfire for we knew we were in great danger, and we had many doubts about the safety of our people, that would soon be decided, and whether for joy or sorrow we could not tell.

From this place, as we walked along, we had a wagon road to follow, in soft sand, but not a sign of a human footstep could we see, as we marched toward this, the camp of the last hope. We had the greatest fears the people had given up our return and started out for themselves and that we should follow on, only to find them dead or dying. My pen fails me as I try to tell the feelings and thoughts of this trying hour. I can never hope to do so, but if the reader can place himself in my place, his imagination cannot form a picture that shall go beyond reality.

We were some seven or eight miles along the road when I stopped to fix my moccasin while Rogers went slowly along. The little mule went on ahead of both of us, searching all around for little bunches of dry grass, but always came back to the trail again and gave us no trouble. When I had started up again I saw Rogers ahead leaning on his gun and waiting for me, apparently looking at something on the ground. As I came near enough to speak I asked what he had found and he said "Here is Captain Culverwell, dead." He did not look much like a dead man. He lay upon his back with arms extended wide, and his little canteen, made of two powder flasks, lying by his side. This looked indeed as if some of our saddest forbodings were coming true. How many more bodies should we find? Or should we find the camp deserted, and never find a trace of the former occupants....

One hundred yards now to the wagons and still no sign of life, no positive sign of death, though we looked carefully for both. We fear that perhaps there are Indians in ambush, and with nervous irregular breathing we counsel what to do. Finally Rogers suggested that he had two charges in this shot gun and I seven in the Coll's rifle, and that I fire one of mine and await results before we ventured any nearer....And now both closely watching the wagons I fired the shot. Still as death and not a move for a moment, and then as if by magic a man came out from under a wagon and stood up looking all around, for he did not see us. Then he threw up his arms high over his head and shouted, "The boys have come," "The boys have come!"

WILLIAM PHIPPS BLAKE

William Phipps Blake was born in New York City in 1826 and went on to become the first geologist to professionally examine southern California's western Sonoran Desert. While working for the Smithsonian Institution, he served as a geologist for Lt. R. S. Williamson's 1853 expedition to survey possible railroad routes from the Pacific Coast eastward across the California desert. Blake is credited with having located the shoreline of Ancient Lake Cahuilla, an extinct inland sea that once extended from the Gulf of California to modern-day Indio, an area now partially filled by the Salton Sea. Blake made many other important contributions to the early geologic understanding of California, and in 1864 he was appointed professor of geology at the new University of California in Berkeley. He died in 1910.

In this excerpt from the "USGS Survey of Railroad—Coachella Valley, 1853," Blake describes meeting some of the area's Cahuilla Indians and visiting several of their village sites.

Journal excerpts, Pacific Railroad Survey, 1853

November 13.—[...]We continued travelling to the southeast, and downwards over the broad slope of the pass, following the shallow bed of a brook in which water was flowing rapidly, but without trees or much vegetation on its banks. It appeared as if it had been entirely dry for the greater part of the summer. On reaching the next spur of San Gorgonio, we encamped on the eastern or lower side, in order to avoid the strong wind which continued to blow without cessation....

We had travelled twenty-two miles from the summit, and were nearly at the base of the pass....

We travelled southeasterly over the new broad plainlike slope of the pass, and continued to descend....

After travelling about seven and a half miles over these long and barren slopes, we saw a green spot in the distance, and soon came to two large springs of water rising in the bare plain, not far from the foot of the mountain. One of these springs is warm, and forms a pool nearly thirty feet in diameter, and three to four feet deep. The cold spring is not quite so large, and is only ten feet distance from the other....

This place was evidently a favorite camping-ground for Indians. When we arrived, many Indian boys and girls were bathing in the warm spring, and a group of squaws were engaged in cooking a meal for a party returning from a great feast held near Weaver's ranch....

A growth of rushes forms a narrow margin of green vegetation around the spring and its outlet. Willows and mezquite bushes grow there also; and I found a young *palm tree* spreading its broad, fanlike leaves among them. The surrounding desert, and this palm tree, gave the scene an Oriental aspect....

November 16.—Hot Spring to Deep Well, 12 miles—A slight dew was deposited on the blankets during the night, but this was probably local, and derived from the warm vapor of the spring. The water was covered with a cloud of condensed vapor, and its temperature at sunrise was only 86°, the air being 46°. It is thus affected by the changes in the temperature of the air, the supply not being very rapid. The barometer indicated an elevation of less than two hundred feet above the sea.

On leaving the green banks of this spring, we again traversed the bare and gravelly surface, and skirted the base of the mountains on the right...

The Indian guide conducted us...to "*Pozo hondo*", or Deep Well, a deep excavation in the clay made by the Indians to obtain water. It was at the base of a high sand-drift, and about twenty-five feet deep, but contained only a little water. It was wide at the top, but became smaller towards the bottom, being a funnel-shaped depression. The water was obtained by means of steps out in the sides of the pit...The opening to the well was shaded by several mezquite trees....

We encamped, and before the mules were satisfied with water, it was all exhausted. It oozed in through the clay very slowly, but appeared to be abundant; about twenty buckets full were obtained in the course of the night.

November 17.—Deep Well to Cohuilla Villages, 13 miles.—Sand hills were observed on the left, or north of the trail, for three or four miles beyond the Deep Well, and formed a succession of low, rolling hills, or long drifts....

On turning around the point, I saw a discoloration of the rocks extending for a long distance in a horizontal line on the side of the mountains....This crust had evidently been deposited under water, and, when seen at a distance of a few yards, its upper margin appeared to form a distinct line, which indicated the former level of the water under which it was deposited...it became evident to every one in the train that we were travelling in the dry bed of a former deep and extended sheet of water, probably an *Ancient Lake* or an extensive bay.

...many small spiral shells were found; they were...very abundant... appearing to have been blown into heaps by the wind. They were so numerous in some places as to whiten the ground. Five or six species of the genera, Planorbis...were soon collected, and showed that the former lake was fresh water....

We passed several Indian trails, and about noon met an Indian family travelling in the opposite direction. The young men came first, carrying bows and arrows and an old flint lock musketoon; an old Indian squaw followed, bearing the burdens. They all stopped with surprise as we came up, and unrolling some rags from a great yellow ball invited us to eat. This proved to be made of the pounded beans and pods of the mezquite, which is an important article of food to them, but prepared in that way, and partly fermented, was not a very agreeable refreshment to us....Up to the time of our arrival their country had never been visited by the whites with a train of wagons. As we approached some of their villages, we passed several holes dug in the clay, two or three feet deep, that contained water, and were evidently springs that the Indians had enlarged. The largest and best of these springs were surrounded by extensive rancherias, or villages of huts, located in thick groves of mezquite trees, which were quite abundant, and grew so thickly together that the Indian huts were completely hid....We encamped at this place and were surrounded by crowds of Indians anxious to trade melons, squashes, corn, and barley, for pork, bacon, or other articles.

The chief, or "captain," and the principal men having collected for a talk with Lieutenant Parke, they learned the object of our visit, and appeared much pleased. When questioned about the shore-line and water marks of the ancient lake, the chief gave an account of a tradition they have of a great water (*agua grande*) which covered the whole valley and was filled with fine fish. There was also plenty of geese and ducks. Their fathers lived in the mountains and used to come down to the lake to fish and hunt. The water gradually subsided...and their villages were moved down from the mountains, into the valley it had left. They also said that the waters once returned very suddenly and overwhelmed many of their people and drove the rest back to the mountains....

November 18.—Cohuilla Villages to Salt Creek—35 miles. The Indians had a grand feast and dance during the night, keeping us awake....

Eight or nine miles from our camp at the villages we stopped at another spring, where the water rose to the surface in abundance, and formed a pool twenty feet or more in diameter, surrounded by an artificial embankment three or four feet high. The water was clear and good; its temperature at noon was 78°, and there appeared to be a never-failing supply....We remained at the spring until three o'clock, to rest the animals and prepare for a long march over the unknown region between us and Carrizo Creek, where the emigrant road from the Gila enters the mountains. None of the Indians could be induced to go with us; they were afraid to venture, saying that there was neither grass nor water, and that we could not take the wagons....

ANONYMOUS

The following anonymous news story was first published in San Francisco's Alta
California *on July 1, 1866. It is included in* The Mojave Road in Newspapers,
*a series of books edited and published by Dennis G. Casebier, a longtime desert
enthusiast and Mojave Road historian.*

*This story, along with other news articles in Casebier's collection, depicts scenes
and activities that existed along the Mojave Road during its frontier heyday. For a
twenty-year period in the mid-nineteenth century, the Mojave Road was an important
east-west wagon route in and out of coastal California. It followed a centuries-old
Indian footpath across the Mojave Desert from the Colorado River near Needles to
the Cajon Pass, and was heavily traveled until the arrival of the railroad.*

from a letter to the *Alta California*, July 1, 1866

Thus far our road has been easterly. We now turn north, and thirteen miles
brings us to the mouth of a cañon, that, winding up through the mountains,
affords an excellent avenue through the otherwise impassable barrier. This
cañon is known as Cajone Pass. Mr. John Brown, one of the oldest pioneers
of the Pacific coast, at a considerable cost has constructed an excellent road
through this pass, and is now enjoying the fruits of his enterprise. Toiling
slowly up through the cañon, occasionally meeting heavy grades, we finally
reached the summit. And a new scenery presented itself. We stood upon the
dividing line of sterility and fertility. Away to the east stretches the desert land;
the cactus and greasewood are lost in the distance in the smoky haze, and then
the low range of barren mountains and sand hills, that mark the course of the
Mohave River, like islands at sea, catch the vision ere it is quite lost in nonentity.
I turned and looked down through the pass, out on the beautiful verdant hills
and valleys, and their inviting beauties and wooing luxuries seemed to present
themselves in more than usual attractiveness, as in my mind's eye I contrasted
all with the desert land ahead of me, and I really detected myself in a sigh of
regret. Our road now runs twenty-two miles to the eastward, where it crossed
the Mohave River, and turning, follows its course to the northeast.

The Mohave

At this point is a beautiful stream of, say, sixty feet wide and about two feet
deep. The valley is irregular in its width; here a mile and a half wide, perhaps;
further down, a few miles. The sand hills crowd in upon it, threatening to cut

the slender thread that connects it with the next larger tract, that spreads out its greensward and continues along thus a few miles, then yielding again to the encroachments of the barren plain. The river, that winds through the valley, among the cottonwood and willow, so sparkling and clear, is lost in the sands, and naught marks its course save an occasional willow or cottonwood, or the rocky channel cut by the torrent flood that the winter rains send down, till suddenly its waters burst forth anew, and then commences bright meadows and green groves again. The traveller sees but little to interest; the road now passing over gravelly masa, then across a bend in the valley, with volcanic hills and barren plains on either hand, their withering desolation relieved only by the bright green belt that for the most part marks the course of the Mohave. This valley is truly an oasis. The soil, though somewhat impregnated with alkali, is rich, and affords an abundance of good grass. Cottonwood and willows are scattered over it, sometimes thickening into dense groves.

Settlers

Scattered along down the valley, at distances of five to twenty miles apart, are settlers (station-keepers), most of whom have ever on hand milk, butter and eggs, and the inevitable black bottle; all of which are highly appreciated by the weary traveller. Camp Cady is the half way point between the Colorado and Los Angeles. It was established in 1858 [1860], under General Carleton's command. There is quite a formidable fortification constructed here in the form of a redoubt; it is about 150 feet square, surrounded by a ditch six feet deep and as many wide. It would be a hard place to take if defended by fifty men. Until within a few months troops have been stationed here, whose presence made the road safe from Indians; but since they have been withdrawn, straggling bands of Pah-Utes have hung along the road, making it unsafe for small parties. But General McDowell having been apprised of the state of affairs, sent twenty troop of cavalry to occupy the post, and it is doubtful if we have any more Indian troubles so long as they remain here.

I must make mention of a work done at this post, by Company C, Fourth California Volunteers, during the few months they were stationed here, which consists in the erection of no less than thirty adobe buildings; among them, three of generous dimensions—I judge twenty by forty feet—the others, neat little buildings, of about twelve by fourteen. They are built a few feet apart, in two rows, a street between, and make a really fine appearance. When it is considered that this work was voluntary, and without other reward than the satisfaction of having done something, at no cost to the Government, we may really say that it is highly creditable to the Volunteers. But I am tarrying too long by the wayside.

Soda and Salt

From Camp Cady to Soda Lake, a distance of forty miles, the road is, for the most part, heavy with sand. "Soda Lake" is, as its name indicates, a lake of soda; here it is about six miles in width. Looking out over its glistening whiteness, were it not for the blistering rays of old "Sol," that seem to come down here with intensified fierceness, one might almost imagine himself in the land of sleigh bells.

A spring of water boils out from beneath a marble mountain, sparkling and bright, and looks so inviting to the thirsty stranger, that he is on his knees, and has gulped down a gill of warm water, seasoned with a little salt and more soda, before he has detected the cheat. I should like much to see the piety of some good Dominie put to the test here. If he did not ejaculate something about that hot place that is supposed to have something to do with warming this water, I should be ready to declare that Pluto's claim was small on him indeed.

Leaving Soda Lake, we have thirty-five miles without water. Most travellers travel in the night. But, arrived at Marl Spring, the water is excellent, grass scarce; twenty-two miles on is Rock Springs, water and grass good and abundant.

EDNA BRUSH PERKINS

Edna Brush Perkins (1880–1930) was born to a life of upper-class privilege in Chicago, Illinois. An early suffragist, she boldly set out in 1920 with her friend Charlotte Hannahs Jordan, bound for Death Valley in a Model T Ford on a primitive road, depending only on an early Auto Club map for directions. The women hoped to escape the constraints of civilization and find "the heart of the California desert." The Mojave at that time was considered desolate, inaccessible, and one of the last remnants of the fading American frontier.

In the following excerpt from the book that chronicles her journey, The White Heart of Mojave, *published in 1922, Perkins describes leaving the Los Angeles Basin and entering the desert at the lip of Cajon Pass.*

from *The White Heart of Mojave*

A good road led through the Cajon Pass to Victorville and thence over sand dotted with groves of Joshua palms to Barstow. A Joshua palm is a grotesque tree-yucca which appears wherever the mesas of the Mojave rise to an elevation of a few thousand feet. It becomes twenty feet high in some places and its ungainly arms stick up into the sky. It has long, dark green, pointed leaves ending in sharp thorns like the yucca. It attains to great age and the dead branches, split off from the trunk or lying on the ground, look as though they were covered with matted gray hair. Charlotte and I never liked them much, they seemed like monsters masquerading as trees; but in that first encounter, when we drove through them mile after mile in a desolation broken only by the narrow ribbon of the gravel road, they were distinctly unpleasant and we were glad when we left them behind at Barstow.

There seemed to be a choice of routes from that town so we had an ice cream soda and interviewed the druggist, having discovered that druggists are among the most helpful of citizens. He proved to be an enthusiast about the desert, the first we had met, and we warmed to him. He brought out an album full of kodak-pictures of the Devil's Playground where the sand-dunes roll along before the wind. He grew almost poetic about them, but when we spread out the map and showed him the proposed route to Death Valley he grew grave. He said the road was so seldom traveled that in places it was obliterated. We would surely get lost. Silver Lake, the next town on it, was eighty-seven miles away. There was one ranch on the road but he was not sure any one was living there. He was not even sure we could get accommodations at Silver Lake. Yes, it was a wonderful country; you went over five mountain ridges. He forgot himself and began to describe it glowingly when a tall man who was looking at the magazines interposed with: "Surely, you would not send the ladies that way!"

The two words "get lost" were what deterred us. We felt we could cope with most calamities, but already, coming through the Joshua palms, we had sensed the size and emptiness of Mojave. At least until we were a little better acquainted with the strange land where even the plants seemed weird, we needed the reassurance of a very definite ribbon of road ahead. We decided to go to Randsburg, then to Ballarat and try to get into Death Valley from there. The druggist doubted if we could get into the valley at all. We began to suspect that it might be difficult.

Randsburg, Atolia and Johannesburg are mining towns close together about forty miles north of Barstow. The road there was no such highway as we had been traveling upon; often it was only two ruts among the sagebrush, but it was well enough marked to follow easily. Great sloping mesas spread for miles on either side of the track, rising to rocky crowns. All the big, open, gradually ascending sweeps are called mesas on the Mojave, though they are in no sense table-lands like the true mesas of New Mexico and Arizona. The groves of Joshua palms had disappeared; we were lower down now where only greasewood and sagebrush grew. The unscientific like us, who accept the word "mesa," lump together all the varieties of low prickly brush as sagebrush. The little bushes grew several feet apart on the white, gravelly ground, each little bush by itself. They smoothed out in the distance like a carpet woven of all shades of blue and green. The occasional greasewood, a graceful shrub covered with small dark green leaves, waved in the wind. Unobstructed by trees the mesa seemed endless. We stopped the car to feel the silence that enveloped it. The place was vast and empty as the stretches we had seen from the railroad, and now we found how still they all had been. Then strong, fresh wind pressed steadily against us like a wind at sea.

Atolia was the first town, golden in the setting sun, on the shoulder of a stern, red mountain. Before it a wide valley fell away in whose bottom gleamed the white floor of a dry lake. All the mountain tops were on fire. The three towns were very close together, separated by the shoulder of the red mountain. Randsburg was the largest, whose one street was a steep hill. It had a score of buildings and two or three stores. Johannesburg, just over the crest, had six buildings, among them an adobe hotel and a large garage. All three towns ornamented the map with big black letters. We thought we were approaching cities and found instead little wooden houses set on the sand with the great simplicity of the desert at their doors.

JOHN STEINBECK

*John Steinbeck (1902–1968) is remembered as one of America's most influential
and beloved writers. Born in Salinas, California, he wrote many novels that
chronicled important and tragic aspects of California life in the first half of the
twentieth century, including* The Grapes of Wrath, *for which he won a Pulitzer
Prize in 1940.*

*In 1960, Steinbeck set out to rediscover America, and part of his journey took
him through the California desert. The following excerpt is taken from his memoir
of the trip,* Travels with Charley, *published in 1961.*

from *Travels with Charley*

I bucketed Rocinante out of California by the shortest possible route—one I
knew well from the old days of the 1930s. From Salinas to Los Banos, through
Fresno and Bakersfield, then over the pass and into the Mojave Desert, a burned
and burning desert even this late in the year, its hills like piles of black cinders in
the distance, and the rutted floor sucked dry by the hungry sun. It's easy enough
now, on the high-speed road in a dependable and comfortable car, with stopping
places for shade and every service station vaunting its refrigeration. But I can
remember when we came to it with prayer, listening for trouble in our laboring
old motors, drawing a plume of steam from our boiling radiators. Then the
broken-down wreck by the side of the road was in real trouble unless someone
stopped to offer help. And I have never crossed it without sharing something
with those early families foot-dragging through this terrestrial hell, leaving the
white skeletons of horses and cattle which still mark the way.

The Mojave is a big desert and a frightening one. It's as though nature
tested a man for endurance and constancy to prove whether he was good
enough to get to California. The shimmering dry heat made visions of water
on the flat plain. And even when you drive at high speed, the hills that mark
the boundaries recede before you. Charley, always a dog for water, panted
asthmatically, jarring his whole body with the effort, and a good eight inches
of his tongue hung out flat as a leaf and dripping. I pulled off the road into a
small gulley to give him water from my thirty-gallon tank. But before I let him
drink I poured water all over him and on my hair and shoulders and shirt. The
air is so dry that evaporation makes you feel suddenly cold.

I opened a can of beer from my refrigerator and sat well inside the shade of
Rocinante, looking out at the sun-pounded plain, dotted here and there with
clumps of sagebrush.

About fifty yards away two coyotes stood watching me, their tawny coats blending with sand and sun. I knew that with any quick or suspicious movement of mine they could drift into invisibility. With the most casual slowness I reached down my new rifle from its sling over my bed—the .222 with its bitter little high-speed, long-range stings. Very slowly I brought the rifle up. Perhaps in the shade of my house I was half hidden by the blinding light outside. The little rifle has a beautiful telescope sight with a wide field. The coyotes had not moved.

I got both of them in the field of my telescope, and the glass brought them very close. Their tongues lolled out so that they seemed to smile mockingly. They were favored animals, not starved, but well furred, the golden hair tempered with black guard hairs. Their little lemon-yellow eyes were plainly visible in the glass. I moved the cross hairs to the breast of the right-hand animal, and pushed the safety. My elbows on the table steadied the gun. The cross hairs lay unmoving on the brisket. And then the coyote sat down like a dog and its right rear paw came up to scratch the right shoulder.

My finger was reluctant to touch the trigger. I must be getting very old and my ancient conditioning worn thin. Coyotes are vermin. They steal chickens. They thin the ranks of quail and all other game birds. They must be killed. They are the enemy. My first shot would drop the sitting beast, and the other would whirl to fade away. I might very well pull him down with a running shot because I am a good rifleman.

And I did not fire. My training said, "Shoot!" and my age replied, "There isn't a chicken within thirty miles, and if there are any they aren't my chickens. And this waterless place is not quail country. No, these boys are keeping their figures with kangaroo rats and jackrabbits, and that's vermin eat vermin. Why should I interfere?"

"Kill them," my training said. "Everyone kills them. It's a public service." My finger moved to the trigger. The cross was steady on the breast just below the panting tongue. I could imagine the splash and jar of angry steel, the leap and struggle until the torn heart failed, and then, not too long later, the shadow of a buzzard, and another. By that time I would be long gone—out of the desert and across the Colorado River. And beside the sagebush there would be a naked, eyeless skull, a few picked bones, a spot of black dried blood and a few rags of golden fur.

I guess I'm too old and too lazy to be a good citizen. The second coyote stood sidewise to my rifle. I moved the cross hairs to his shoulder and held steady. There was no question of missing with that rifle at that range. I owned both animals. Their lives were mine. I put the safety on and laid the rifle on the table. Without the telescope they were not so intimately close. The hot blast of light tousled the air to shimmering.

Then I remembered something I heard long ago that I hope is true. It was unwritten law in China, so my informant told me, that when one man saved another's life he became responsible for that life to the end of its existence. For, having interfered with a course of events, the savior could not escape his responsibility. And that has always made good sense to me.

Now I had a token responsibility for two live and healthy coyotes. In the delicate world of relationships, we are tied together for all time. I opened two cans of dog food and left them as a votive.

I have driven through the Southwest many times, and even more often have flown over it—a great and mysterious wasteland, a sun-punished place. It is a mystery, something concealed and waiting. It seems deserted, free of parasitic man, but this is not entirely so. Follow the double line of wheel tracks through sand and rock and you will find a habitation somewhere huddled in a protected place, with a few trees pointing their roots at under-earth water, a patch of starveling corn and squash, and strips of jerky hanging on a string. There is a breed of desert men, not hiding exactly but gone to sanctuary from the sins of confusion.

At night in this waterless air the stars come down just out of reach of your fingers. In such a place lived the hermits of the early church piercing to infinity with unlittered minds. The great concepts of oneness and of majestic order seem always to be born in the desert. The quiet counting of the stars, and observation of their movements, came first from desert places. I have known desert men who chose their places with quiet and slow passion, rejecting the nervousness of a watered world. These men have not changed with the exploding times except to die and be replaced by others like them.

And always there are mysteries in the desert, stories told and retold of secret places in the desert mountains where surviving clans from an older era wait to re-emerge. Usually these groups guard treasures hidden from the waves of conquest, the golden artifacts of an archaic Montezuma, or a mine so rich that its discovery would change the world. If a stranger discovers their existence, he is killed or so absorbed that he is never seen again. These stories have an inevitable pattern untroubled by the question, If none return, how is it known what is there? Oh, it's there all right, but if you find it you will never be found.

And there is another monolithic tale which never changes. Two prospectors in partnership discover a mine of preternatural richness—of gold or diamonds or rubies. They load themselves with samples, as much as they can carry, and they mark the place in their minds by landmarks all around. Then, on the way out to the other world, one dies of thirst and exhaustion, but the other crawls on, discarding most of the treasure he has grown too weak to carry. He comes at last to a settlement, or perhaps is found by other prospecting men. They examine his samples with great excitement. Sometimes in the story the survivor dies after leaving directions with his rescuers, or again he is nursed back to strength. Then a well-equipped party sets out to find the treasure, and it can never be found again. That is the invariable end of the story—it is never found again. I have heard this story many times, and it never changes. There is nourishment in the desert for myth, but myth must somewhere have its roots in reality.

And there are true secrets in the desert. In the war of sun and dryness against living things, life has its secrets of survival. Life, no matter on what level, must be moist or it will disappear. I find most interesting the conspiracy of life in the desert to circumvent the death rays of the all-conquering sun. The beaten earth appears defeated and dead, but it only appears so. A vast and inventive organization of living matter survives by seeming to have lost. The gray and dusty sage wears oily armor to protect its inward small moistness. Some plants engorge themselves with water in the rare rainfall and store it for future use. Animal life wears a hard, dry skin or an outer skeleton to defy the desiccation. And every living thing has developed techniques for finding or creating shade. Small reptiles and rodents burrow or slide below the surface or cling to the shaded side of an outcropping. Movement is slow to preserve energy, and it is a rare animal which can or will defy the sun for long. A rattlesnake will die in an hour of full sun. Some insects of bolder inventiveness have devised personal refrigeration systems. Those animals which must drink moisture get it at second hand—a rabbit from a leaf, a coyote from the blood of a rabbit.

One may look in vain for living creatures in the daytime, but when the sun goes and the night gives consent, a world of creatures awakens and takes up its intricate pattern. Then the hunted come out and the hunters, and hunters of the hunters. The night awakes to buzzing and to cries and barks.

When, very late in the history of our planet, the incredible accident of life occurred, a balance of chemical factors, combined with temperature, in quantities and in kinds so delicate as to be unlikely, all came together in the retort of time and a new thing emerged, soft and helpless and unprotected in the savage world of unlife. Then processes of change and variation took place in the organisms, so that one kind became different from all others. But one ingredient, perhaps the most important of all, is planted in every life form— the factor of survival. No living thing is without it, nor could life exist without this magic formula. Of course, each form developed its own machinery for survival, and some failed and disappeared while others peopled the earth. The first life might easily have been snuffed out and the accident may never have happened again—but, once it existed, its first quality, its duty, preoccupation, direction, and end, shared by every living thing, is to go on living. And so it does and so it will until some other accident cancels it. And the desert, the dry and sun-lashed desert, is a good school in which to observe the cleverness and the infinite variety of techniques of survival under pitiless opposition. Life could not change the sun or water the desert, so it changed itself.

The desert, being an unwanted place, might well be the last stand of life against unlife. For in the rich and moist and wanted areas of the world, life pyramids against itself and in its confusion has finally allied itself with the enemy non-life. And what the scorching, searing, freezing, poisoning weapons of non-life have failed to do may be accomplished to the end of its destruction and extinction by the tactics of survival gone sour. If the most versatile of living forms, the human, now fights for survival as it always has, it can eliminate not only itself but all other life. And if that should transpire, unwanted places like the desert might be the harsh mother of repopulation. For the inhabitants of the desert are well trained and well armed against desolation. Even our own misguided species might re-emerge from the desert. The lone man and his sun-toughened wife who cling to the shade in an unfruitful and uncoveted place might, with their brothers in arms—the coyote, the jackrabbit, the horned toad, the rattlesnake, together with a host of armored insects—these trained and tested fragments of life might well be the last hope of life against non-life. The desert has mothered magic things before this.

VICTOR VILLASEÑOR

Victor Villaseñor was born in 1940 in Carlsbad, California. Among his many works is Burro Genius: A Memoir *(2004), which was nominated for a Pulitzer Prize; the best-seller* Rain of Gold *(1991); and* Thirteen Senses: A Memoir *(2001), which chronicles his family's history for the past century, beginning with their immigration from Mexico to southern California after the Mexican Revolution of 1910 and continuing into the present. Villaseñor also wrote the screenplay for the movie* The Ballad of Gregorio Cortez, *produced in 1982, and more recently he served as the first Steinbeck Chair at Hartnell College and at the National Steinbeck Center in Salinas, California, from 2003-2004.*

The following excerpt is from Villaseñor's first novel, Macho!, *published in 1973. Set in the turbulent 1960s, the story follows the life of seventeen-year-old Roberto, a young man from Mexico, and his illegal crossing into the United States. Roberto and a small group of men face many perils as they travel across the sparsely inhabited and brutal desert between the Mexico–U.S. border and Palm Springs, California.*

from *Macho!*

He too fell down. Exhausted. He was terrified. He swallowed and remained still. They, a *camarilla* of twenty-some men, were at the border. It was midnight, and they were by the American canal about seven miles west of Mexicali. A U.S. Border Patrol car had just driven by, and they were belly-down on the bank. They waited. Hearts pounding against the earth, and then, on the word of Aguilar, they jumped up and began their slow lope along the canal. There were twenty-some men, and most of these Roberto did not know.

A few miles farther into California they came to a road. Highway 98. They left the canal and went across the flat fields of sandy dirt, climbing over fences of barbed wire. They saw lights coming. One man panicked and yelled as he ripped his leg on the barbed wire. Another man hit him.

All was silent.

They waited. Hearts against the earth. The lights passed by. They got up and began running. The man with the cut leg began dragging behind. No one waited for him. They crossed the highway and came to another fence. One wire. There were cattle in the field. One man touched the wire. He screamed. This wire was electric. Suddenly lights came on from over there, and a bullhorn yelled in English, then in Spanish, to stop! Men scattered, running every which way. Juan gripped Roberto and they went back toward Mexico. A few others followed them. The patrol ran after the ones who raced toward the north. Juan Aguilar turned, going east, and ran with all his might, and Roberto was right beside him. Later they, a *camarilla* of about fifteen men, turned north once more, and they ran until they fell down gasping.

Juan coughed and coughed, and then got up and said, "Boy, now your power of youth starts paying its way. Give me a hand, and let's go!"

Roberto, tired and sweaty, jumped up and gripped Juan by the belt and began running.

"Faster!" said Juan Aguilar. "Pull faster and never stop until my legs fold or I drop."

Roberto gripped and pulled, and they went on. Mile after mile. There were only ten others with them now, and they were all blowing hard like dogs. Suddenly Roberto gasped and choked and fell down spitting and heaving.

"What is it?" asked Juan between his own heaves for air.

"I swallowed! I swallowed!" And he pointed to this great mass of insects which they had stirred up. "I swallowed a mouth full of 'em."

"Oh." And Juan took out their plastic milk bottle. It was filled with water. "Drink! Wash and spit out these damn American bugs. Quickly! They're probably sprayed with poison." Roberto drank and coughed. "Here, one more swallow and...okay. Okay! No more. Let's go!"

They were off once more at a run, and like that they ran and walked most of the night. Through bugs and insects and over electric fences and across fields of produce. Then up ahead through an area of granite and brush, and worst of all, little slopes of deep barren sand. They went on. Escaping into the land of plenty. They came to a freeway. Such a road that Roberto stopped, forgot he was hiding, and just stood upright and looked in wonder. Such a fantastic sight. And all man made. Juan grabbed him, and they crossed and went on and on, and then Juan said, "Okay, see those lights over there? That mountain of lights? That's the plant where they make plaster. So we go this way..." He pointed northwest. "I know a place where we can get more water." He turned to Roberto. "How much water we got left?"

"Not much."

"Well, come on. We must go far before daybreak."

There were just ten of them in total now. More had fallen. And so they, the able, now headed for a small light in the distance, which Aguilar said he knew was a farmhouse with water, and only a few miles away. For hours and hours they traveled. Up the rich Imperial Valley at the southeast end of California. And the light in the distance never seemed to get any closer.

Finally one man, a stranger, said, "Look...I don't think you know where we're going. I think that light might be forty miles away."

"Then don't follow. Go where you damn please," said Aguilar, and he went on toward the light in the distance.

Daylight found them no closer to the distant light. They were in a field of produce, and an airplane was coming. It was coming low and loud, so they hit the ground, chest to the good earth between the rows of green, and there they lay in the misty dawn of the first light of day. Hearts pounding and nostrils raging as the plane came down in a sweep and sprayed a blue-green cloud of chemicals all about them.

Juan Aguilar began coughing. Quickly he covered his mouth and nose with his shirt sleeve, and poked Roberto to do the same, and the plane swept upward in a huge roar of power and then circled and came by two more times, and each time it swept up and the pilot couldn't see them, they would get up and run. Leaping over rows of green, and then hitting the good earth, trying to rest as the plane came down again with its deadly spray of man-made chemicals. They breathed. Deeply. Then raced on, escaping farther into the land of plenty as the plane clouded the dawn with spray.

By sunrise they were coming out of the fields of produce and going into dry desert. They came across an irrigation pump. It had a big tank on posts some ten feet high, and in a pile on the ground were huge empty bags; and water, or what seemed to be water, was dripping down one side of the tank. The stranger who had earlier talked back to Aguilar now felt this dripping with his fingertips and then smelled it and looked at the pile of empty bags and read their labels and said, "No, don't drink. It's bad."

"You can read English?" asked Aguilar.

"A little," said the stranger. His name was Luis.

"Oh," said Aguilar, and nodded. "Well, then, I'll just rinse my mouth a little."

"Don't," said Luis. He was tall and thin, and his face was all covered with pock marks. He had a big moustache and large open eyes.

"What are you saying to me!" said Aguilar. "No one tells me what to do." He began gathering the dripping liquid in his hand.

"I wouldn't fool with it," said Luis. "It may be poison."

"It may not be. Hell! I'm thirsty! I'll just rinse out. It can do me no harm if I don't drink."

Roberto went to join Aguilar, who was catching the dripping water, sipping it, slapping it about his face, and refreshing himself. The stranger gripped Roberto.

"No, don't. Believe me."

Roberto stopped, and he and Luis watched as the other men freshened themselves, and then they all went on. They went out of the cultivated land and into the barren desert of sand and scattered brush. Soon the heat began to sing and dance with glistening bright nothingness.

The flat sand would reflect heat waves of dancing whiteness. The insects would sing in eery outer-space screeches. The sun rose higher and higher, and soon the insects screeching and the heat waves dancing all began numbing the eyes and ears, until all was not here. But over there. Up ahead. And here, all around, was nothing. *Nada. Nada. Nada. Nada.* And Roberto instinctively began to pray memorized prayers as he walked on, numb. All alone. Separated from all men, all life, he felt his human awareness evaporating into the huge infinite nothingness of all sealess deserts since before recorded time. He continued to mumble prayers as they walked on. It was only mid-morning, and men were beginning to stagger. They walked over to a pile of rocks. They kicked some rattlesnakes out of the shade and lay down like dogs. Panting and hurting and trying to gain moisture from deep within their throats. The men who had drunk from the tank had a little whiteness about their lips. Some began to babble nada words. The others all rested. Not saying one word. Roberto, being strong and young, was able to go right off to sleep and truly gain rest for both body and mind, and he dreamed. He was killing and robbing and gathering money, but his father was drunk and begging a drink from a man in a fancy shirt.

He awoke. His tongue felt dry and coarse as sandpaper. He glanced about. He saw Pedro. And he felt good that Pedro was still along, and Roberto's heart uplifted with the good feeling of revenge. He stood up. Aguilar saw and smiled.

"Well, one of us is still strong. Come. Let's go a little farther." Aguilar was trying to act strong, but he looked sick. "Find a place with better shade."

"No," said one man. "I stay here. My head. My stomach. Oh, God."

"I told you not to drink," said Luis, getting up.

"Telling us of the past does not help. Why don't you go back there to the cultivated fields and bring us help! Oh, for the love of God, are we not all brothers?"

All the men who had drunk were yellowish-looking, but only two were truly complaining and getting sick in the head.

"No help from me," said Juan Aguilar. "I drank, but I don't cry. Stay and die if you're so cowardly."

"My friend?" said one sick man. "We're all of *la Raza*. Please, don't leave us."

"All my friends died," said Aguilar, and he began getting up, but he groaned with pain.

Luis shook his head. "No, I don't think any of us are going to make it, especially not you people who drank."

"Bullshit!" yelled Aguilar. He was sick, but he still had gut concentration. You could see it in his eyes. "Me and my boy are going on." He got up. "Roberto, give me a hand."

Roberto obeyed.

Luis glanced around. "Well, at least we got to try."

Some men yelled not to leave them. Others began scrambling to get up and give it a try. Aguilar leaned on Roberto, and they began. Men cried out in fear of being left.

Luis stopped and said, "Don't be fools. Stay here in this shade, and I promise you, if I make it out, I'll send you people help."

Luis turned, going after Roberto and Aguilar and Pedro, and little by little, the cries of the ones left behind began to blend into the great numbing silence of the eery screeching bugs and insects, and then were gone. Not here. But over there. Ahead. Everywhere. And in the next few miles two others fell, but still they, the strong, walked on. Into the dancing, singing, eery bright nothingness of *nada nada* desert.

Near the Salton Sea on Highway 78 there is a restaurant at Ocotillo Wells with a historical landmarker, and the landmarker says that six skeletons were found there. They were Mexicans. No doubt trying to make it across the desert border to get work in the land of plenty.

Also, in a two-month period in the late 1960's, seventeen dead were found in Southern California by the Border Patrol and local law enforcements. And one old border patrolman said, "That's nothing. Imagine how many are never found."

ALDOUS HUXLEY

Aldous Huxley (1894–1963) was born in England and spent the latter part of his life in the United States. He was the author of the landmark dystopian novel Brave New World *(1932), the apocalyptic* Ape and Essence *(1948), the optimistic* Island, *and the psychedelic exploration* The Doors of Perception *(1954). Toward the end of his life, Huxley spent much time in his small house in Llano, in the Mojave Desert, devoting hours to contemplation and mysticism and experimenting with hallucinogenic drugs.*

The following excerpt is from the essay "Tomorrow and Tomorrow and Tomorrow," published in 1956.

from **"Tomorrow and Tomorrow and Tomorrow"**

The facts of silence and emptiness are traditionally the symbols of divine immanence—but not, of course, for everyone, and not in all circumstances. "Until one has crossed a barren desert, without food or water, under a burning tropical sun, at three miles an hour, one can form no conception of what misery is." These are the words of a gold seeker, who took the southern route to California in 1849. Even when one is crossing it at seventy miles an hour on a four-lane highway, the desert can seem formidable enough. To the forty-niners it was unmitigated hell. Men and women who are at her mercy find it hard to see in Nature and her works any symbols but those of brute power at the best and, at the worst, of an obscure and mindless malice. The desert's emptiness and the desert's silence reveal what we may call their spiritual meanings only to those who enjoy some measure of physiological security. The security may amount to no more than St. Anthony's hut and daily ration of bread and vegetables, no more than Milarepa's cave and barley meal and boiled nettles—less than what any sane economist would regard as the indispensable minimum, but still security, still a guarantee of organic life and, along with life, of the possibility of spiritual liberty and transcendental happiness.

But even for those who enjoy security against the assaults of the environment, the desert does not always or inevitably reveal its spiritual meanings. The early Christian hermits retired to the Thebaid because its air was purer, because there were fewer distractions, because God seemed nearer there than in the world of men. But, alas, dry places are notoriously the abode of unclean spirits, seeking rest and finding it not. If the immanence of God was sometimes more easily discoverable in the desert, so also, and all too frequently, was the immanence of the devil. St. Anthony's temptations have become a legend, and Cassian speaks of "the tempests of imagination" through which every newcomer to the eremitic life had to pass. Solitude, he writes, makes men feel "the many-winged folly of their souls[...]; they find the perpetual silence intolerable, and those whom no labor on the land could weary, are vanquished by doing nothing and worn out by the long duration of their peace." *Be still, and know that I am God;* be still, and know that *you* are the delinquent imbecile who snarls and gibbers in the basement of every human mind. The desert can drive men mad, but it can also help them to become supremely sane.

The enormous drafts of emptiness and silence prescribed by the eremites are safe medicine only for a few exceptional souls. By the majority the desert should be taken either dilute or, if at full strength, in small doses. Used in this way, it acts as a spiritual restorative, as an anti-hallucinant, as a de-tensioner and alterative.

In his book *The Next Million Years*, Sir Charles Darwin looks forward to thirty thousand generations of ever more humans pressing ever more heavily on ever dwindling resources and being killed off in ever increasing numbers by famine, pestilence and war. He may be right. Alternatively, human ingenuity may somehow falsify his predictions. But even human ingenuity will find it hard to circumvent arithmetic. On a planet of limited area, the more people there are, the less vacant space there is bound to be. Over and above the material and sociological problems of increasing population, there is a serious psychological problem. In a completely homemade environment, such as is provided by any great metropolis, it is as hard to remain sane as it is in a completely natural environment such as the desert or the forest. O Solitude, where are thy charms? But, O Multitude, where are *thine?* The most wonderful thing about America is that, even in these middle years of the twentieth century, there are so few Americans. By taking a certain amount of trouble you might still be able to get yourself eaten by a bear in the state of New York. And without any trouble at all you can get bitten by a rattler in the Hollywood hills, or die of thirst, while wandering through an uninhabited desert, within a hundred and fifty miles of Los Angeles. A short generation ago you might have wandered and died within only a hundred miles of Los Angeles. Today the mounting tide of humanity has oozed through the intervening canyons and spilled out into the wide Mojave. Solitude is receding at the rate of four and a half kilometers per annum.

And yet, in spite of it all, the silence persists. For this silence of the desert is such that casual sounds, and even the systematic noise of civilization, cannot abolish it. They coexist with it—as small irrelevances at right angles to an enormous meaning, as veins of something analogous to darkness within an enduring transparency. From the irrigated land come the dark gross sounds of lowing cattle, and above them the plovers trail their vanishing threads of shrillness. Suddenly, startlingly, out of the sleeping sagebrush there bursts the shrieking of coyotes—Trio for Ghoul and Two Damned Souls. On the trunks of cottonwood trees, on the wooden walls of barns and houses, the woodpeckers rattle away like pneumatic drills. Picking one's way between the cactuses and the creosote bushes one hears, like some tiny whirring clockwork, the soliloquies of invisible wrens, the calling, at dusk, of the nightjays and even occasionally the voice of Homo sapiens—six of the species in a parked Chevrolet, listening to the broadcast of a prize fight, or else in pairs necking to the delicious accompaniment of Crosby. But the light forgives, the distances forget, and this great crystal of silence, whose base is as large as Europe and whose height, for all practical purposes, is infinite, can coexist with things of a far higher order of discrepancy than canned sentiment or vicarious sport. Jet planes, for example—the stillness is so massive that it can absorb even jet planes. The screaming crash mounts to its intolerable climax and fades again, mounts as another of the monsters rips through the air, and once more diminishes and is gone. But even at the height of the outrage the mind can still remain aware of that which surrounds it, that which preceded and will outlast it.

HUNTER S. THOMPSON

Hunter S. Thompson was born in 1937 in Louisville, Kentucky. A journalist and author, he is credited with innovating the popular, renegade style of news reporting known as Gonzo journalism, and he is best remembered for his book Fear and Loathing in Las Vegas, *published in 1971. He died at his home in Woody Creek, Colorado, in 2005.*

The following excerpt, taken from Fear and Loathing in Las Vegas, *is an infamous and partially fictionalized account of Thompson's drug-addled, mind-boggling drive along remote Interstate 15 from Los Angeles to Las Vegas on his way to report on a motorcycle race for a New York sports magazine.*

from *Fear and Loathing in Las Vegas*

We were somewhere around Barstow on the edge of the desert when the drugs began to take hold. I remember saying something like "I feel a bit lightheaded; maybe you should drive...." And suddenly there was a terrible roar all around us and the sky was full of what looked like huge bats, all swooping and screeching and diving around the car, which was going about a hundred miles an hour with the top down to Las Vegas. And a voice was screaming: "Holy Jesus! What are these goddamn animals?"

Then it was quiet again. My attorney had taken his shirt off and was pouring beer on his chest, to facilitate the tanning process. "What the hell are you yelling about?" he muttered, staring up at the sun with his eyes closed and covered with wraparound Spanish sunglasses. "Never mind," I said. "It's your turn to drive." I hit the brakes and aimed the Great Red Shark toward the shoulder of the highway. No point mentioning those bats, I thought. The poor bastard will see them soon enough.

It was almost noon, and we still had more than a hundred miles to go. They would be tough miles. Very soon, I knew, we would both be completely twisted. But there was no going back, and no time to rest. We would have to ride it out. Press registration for the fabulous Mint 400 was already underway, and we had to get there by four to claim our sound-proof suite. A fashionable sporting magazine in New York had taken care of the reservations, along with this huge red Chevy convertible we'd just rented off a lot on the Sunset Strip... and I was, after all, a professional journalist; so I had an obligation to *cover the story,* for good or ill.

The sporting editors had also given me $300 in cash, most of which was already spent on extremely dangerous drugs. The trunk of the car looked like a mobile police narcotics lab. We had two bags of grass, seventy-five pellets of mescaline, five sheets of high-powered blotter acid, a salt shaker half full of cocaine, and a whole galaxy of multi-colored uppers, downers, screamers, laughers…and also a quart of tequila, a quart of rum, a case of Budweiser, a pint of raw ether and two dozen amyls.

All this had been rounded up the night before, in a frenzy of high-speed driving all over Los Angeles County—from Topanga to Watts, we picked up everything we could get our hands on. Not that we *needed* all that for the trip, but once you get locked into a serious drug collection, the tendency is to push it as far as you can.

The only thing that really worried me was the ether. There is nothing in the world more helpless and irresponsible and depraved than a man in the depths of an ether binge. And I knew we'd get into that rotten stuff pretty soon. Probably at the next gas station. We had sampled almost everything else, and now—yes, it was time for a long snort of ether. And then do the next hundred miles in a horrible, slobbering sort of spastic stupor. The only way to keep alert on ether is to do up a lot of amyls—not all at once, but steadily, just enough to maintain the focus at ninety miles an hour through Barstow.

"Man, this is the way to travel," said my attorney. He leaned over to turn the volume up on the radio, humming along with the rhythm section and kind of moaning the words: "One toke over the line, Sweet Jesus…One toke over the line…"

One toke? You poor fool! Wait till you see those goddamn bats. I could barely hear the radio…slumped over on the far side of the seat, grappling with a tape recorder turned all the way up on "Sympathy for the Devil." That was the only tape we had, so we played it constantly, over and over, as a kind of demented counterpoint to the radio. And also to maintain our rhythm on the road. A constant speed is good for gas mileage—and for some reason that seemed important at the time. Indeed. On a trip like this one *must* be careful about gas consumption. Avoid those quick bursts of acceleration that drag blood to the back of the brain.

My attorney saw the hitchhiker long before I did. "Let's give this boy a lift," he said, and before I could mount any argument he was stopped and this poor Okie kid was running up to the car with a big grin on his face, saying, "Hot damn! I never rode in a convertible before!"

"Is that right?" I said. "Well, I guess you're about ready, eh?"

The kid nodded eagerly as we roared off.

"We're your friends," said my attorney. "We're not like the others."

O Christ, I thought, he's gone around the bend. "No more of that talk," I said sharply. "Or I'll put the leeches on you." He grinned, seeming to understand.

Luckily, the noise in the car was so awful—between the wind and the radio and the tape machine—that the kid in the back seat couldn't hear a word we were saying. Or could he?

How long can we *maintain?* I wondered. How long before one of us starts raving and jabbering at this boy? What will he think then? This same lonely desert was the last known home of the Manson family. Will he make that grim connection when my attorney starts screaming about bats and huge manta rays coming down on the car? If so—well, we'll just have to cut his head off and bury him somewhere. Because it goes without saying that we can't turn him loose. He'll report us at once to some kind of outback nazi law enforcement agency, and they'll run us down like dogs.

Jesus! Did I *say* that? Or just think it? Was I talking? Did they hear me? I glanced over at my attorney, but he seemed oblivious—watching the road, driving our Great Red Shark along at a hundred and ten or so. There was no sound from the back seat.

Maybe I'd better have a chat with this boy, I thought. Perhaps if I *explain* things, he'll rest easy.

Of course. I leaned around in the seat and gave him a fine big smile... admiring the shape of his skull.

"By the way," I said. "There's one thing you should probably understand."

He stared at me, not blinking. Was he gritting his teeth?

"Can you *hear* me?" I yelled.

He nodded.

"That's good," I said. "Because I want you to know that we're on our way to Las Vegas to find the American Dream." I smiled. "That's why we rented this car. It was the only way to do it. Can you grasp that?"

He nodded again, but his eyes were nervous.

"I want you to have all the background," I said. "Because this is a very ominous assignment—with overtones of extreme personal danger....Hell, I forgot all about this beer; you want one?"

He shook his head.

"How about some ether?" I said.

"What?"

"Never mind. Let's get right to the heart of this thing. You see, about twenty-four hours ago we were sitting in the Polo Lounge of the Beverly Hills Hotel—in the patio section, of course—and we were just sitting there under a palm tree when this uniformed dwarf came up to me with a pink telephone and said, 'This must be the call you've been waiting for all this time, sir.'"

I laughed and ripped open a beer can that foamed all over the back seat while I kept talking. "And you know? He was right! I'd been *expecting* that call, but I didn't know who it would come from. Do you follow me?"

The boy's face was a mask of pure fear and bewilderment.

I blundered on: "I want you to understand that this man at the wheel is my *attorney!* He's not just some dingbat I found on the Strip. Shit, *look* at him! He doesn't look like you or me, right? That's because he's a foreigner. I think he's probably Samoan. But it doesn't matter, does it? Are you prejudiced?"

"Oh, hell *no!*" he blurted.

"I didn't think so," I said. "Because in spite of his race, this man is extremely valuable to me." I glanced over at my attorney, but his mind was somewhere else.

I whacked the back of the driver's seat with my fist. "This is *important,* goddamnit! This is a *true story!*" The car swerved sickeningly, then straightened out. "Keep your hands off my fucking neck!" my attorney screamed. The kid in the back looked like he was ready to jump right out of the car and take his chances.

Our vibrations were getting nasty—but why? I was puzzled, frustrated. Was there no communication in this car? Had we deteriorated to the level of *dumb beasts?*

Because my story *was* true. I was certain of that. And it was extremely important, I felt, for the *meaning* of our journey to be made absolutely clear. We had actually been sitting there in the Polo Lounge—for many hours—drinking Singapore Slings with mescal on the side and beer chasers. And when the call came, I was ready.

The Dwark approached our table cautiously, as I recall, and when he handed me the pink telephone I said nothing, merely listened. And then I hung up, turning to face my attorney. "That was headquarters," I said. "They want me to go to Las Vegas at once, and make contact with a Portuguese photographer named Lacerda. He'll have the details. All I have to do is check into my suite and he'll seek me out."

My attorney said nothing for a moment, then he suddenly came alive in his chair. "God *hell!*" he exclaimed. "I think I see the *pattern.* This one sounds like real trouble!" He tucked his khaki undershirt into his white rayon bellbottoms and called for more drink. "You're going to need plenty of legal advice before this thing is over," he said. "And my first advice is that you should rent a very fast car with no top and get the hell out of L.A. for at least forty-eight hours." He shook his head sadly. "This blows my weekend, because naturally I'll have to go with you—and we'll have to arm ourselves."

"Why not?" I said. "If a thing like this is worth doing at all, it's worth doing right. We'll need some decent equipment and plenty of cash on the line—if only for drugs and a super-sensitive tape recorder, for the sake of a permanent record."

"What kind of a story is this?" he asked.

"The Mint 400," I said. "It's the richest off-the-road race for motorcycles and dune-buggies in the history of organized sport—a fantastic spectacle in honor of some fatback *grossero* named Del Webb, who owns the luxurious Mint Hotel in the heart of downtown Las Vegas...at least that's what the press release says; my man in New York just read it to me."

"Well," he said, "as your attorney I advise you to buy a motorcycle. How else can you cover a thing like this righteously?"

"No way," I said. "Where can we get hold of a Vincent Black Shadow?"

"What's that?"

"A fantastic bike," I said. "The new model is something like two thousand cubic inches, developing two hundred brake-horsepower at four thousand revolutions per minute on a magnesium frame with two styrofoam seats and a total curb weight of exactly two hundred pounds."

"That sounds about right for this gig," he said.

"It is," I assured him. "The fucker's not much for turning, but it's pure hell on the straightaway. It'll outrun the F-111 until takeoff."

"Takeoff?" he said. "Can we handle that much torque?"

"Absolutely," I said. "I'll call New York for some cash."

JEANETTE CLOUGH

Jeanette Clough was born in Paterson, New Jersey. She received a master's degree from the University of Chicago and currently works for the Getty Research Institute in Los Angeles. Her poetry collections include Island *(2007),* Cantatas *(2002), and two limited editions,* Celestial Burn *and* Dividing Paradise. *Along with poet Jim Natal, she has taught the Plein Air poetry workshop in Joshua Tree National Park since 1999.*

Clough's poem "Leaving Palm Springs" emerged from an early evening drive on Indian Avenue through the wind farms to Twentynine Palms and Joshua Tree National Park.

Leaving Palm Springs

Hot blasts lift a scarf of dust,
and fault zones shift their ribs.

The ground sighs and rolls over,
restless as a desert saint
whose ridges and misalignments
are scourged bare.

At these bends in the road the earth's spine
slings back on itself like shoelaces

then with powdery softness goes forward, twined.

I like the look of the windmill bones
backlit by the gasoline sun.
All skeleton and no meat. My elbow

juts over the rolled-down car window,
knobbing at an important hinge.

Another knobbed bone at the wrist
makes the delicate transition to hand.

My cascading bones stay in the dark
doing their linear work, then splay at the end

into five stalks where one licked finger,
surely as thread pulled through a needle's eye,
reports the direction of every wind.

My staccato thoughts gap.
Again I fail the test of coherence
by assuming sense is a chain link fence
or that it buttons to the neck. Instead

a Hindu god's windmill arms circle completely
like my elbow fancies to proceed, whirling.

The desert's origami blue
folds around itself,
and matter is brought to the threshold.

The truth of windmills allows space
between the shafts. Solidity is illusion

viewed from a long way off.

The distance between this and the wind farm
increases. Take in the scenery, ignore the gloss.

Tonight when constellations rotate overhead
I'll connect their star-dots and ride them across the sky.

The stars perceive my weight, turn their heads
and nip. *Wake up. It isn't night at all.* Only dusk,

and I'm driving across white sand
that reflects windmills brightly
as if they stood in bright water. Mirage

is a desire so strong
I create out of thin air what is not.

Here is emptiness. My bones and yours
are disposable, if ever we were here.

REFUGE AND EXILE

DOROTHY RAMON and ERIC ELLIOTT

Dorothy Ramon was the last person for whom Serrano, a California Indian language, was a first language. Born in 1909 on the Morongo Indian Reservation, Ramon was in her eighties when she began working with linguist Eric Elliott to record stories of her life and the Serrano culture. She died in 2002. The result of Ramon and Elliott's collaboration is the two-volume bilingual collection Wayta' Yawa': "Always Believe," *published by Malki Museum Press in 2000. Elliott has coauthored similar works with Luiseño speaker Villiana Hyde and Cahuilla speaker Katherine Siva Sauvel; he has also revised translations of Cupeño texts.*

"Massacre at Desert Hot Springs" is an excerpt from Wayta' Yawa': "Always Believe." *"Problems with Windmills," a passage from Elliott's collaboration with Katherine Siva Sauvel, can be found in the "Changing Desert" section of this anthology.*

Massacre at Desert Hot Springs

'Aam taaqtam kiikam xhayp: 'uviht 'ipyu' nenyam qatt Maarrênga'yam.
I don't know what tribe (it was): long ago my Serrano ancestors were from there.

'Aam tewa'nkin 'amay terva'tti' *Paatt Wirrarrka'*.
They call that land *Paatt Wirrarrka'* (which means 'Swirling Water').

Xhinyim Wanepuhpa'yam warre' kiti' huwa'i' peewerrafka'.
They were *Wánikik* Pass Cahuilla but they spoke a little differently.

Qayn 'enan peekichawani'. 'Uvihtim puuyu' chewp. Kwenemu' qatt 'at.
I don't know what tribe they were. They all disappeared long ago. They lived there.

Peenyu' tervatt xhiit Peenepvu' 'amaqa' kwenevu': 'animu' 'ap qatte'.
Their land was given to them by their Lord: they lived there.

'Anivu' 'ap qatt paatt waha'. 'Ercert, 'ami' 'ecet waha'.
There was water there. Hot water, and cold water too.

'Apyu' wangatk paatt tervanu'. Weerv qatte'.
Water comes out of the ground there. It comes out at two places.

91

Kwan paatt paca'xq 'apyu' paatt 'ercert. 'Ipyu' huwatt kwan paca'xq paatt 'ecet.
People say hot water spews from there. On the other side cold water flows out.

'Api'a' kwenemu' qatt taaqtam Wanepuhpa'yam.
That's where the Pass Cahuilla lived.

Peenyuf kwenemu' 'ap hawayt qatt tervav. Kwenemu' Taaqtam 'uvihanu' kiikam 'ap.
They always lived there on their land. The Indians had lived there all their lives.

'Ap kwenemu' qatte' tervav mutu' 'amayv. 'Aam 'animu' qatte'. Peenyu' tervatt 'ayee'.
They lived there since the world began. They lived there still. It was just their land.

Xhinyim kwenemu' wêan 'angkwa' xhinyim parckwam, xhinyim yarra'nkam parckwam.
They sent some others there, some soldiers, some white soldiers.

Kwenemu' wêan 'angkwa' mita' hinyim haym.
Some people sent them there.

Peenyu' *government*: hiit hami' kwenevu' wêan.
Their government, or whatever, sent them.

Hiit hami' kwenevu' wêan 'angkwa' mita' hamin nyaawnk.
Someone, whoever it was, sent them there, for some reason.

Kwenemu' puuyu' qeerncu': 'angkwa' kwenemu' muum.
They went there to kill them all: they shot them.

'Aam Taaqtam 'ap kiikam kwenemu' qaymu' wi'wan peerqriktti'.
The Indians residing there did not wish to die.

Kwenemu' pemeka' heterheterk parckwam peepamkw.
They knelt down before the soldiers.

Kwenemu' pemeka' werra'n 'enaac, "Qaycheme' qeern," key kwenemu'.
They beseeched them, saying, "Please don't kill us."
Warrêngk kwenemu' puuyu' qeern. Puuyu' kwenemu' 'empkin 'angkwa'.
But they killed them all. They wiped them all out there.

'Apyu' kwenevu' 'ayee' tervatt pana' 'erner'erk. Qayvu' hami' 'ap qatt.
The land was just lying flat on its back. There was no one living there then.

'Amay' chevêk wuuwerham 'ap qatt. 'Amêrrikaanu'yam puuyu' qatt.
Today there are many living there. They are all Americans.

'Amay'm paatti' 'ercer'ti' mutu' yaanam Xay'ku'yam.
The white people still own the hot springs.

'Ani' 'atewan 'amay' *Desert Hot Springs*, keym 'Amêrrikaanu'yam, Xay'ku'yam
xhinyim.
Its name is now 'Desert Hot Springs', as the Americans, the white people, say.

'Ani' pana' nyaawnk Taaqtam huwam chevêk txamu' xhaynkwa' hwinhwink.
But some Indians escaped.

'Ipvu' qatt mutu' neert.
There was one woman left (who died in the 1940's).

Mu' 'ip qatt: 'ip mutu' niihtavec 'ayee': 'ama' 'ana' waha'vu' qatt wehtivec.
She survived: an old woman and her father survived.

'Aninu' hihii: 'uvya'vu' 'atuchini' wehtivec. 'Atewanvu' *Tomás (Cisco)*.
He was already an old man when I saw him. His name was *Tomás (Cisco)*.

'Icham tewa'nkin 'atewani' Chiinarrup. *Tomás* keychemu'. 'Ama'vu' qatt.
We called him by his Spanish name. Tomás we called him. He survived.

'Ama'vu' 'ana' 'ama' 'apyuch, 'ama' kwenevu' 'apyuch 'ama' *Paatt
Wirrarrkanu'*.
His father was apparently from there, from Desert Hot Springs.

'Ama' wehtivec 'ani' kwenevu' hunuk 'ingkwa'. 'Ipvu' meme'.
And then the old man moved over here. He died here (at Morongo).

'Ani' 'ashuung 'ip meme' waha'. 'Atewanvu' *Polly (Cisco)*.
Then his daughter died here as well. Her name was Polly *(Cisco)*.

'Ani' na'u' wecerc peyika': wêher' kwenevu' na'u'.
She got married twice.

'Ama' kwenevu' huwatt 'awercehaf qatt. 'Ani' pana' 'empkcu' 'ip.
She had a second husband. And that's how it ended here.

'Ap kiikam: kwenemu' terva'tti' tewa'nkin 'amay. *Paatt Wirrarrka'* keym.
The people that lived there named it that. They call it 'Swirling Water'.

'Ama'vu' 'atewan 'ama'. 'Ani' kwenemu' 'ap qatt Wanepuhpa'yam xhinyim.
That was its name. The Pass Cahuilla were the ones who lived there.

'Ama' 'ayee'.
That's all.

FRANK NORRIS

Frank Norris was born in 1870 in Chicago, Illinois, and moved to San Francisco as a young man. His books include The Octopus *(1901),* The Pit *(1902), and* McTeague, *published in 1899 and made into the film* Greed *in 1924. Reflecting the author's idealization of the socialist/progressive movement in the United States in the early twentieth century,* McTeague *is a criticism of capitalism, and illustrates its pitfalls. Norris died in San Francisco in 1902.*

McTeague *is the story of an urban dentist at the bottom of his life's downward spiral, and the excerpt that follows is his final life-and-death struggle in Death Valley, California, where he fled following a murder in San Francisco.*

from *McTeague*

To the Bitter End

Far behind [McTeague], the Panamint hills were already but blue hummocks on the horizon. Before him and upon either side, to the north and to the east and to the south, stretched primordial desolation. League upon league the infinite reaches of dazzling white alkali laid themselves out like an immeasurable scroll unrolled from horizon to horizon; not a bush, not a twig relieved that horrible monotony.

Even the sand of the desert would have been a welcome sight; a single clump of sagebrush would have fascinated the eye; but this was worse than the desert. It was abominable, this hideous sink of alkali, this bed of some primeval lake lying so far below the level of the ocean. The great mountains of Placer County had been merely indifferent to man; but this awful sink of alkali was openly and unreservedly iniquitous and malignant.

McTeague had told himself that the heat upon the lower slopes of the Panamint had been dreadful; here in Death Valley it became a thing of terror. There was no longer any shadow but his own. He was scorched and parched from head to heel. It seemed to him that the smart of his tortured body could not have been keener if he had been flayed.

"If it gets much hotter—" he muttered, wringing the sweat from his thick fell of hair and mustache, "if it gets much hotter, I don' know what I'll do." He was thirsty, and drank a little from his canteen. "I ain't got any too much water," he murmured, shaking the canteen. "I got to get out of this place in a hurry, sure."

By eleven o'clock the heat had increased to such an extent that McTeague could feel the burning of the ground come prickling and stinging through the soles of his boots. Every step he took threw up clouds of impalpable alkali

95

dust, salty and choking, so that he strangled and coughed and sneezed with it. "*Lord!* what a country!" exclaimed the dentist.

An hour later the mule stopped and lay down, his jaws wide open, his ears dangling. McTeague washed his mouth with a handful of water and for a second time since sunrise wetted the flour sacks about the bird cage. The air was quivering and palpitating like that in the stokehold of a steamship. The sun, small and contracted, swam molten overhead. "I can't stand it," said McTeague at length. "I'll have to stop and make some kinda shade."

The mule was crouched upon the ground, panting rapidly, with half-closed eyes. The dentist removed the saddle and unrolling his blanket, propped it up as best he could between him and the sun. As he stooped down to crawl beneath it, his palm touched the ground. He snatched it away with a cry of pain. The surface alkali was oven-hot; he was obliged to scoop out a trench in it before he dared to lie down.

By degrees the dentist began to doze. He had had little or no sleep the night before, and the hurry of his flight under the blazing sun had exhausted him. But his rest was broken; between waking and sleeping, all manner of troublous images galloped through his brain. He thought he was back in the Panamint hills again....Something was following him...something dark crawling upon the ground, an indistinct gray figure, man or brute, he did not know. Then he saw another and another; then another. A score of black, crawling objects were following him, crawling from bush to bush, converging upon him. "*They*" were after him, were closing in upon him, were within touch of his hand, were at his feet—*were at his throat.*

McTeague jumped up with a shout, oversetting the blanket. There was nothing in sight. For miles around the alkali was empty, solitary, quivering and shimmering under the pelting fire of the afternoon's sun. But once more the spur bit into his body, goading him on. There was to be no rest, no going back, no pause, no stop. Hurry, hurry, hurry on....."I *can't* go on," groaned McTeague, his eyes sweeping the horizon behind him. "I'm beat out. I'm dog tired. I ain't slept any for two nights." But for all that he roused himself again, saddled the mule, scarcely less exhausted than himself, and pushed on once more over the scorching alkali and under the blazing sun.

From that time on, the fear never left him, the spur never ceased to bite, the instinct that goaded him to flight never was dumb; hurry or halt, it was all the same. On he went, straight on, chasing the receding horizon, flagellated with heat, tortured with thirst, crouching over, looking furtively behind, and at times reaching his hand forward, the fingers prehensile, grasping as it were toward the horizon that always fled before him.

The sun set upon the third day of McTeague's flight; night came on; the stars burned slowly into the cool dark purple of the sky. The gigantic sink of white alkali glowed like snow. McTeague, now far into the desert, held steadily on, swinging forward with great strides. His enormous strength held him doggedly to his work. Sullenly, with his huge jaws gripping stolidly together, he pushed on. At midnight he stopped. "Now," he growled with a certain desperate defiance, as though he expected to be heard, "now, I'm going to lay up and get some sleep. You can come or not."

He cleared away the hot surface alkali, spread out his blanket, and slept until the next day's heat aroused him. His water was so low that he dared not make coffee now, and so he breakfasted without it. Until ten o'clock he tramped forward, then camped again in the shade of one of the rare rock ledges, and "lay up" during the heat of the day. By five o'clock he was once more on the march.

He traveled on for the greater part of that night, stopping only once toward three in the morning to water the mule from the canteen. Again the red-hot day burned up over the horizon. Even at six o'clock it was hot. "It's going to be worse than ever today," he groaned. "I wish I could find another rock to camp by. Ain't I ever going to get out of this place?"

There was no change in the character of the desert. Always the same measureless leagues of white-hot alkali stretched away toward the horizon on every hand. Here and there the flat, dazzling surface of the desert broke and raised into long low mounds, from the summit of which McTeague could look for miles and miles over its horrible desolation. No shade was in sight. Not a rock, not a stone broke the monotony of the ground. Again and again he ascended the low unevenness, looking and searching for a camping place, shading his eyes from the glitter of sand and sky.

He tramped forward a little farther, then paused at length in a hollow between two breaks, resolving to make camp there.

Suddenly there was a shout. "Hands up. By damn, I got the drop on you!"

McTeague looked up. It was Marcus [, his former best friend].

"Hands up!" shouted Marcus a second time. "I'll give you three to do it in. One, two—"

Instinctively McTeague put his hands above his head.

Marcus rose and came toward him over the break. "Keep 'em up," he cried. "If you move 'em once I'll kill you, sure." He came up to McTeague and searched him, going through his pockets; but McTeague had no revolver, not even a hunting knife. "What did you do with that money, with that five thousand dollars?"

"It's on the mule," answered McTeague sullenly.

Marcus grunted and cast a glance at the mule, who was standing some distance away, snorting nervously and from time to time flattening his long ears.... "Got any water?" he demanded.

"There's a canteen of water on the mule."

Marcus moved toward the mule and made as if to reach the bridle rein. The mule squealed, threw up his head and galloped to a little distance, rolling his eyes and flattening his ears. Marcus swore wrathfully.

"He acted that way once before," explained McTeague, his hands still in the air. "He ate some locoweed back in the hills before I started."

For a moment Marcus hesitated. While he was catching the mule McTeague might get away. But where to, in heaven's name? A rat could not hide on the surface of that glistening alkali, and besides, all McTeague's store of provisions and his priceless supply of water were on the mule. Marcus ran after the mule, revolver in hand, shouting and cursing. But the mule could not be caught. He acted as if possessed, squealing, lashing out and galloping in wide circles, his head high in the air....McTeague came up. "He's eatun some locoweed," he repeated. "He went kinda crazy once before."

"If he should take it into his head to bolt and keep on running—" Marcus did not finish. A sudden great fear seemed to widen around and enclose the two men. Once their water was gone, the end would not be long.

Already the sense of enmity between the two had weakened in the face of a common peril. Marcus let down the hammer of his revolver and slid it back into the holster. The mule was trotting ahead, snorting and throwing up great clouds of alkali dust....."He's clean crazy," fumed Marcus, panting and swearing....

Then began an interminable pursuit. Mile after mile under the terrible heat of the desert sun the two men followed the mule, racked with a thirst that grew fiercer every hour. A dozen times they could almost touch the canteen of water, and as often the distraught animal shied away and fled before them. At length Marcus cried: "It's no use; we can't catch him, and we're killing ourselves with thirst. We've got to take our chances." He drew his revolver from his holster, cocked it and crept forward.

"Steady, now," said McTeague; "it won't do to shoot through the canteen."

Within twenty yards Marcus paused, made a rest of his forearm and fired. "You've *got* him," cried McTeague. "No, he's up again. Shoot him again. He's going to bolt."

Marcus ran on, firing as he ran. The mule, one foreleg trailing, scrambled along, squealing and snorting. Marcus fired his last shot. The mule pitched forward upon his head, then rolling sideways, fell upon the canteen, bursting it open and spilling its entire contents into the sand. Marcus and McTeague ran up....There was no water left...."We're dead men," said Marcus.

McTeague looked from him out over the desert. Chaotic desolation stretched from them on either hand, flaming and glaring with the afternoon heat. There was the brazen sky and the leagues upon leagues of alkali, leper white. There was nothing more. They were in the heart of Death Valley. "Not a drop of water," muttered McTeague, "not a drop."...

Marcus ground his teeth. "Done for," he muttered; "done for....Well, we can't stop here; we got to go somewhere....Anything we want to take along with us from the mule? We can—"

Suddenly he paused. In an instant the eyes of the two doomed men had met as the same thought simultaneously rose in their minds. The canvas sack with its five thousand dollars was still tied to the horn of the saddle. Marcus had emptied his revolver at the mule, and though he still wore his cartridge belt, he was for the moment as unarmed as McTeague. "I guess," began McTeague, coming forward a step, "I guess, even if we are done for, I'll take—some of my truck along."

"Hold on," exclaimed Marcus with rising aggressiveness. "Let's talk about that....You're my prisoner, do you understand? You'll do as I say." Marcus had drawn the handcuffs from his pocket and stood ready with his revolver held as a club. "...Don't you lay your finger on that sack."

Marcus barred McTeague's way, white with passion. McTeague did not answer. His eyes drew to two fine, twinkling points, and his enormous hands knotted themselves into fists, hard as wooden mallets. He moved a step nearer to Marcus, then another. Suddenly the men grappled and in another instant were rolling and struggling upon the hot white ground. McTeague thrust Marcus backward until he tripped and fell over the body of the dead mule. The little bird cage broke from the saddle with the violence of their fall and rolled upon the ground, the flour bags slipping from it. McTeague tore the revolver from Marcus's grip and struck out with it blindly. Clouds of alkali dust, fine and pungent, enveloped the two fighting men, all but strangling them.

McTeague did not know how he killed his enemy, but all at once Marcus grew still beneath his blows. Then there was a sudden last return of energy. McTeague's right wrist was caught; something clicked upon it; then the struggling body fell limp and motionless with a long breath.

As McTeague rose to his feet, he felt a pull at his right wrist; something held it fast. Looking down, he saw that Marcus in that last struggle had found strength to handcuff their wrists together. Marcus was dead now; McTeague was locked to the body. All about him, vast, interminable, stretched the measureless leagues of Death Valley.

McTeague remained stupidly looking around him, now at the distant horizon, now at the ground, now at the half-dead canary chittering feebly in its little gilt prison.

HARRY LAWTON

*Harry Lawton was born in 1927 in Long Beach, California. He attended the
University of California, Riverside, where he later taught creative writing and
founded UCR Writers Week, which is still held every year. A journalist and
historian, Lawton worked for the* Riverside Press-Enterprise *and wrote* Willie
Boy: A Desert Manhunt, *a novelized version of a 1909 headline story, the facts of
which have been the subject of academic dispute in recent years.*

In the novel Willie Boy, *a Chemehuevi/Paiute Indian, with his lover, Carlota
(called Lolita in Lawton's book), evades a posse of trackers in the desert after
murdering the girl's father, Old Mike Boniface, another Chemehuevi. The following
excerpt from Chapter 10 of* Willie Boy, *"The Long Chase," describes the period of
time after Carlota is found dead of a gunshot wound and Willie Boy is successfully
outrunning the law.*

from *Willie Boy*

Tuesday, Oct. 5

Willie Boy had been running since daybreak, darting along at a slow, even pace
westward through the brush. Occasionally he startled a jackrabbit, and then
both of them would run together a short way, until suddenly the rabbit would
bound off abruptly at right angle and he would once again be racing on alone.

He was exhausted from the previous day's run, and although he had slept in the
warm sand beneath a stand of mesquite the night winds and colder temperatures had
numbed his body. His cramped legs throbbed as they jarred against the ground.

He kept turning his head to glance back to the east, where he believed the
posse had made camp at Surprise Springs. Once he thought he detected a
group of riders moving through the brush far away, but then they disappeared
into the rolling plain and the interval between their appearance and departure
was so short he wondered if he had imagined them.

He had eaten nothing since the afternoon before. There was a steady,
persistent ache in his belly, and his tongue was thick.

Eventually the jogging motion of running quelled sensations of hunger and
thirst and tiredness. He was running beyond the limits of human endurance,
where stopping becomes too painful to contemplate.

The Mohave plain gave an illusion of flatness, unending, since from any
point in the desert bowl the observer sees only the long vistas of extending
flatland and not the intervening depressions.

Running always westward across this saucer country, Willie Boy began to
feel as if the land was rising and falling beneath his feet, so that he would dart

down a grade, twist in and out of brush at the bottom, and then stride up the other side, and once more look out across endless desert, surrounded on all sides by distant mountain ranges.

As he ran there came a fear he was trapped. Danger threatened from all sides. He might run on and on within the confines of the saucer, but eventually they would press in on him, closer and closer, until there would be no space left in which to run. He shook this fear off. There was a break far ahead where the Victor Road wound up the Lucerne Valley to Victorville. He might find both food and water at Rock Corral or Old Woman Springs about thirty miles to the west.

He drove his legs forward, suppressing the knowledge that such a distance was impossible. By midday a withering heat would bake the basin. He might find shade, but without water he could not continue on that night. He ran and there was no spring in his legs, only a dull movement of muscle.

The morning sun topped the Bullions behind him, throwing his shadow ahead as he ran. He came over a rise and then ran on in shadow. The sky gathered overhead, leaden and gray from a small thunderhead that had swiftly moved down from the western slopes of the Coffin Mountains. He strode on beneath the clouds and occasionally a globule of rain spattered his face. The scattered drops reminded him of his thirst, but offered no chance of quenching it.

The clouds rolled on and sunlight bathed the plain once more. The sun's heat pressed down on him, shriveling his thoughts, and his pace slackened. His rifle butt dragged the sand, leaving a snake-like trail. It cost an effort to raise the gun. His arm was numb from carrying it. He shifted the weapon to his left hand.

His eyes blurred. His legs stumbled. He ran jerkily, crashing through brush. He became conscious only of the expiring gasps of each breath he drew and a pounding in his head that seemed synchronized with breathing. He was running downhill toward a dry lakebed, a level sea of sand stretching out a mile or more and then it began wavering in front of him. He tripped and fell forward, overcome by dizziness, tried to gain his balance, and then toppled, falling face down, sprawled on the ancient shoreline.

He knew only the pressure of the sand against his chest and the deep pain that accompanied each movement of his ribs. He lay on the bare sand a long time, until once more he became conscious of the sun's heat and his tongue swollen like a balloon. Again the prodding fears began and the tiny sounds of wind rattling the greasewood became transformed into the plodding beat of hooves in the distance. He was certain if he were to sit up and turn his head he would see horsemen staring down at him. He heard their voices, faint and indistinct. They had water. They were passing a canteen back and forth and laughing at him. He made a sudden clutch for his rifle, lying beside him, rolled over and came up half crouched, his gun aimed at the top of the ridge. There was no one there.

He spun about to look behind. There were no riders behind him either, only sunlight sparkling on the bottom of the old lakebed, a few yards out from shore, reflecting from a dark, metallic expanse that gleamed over the entire lakebed. It took him a moment to understand what he was looking at. The thundercloud had dumped its load on the dry lake.

He lurched toward the water and ran into it, his feet splashing and throwing up a muddy spray. He hurled himself face forward into the water.

Late Tuesday afternoon six posse members tracked north of Giant Rock on a plain that slanted gradually upward into the Lucerne Valley. They included four members of the party that had trailed Willie Boy until darkness on Monday: Toutain, Reche, Wal de Crevecouer, and Nowlin. Wallace Evans had been replaced by John Hyde and Segundo Chino had joined the group on his black stallion.

Segundo had ridden in to Surprise Springs the night before to report that Ben de Crevecouer was raising a fresh posse and would be at Warren's Well on Wednesday. Sheriff Ralphs had ordered the base camp shifted to Warren's Well.

Unless they found fresh tracks, the six trackers were to join Ralphs at the new base camp on Tuesday evening.

Tracking was slow, mostly on foot, but by mid-afternoon the possemen were satisfied they were once more only a few hours behind the fugitive. The land tilted steadily upward through a country of yucca trees.

"He's starting to drag his rifle again," observed Reche, leaning over the pommel of his horse to study a wavering line in the sand.

"He's sure got endurance enough for five men," de Crevecouer muttered.

"That cloudburst saved him," said Toutain. "If it hadn't been for that rain we'd have him."

They rode in silence and Reche cantered ahead to an outcropping of granite boulders. He dropped quickly from the saddle.

"Hey, feller!" he whispered. The other five spurred up to Reche and noted a grimace on his mouth. They glanced at his feet.

Willie Boy had attempted to start a fire. He had lacked kindling and been unable to keep it going. Apparently he had run out of matches. A dead chuckawalla, singed in places, mostly raw, lay in the sand, the flesh of its back torn in strips. A few feet away were the chunks of lizard Willie Boy had retched up.

His tracks went a few yards further in the sand, where he had lain down, retching again.

"Couldn't keep it down," said Reche. "And yet I'll bet his own father could eat a chuckawalla raw and consider himself well fed."

"That right, Segundo?" asked de Crevecouer.

"Maybe," said the Indian. He did not like this type of conversation.

The horsemen rode on to Rock Corral. They watered the horses and discovered Willie Boy had been there before them and raided a cache left by Swarthout's cowboys. The tarp had been slashed open and the Indian had helped himself.

"He's got a decent meal under his belt by now," Reche said tiredly. "Maybe we better get one under ours. You boys want to turn down to Warren's or shall we ride on to Old Woman Springs and rustle up some grub?"

The men were weary and the idea of a long ride didn't sound appealing, particularly when there was a chance of finding food at Old Woman Springs. They followed the Victor Road up the valley. Toward sunset they rode in to the springs, only to discover the ranch-house in ashes. It had burned down several days before.

They turned their horses loose in the cienega and opened the few cans of kidney beans they were carrying. They found a gnarled apple tree—its fruit sour and acid—but they munched on the apples. Then they bedded down for the night.

Wednesday, Oct. 6

Wednesday morning they decided they were too close to Willie Boy to turn back for Warren's Well and could only hope that Ralphs would send out a chuck wagon to catch up with them. They had one can of kidney beans apiece, and they ate half a can each at noon.

They tracked Willie Boy south into the foothill country of the San Bernardino Mountains. Sometimes they lost the trail for hours and then encountered it again by doubling back and forth across the canyons. Food had revived Willie Boy's cunning and he was covering his tracks well.

Several times Reche sent one of the men back to Victor Road for some sign of the expected chuck wagon. Reche was determined that only a long, hard push after Willie Boy—before he had full opportunity to regain his strength—could end the hunt quickly. Toward sunset he asked Toutain to ride back to base camp and get the chuck wagon.

"Not me," Toutain argued. "I want to see what happens."

But the men persuaded Toutain he was the logical relief man. He was the only posse member who had been in the saddle constantly since Sunday, even refusing to relinquish his horse to any other rider.

"All right, boys," Toutain said. "Tell Willie Boy hello for me."

WAKAKO YAMAUCHI

Wakako Yamauchi was born in 1924 in the agricultural Imperial Valley town of Westmoreland, south of the Salton Sea. Her parents were Japanese immigrants and farmers in the area, and during World War II, Yamauchi and her family were interned at Poston, Arizona.

Yamauchi's play And the Soul Shall Dance *won the Los Angeles Drama Critics Circle Award for Best New Play of 1977. The following short story, "Songs My Mother Taught Me," takes place during the 1930s.* Songs My Mother Taught Me *is also the title of a collection of Yamauchi's work published in 1994.*

"Songs My Mother Taught Me"

I was eleven in the summer of 1935. My father was farming then in the valley of the Santa Rosa, San Jacinto, and Chocolate Mountain Ranges called Imperial Valley near the Mexican border of California. The land was fertilized with tons of chicken manure and irrigated by the All American Canal that flowed out of the Colorado River. Planting started in late September: tomatoes, squash, cantaloupe. All during the winter months, it was thinning the seedlings, weeding, building brush covers for them, repairing the covers after a storm, and starting the smudge pots to ward off the frost. In early spring, it was harvesting the crops. By May, after the broiling sun had reduced the plants to dry twigs, the plowing began. Then the land was flooded to start the weeds and fallen seeds growing, after which there was another plowing to destroy these sprouts, and once again fertilizing, furrowing, and preparing the land for late September planting.

At best, the entire process kept us alive and clothed. There were five of us: my father, Junsaburo Kato; mother, Hatsue; sister, Nami, fourteen; brother, Tetsuo, thirteen; and myself, Sachiko. Why my father kept at this unrewarding work, I never knew or questioned or thought about. Maybe there was nothing else he could do. Maybe he worked in hope that one day that merciful God (to whom we prayed, before whose lacquered shrine we burned our incense) would provide the miracle crop that would lift us to Japan, rich and triumphant.

My father was a quiet man. If he suspected this was the whole cloth of his life, that he would live and die this way, he never showed it. He never revealed his dream for us, his children, in America, or articulated his own dreams. But sometimes when he was in his cups, when the winter rain drowned his hopes for a good harvest, he would sing the saddest and loneliest Japanese songs. My mother spoke often of returning to Japan, of smelling again the piney woods, tasting the exquisite fruits, of seeing her beloved sisters. The stories of her

Japan came on like a flashback in the movies—misty, wavering, ethereal, and her beautiful eyes would grow soft.

Summer is unbearable in the valley. The sun beats on miles of scrub prairie and the air is deathly still. One listless day succeeds another, without incident, without change. The farmers who had had some luck with their harvest usually spent their summer north along the coast in San Pedro where a colony of Japanese resided permanently, occupied with fishing and canning. On a small island off the mainland called Terminal Island, most of the Valley people rented cheap houses and spent the summer fishing off the scummy San Pedro wharf, clamming, crabbing, and swimming in the Pacific. My mother said it was hard to believe these same waters broke on the shores of Japan.

The harvest of 1935 was a good one. That summer we started north in early July, after we popped off the firecrackers my father had bought for Tets. They loved firecrackers; I can't remember a year that we didn't have some, no matter how poor we were. Usually, though, they couldn't wait until the 4th, and they'd fire off a few each night, until there was nothing left for the 4th. So on Independence Day we often sat outside and watched the neighbors' fireworks a quarter mile away. Sometimes my mother made iced tea with sugar and lemons while we waited for the dark. That year we had enough to string out until the actual celebration.

After the 4th, we packed our old gray Chevrolet and drove 300 miles north to Terminal Island. It was always a hot tiresome trip and when we finally drove up to the drawbridge that spans the narrow channel, and recognized the Ford plant to the left, we let out whoops of delight. My father quieted us with a grunt. He hated noise.

There was already a group of people from the Valley, townspeople we knew from the Buddhist Church. We rented a small four-room house for the summer.

The days were filled with vacation fun. We trapped crab at the piling of a demolished pier. When the sun went down, we boiled the catch in large galvanized tubs and ate the sweet meat, our skin chilled on one side by the night air, and seared on the other by the campfire. There were roasts; I can still taste the succulent meat and the bitter ash. There were hot days in the sun, blistered backs, and the good pain. In the evening, the older folks sat on front stoops or wandered from house to house looking to start a card game or a session of *go*, and we were allowed to play outside until darkness blurred our faces and hide-and-seek became too hazardous. We went everywhere with friends, huge contingents of raucous children and adults. Every day was a holiday, far removed from the isolated farm and its chores and the terror of another blighted year.

But there was a subtle change in the family. I don't know if Nami or Tets felt it. If they did, they said nothing. Or maybe I too would not have noticed if Sayo, my summer friend, had not called attention to it. Sayo was two years older, and wiser, more perceptive and knowledgeable than I. She wrote poetry and read masses of books and knew all the latest Japanese songs and the meaning of the lyrics. She said, "Your mother plays the same record again and again. Why is she so sad?"

I said, "Maybe she likes the song. Maybe she wants to learn it."

It was true. My mother joined our excursions less and less. Often I found her sitting at the kitchen table writing, nibbling at the stub of a pencil and looking past the window. It was also true that she played the Victrola and usually the same record over and over: "*Mujo No Tsuki*," which Sayo said meant transient moon. She translated the lyrics for me:

Samishiku kyo mo	Today too
Kurete uki	Passes in solitude
Yuhi wa toku	The evening sun is distant
Umi no hate	Beyond the rim of the sea
Dare ga yobuno ka	Who is it that calls
Sagasu no ka yo?	Who, that seeks me?
Namima ni sakebu	The sea birds that
Hama chidori	Cry from the shore

It was all true. But I couldn't always remember to think about it while everyone else appeared so happy. I laughed with them and fought with them and pretended, like Nami and Tets, that nothing was wrong.

After one of those quarrels at the beach, I returned home alone. Nami and Tets always sided against me, and though I had a well-honed tongue for my age, I couldn't match the two together. I begged my father to come home with me, but he was playing cards with some cronies and he waved me off like an annoying fly. I walked home alone, sulky and morose, and went to the rear of the house to brush the sand off my feet. Mrs. Griffin, the landlady, lived within eyeshot and she was very particular about sand in her threadbare rug. We called her *Okorimbo*, which is a diminutive for Angry One. *Okorimbo's* face was permanently creased in a pattern of rage. She spent the daylight hours at her window, and when she caught us heading toward the front door, she rushed from her house, her ragged slippers flapping, screaming, "Go to the back, to the back."

I was rubbing sand from my legs when I heard soft voices. My mother had company. Not wanting to be seen, I crouched behind a privet hedge. There's a humiliating routine a Japanese child must endure from adults: the pat on the head, the examination, the appraisal, "How tall she's growing! She will be beautiful some day. She looks very bright!" Then: "Not at all, very small for her age, homely like her mother, not very smart, no help at all around the house, lazy…"

Within fifteen minutes, the visitor left through the rear door and passed very close to the hedge where I was hiding. I could smell the polish on his shoes. I recognized him as Yamada-san, the young man from northern California who earlier that spring had come to the Valley with a group of laborers. He was a *Kibei*, a Japanese born in America and reared in Japan. My father had hired him to help us with our good harvest. Yamada-san was different from my father—handsome and younger. He came to the dinner table in a fresh shirt, his hair combed back with good-smelling oil. My father, like most farmers, wore his hair so short that neither comb nor rain (nor sun) changed its shape. Yamada-san had eyes that looked at you. When you talked he committed himself to you. My father's eyes were squinty from the sun and he hardly saw or heard you.

My father had been dissatisfied with Yamada-san. For a man of few words, I was surprised how strongly he spoke.

"Someone like that I wouldn't want around here long. He's a bad example for our Tetsuo," he said.

"How can we finish harvesting without him?" my mother had asked. "He works hard, Kato, you know that."

"You see how he is—plucking at that frivolous mandolin all the time—like he was practicing for the pictures or something. The clothes he wears—those two-tone shoes, that downtown haircut, the fancy cigarettes. Like a woman! Do you want Tetsuo to get like that? So he works hard. So? He stays only until harvest is over. I won't tolerate him after May! I don't want him in the house except at meal time. Tell him."

"Kato, I can't tell him this."

Some of what my father said was true. But I watched Yamada-san's sure body move effortlessly and I knew he worked hard. He was not like a woman. He had a quality of the orient my father did not have. He was the affirmation of my mother's Japan—the haunting flutes, the cherry blossoms, the poetry, the fatalism. My mother changed when he was around. Her smile was softer, her voice more gentle. I suspect this was what my father disliked more than the man himself—the change he brought over the rest of us.

Though we loved to listen to Yamada-san's mandolin, loved to touch the pearl frets and pluck the strings, we began to avoid him. He sensed the change in us, and thereafter confined his visits to mealtimes. He lived in the cabin my father had built for Tets, but Tets didn't like sleeping away from the rest of us, so it had stayed unused until now. I pictured Yamada-san sitting on his cot reading by lamplight, or mending his clothes, or singing his songs alone.

We pretended not to hear the lonely music that came from his quarters; we pretended not to notice how each missed him.

Immediately after the harvest, Yamada-san packed his things and shook hands with us. My father drove him to the bus station. I went to the vacant cabin. The cot was in the corner with a naked mattress on top and the bedclothes neatly folded at one end. He'd left an empty jar of pomade on the table.

I saw Yamada-san one more time before summer was over.

One evening Sayo and I crept away from the other children and walked to the dunes where the sea figs sprawled to the edge of the surf. The sky was still pinkish from the dying sun and a warm wind blew in from the west. I only saw his back. He was standing against the wind, facing the sea. When he heard us, he walked away without turning, flicking the stub of his fancy cigarette to the wind.

As summer drew to a close, my mother's malaise would not be ignored. She spent a lot of time in bed. Nami and my father did the cooking because the smell of food sent her retching to the bathroom. My father finally called Dr. Matsuno. It was a frightening time for me because I remembered years before when my mother was very sick and had to stay with friends in town. They had recommended the blood of carp, so a live carp was kept in a tub on the back porch for her. My mother told me she put her mouth directly to the living fish. When she didn't improve, a white doctor was called. He diagnosed the illness as typhoid and we all went to his office to get shots. I was very frightened. I thought often of her dying and my heart would contract.

We waited disconsolately on *Okorimbo's* ratty couch not saying a word to one another. The doctor had closed the door behind him. When he finally emerged, he told us there was nothing to worry about. "She will give birth in late March," he said.

The next day we strapped the luggage to the top of our car and started the dismal trip back to the Valley.

The baby grew steadily in my mother's belly, distending and misshaping her body. The black hair she wore in a smooth coil at her neck grew crisp and faded. Broken strands hung from her temples like dry summer grass and brown splotches appeared on her skin.

She withdrew from us. I couldn't coax a smile from her. My father was kinder than he'd ever been, but everyone seemed to retreat to a separate place. I wanted someone to make room for me, someone to tell me what was going on, to assure me everything was all right. But things were far from right; no one would let me in. I was too afraid to ask. It was as though each prepared to fight this monstrous condition alone, with silence, and will it back to its dark origin.

I looked for someone to turn to. I chose the highest source.

In December, the Jodo Shinshu sect of the Buddhist church headquartered in Los Angeles sent down a new minister to stimulate religious life in the Valley. The previous man was a bachelor, too handsome and too popular for his own good. Many members of the Ladies' Auxiliary desired him for their daughters. He may have asked for his own transfer.

There was something disconcerting about the new minister, Reverend Umino, his eczema, his full lips, his obesity (in contrast to ideals of austerity and self denial), but he appealed to me just because of these very human traits. I thought if God could accept so ordinary a man for His emissary, there was a good chance He would listen to *me*. I thought if prayer and faith could heal the sickness in our house, I would build a faith matched by no one, and I would pray with every breath I drew. So with limited knowledge of ritual, I prayed to my eastern deity in western methods. The request was always the same: Sweet Buddha, let us be happy again. Let there be six living people in our family. I was afraid something terrible and unmentionable would happen to my mother or the baby. After a while the prayers became so abbreviated, only Buddha could have understood them. I slipped them in all the activities of school and home—while turning pages in my reader, while cleaning lamp chimneys, "Six, six, six…"

On the morning of the birth Nami shook me awake: "Sachiko, wake up! Something's happening!" We ran to my mother's room. She lay twisting, clutching the metal bars of the bed and breathing in short gasps. Tets stood beside her, his head low, his tears dropping silently to the floor.

"Papa has gone to get Mrs. Nakagawa," she said. Mrs. Nakagawa was the midwife who delivered most of the Japanese babies. She lived in town, her husband was an accountant, and they were both religious Buddhists. "Go quickly to school, be good children and take care of the little one," she said. She touched my face.

It was the first time in a long time she had touched me. I screamed, "Mama, don't die!"

"Don't worry," she said.

At school I was unable to concentrate. I loitered behind everyone so I could whisper privately into my joined palms, "Six, six, six…" I spoke to no one. Two or three times the teacher peered into my face and shook her head.

Once home, I went directly to my mother's room and watched until I caught the movement of my mother's breathing. The baby was lying next to her.

Buddha had saved this infant for me, Kenji was very special. I loved him, cared for him, changed his diapers, pretended he was my own. I spoiled him. But he took only my mother's milk and when it was gone, he would refuse a bottle. Then he'd fret and cry and Nami and I took turns carrying him, rocking and teasing him until her milk returned. Times when we were already in bed, my mother would get up and carry him off to the desert. I listened to her footsteps on the soft sand, terrified they would not return. The more he cried, the less she liked him; the less she liked him, the more he cried.

She grew despondent from this perpetual drain on her body and her emotions, and slipped farther and farther from us, often staring vacantly into space. Every day I rushed home from school to take care of the baby so he wouldn't start my mother's terrible strangled sighs.

One evening after supper, Kenji began his crying again. My mother snatched him from his bed and walked rigidly away, bouncing him erratically and vigorously. I followed her, begging her to let me carry him, assuring her that I could quiet him.

"Go back!" she snarled. I was frightened but I would not turn back, and half running, I followed her to the packing shed. By this time Kenji, shocked by the unnatural motion, had stopped his crying. My mother too, had calmed down, but she saw a hatchet my father had left outside his toolbox, and she lifted it. While I stood there watching, my blood chilling, she lay the blade on the baby's forehead. I began to cry, and she pulled me to her and cried too. She said, "My children are all grown. I didn't want this baby—I didn't want this baby…"

I know now what she meant: that time was passing her by, that with the new baby she was irrevocably bound to this futile life, that dreams of returning to Japan were shattered, that through the eyes of a younger man she had glimpsed what might have been, could never and would never be.

I turned to prayer again.

Sunday service was largely a congregation of children and adolescents. Adults were at the fields preparing for Monday's market. Whereas the previous minister geared his sermons to the young, using simple parables and theatrical gestures, Reverend Umino never talked down to us. He spoke on abstract philosophies in rhetoric too difficult to follow. Peculiarly responsive to shifting feet and rustling hymn books, he often stopped abruptly, leaving the lectern massaging the festering eczema on his wrists. The church people who lived in the city and had the most influence didn't like him, but I didn't care. I only hoped he'd sense my need and put in a special word for me.

The spring harvest was meager and market prices low. My father couldn't afford the hired help of the year before. Tets and Nami helped wherever and whenever they could. My father would rise before daylight and work on until dark. He grew morose and short with us.

That summer there wasn't enough money for a vacation. Sometimes my father would drive us out to town and give each of us fifteen cents for the movies: ten cents for admission and a nickel for popcorn. I loved the movies, but I couldn't enjoy them, I was so worried about my mother and Kenji. Most of the summer we played cards. The baby grew rapidly in spite of the lack of breast milk and mother love. He responded to his name, slouching like a giant doll in an attempt to sit up. My father went to the burning fields to parcel out the land for flooding.

Then one day toward the end of August, it happened. As it had its beginning in August, so it found its end then.

A hot wind blew in from the desert, drying the summer grass and parching throat and skin. Tets had gone out to help my father with flooding, controlling the flow of water at the water gate. My father was at the far end of the field repairing the leaks.

I was playing with Kenji who was sitting in a shallow tub of water. He loved water. It was one of the few ways to keep him happy. With one hand he grasped the rim of the tub and with the other he pawed at a little red boat I pushed around for him. My father bought the boat for Kenji shortly after he was born. Nami was reading an old Big-Little book, carefully turning the brittle pages. My mother had prepared some rice balls and a Mason jar of green tea for my father's lunch and since Tets was also in the fields, she divided the lunch in two separate cloths and sent Nami and me to take them to the men. We went together, first to Tets, then to my father.

My father had seen us coming and waited for us by the scant noonday shade of a cottonwood tree. We sat together quietly while he ate his rice balls. A rice ball is to a Japanese what a sandwich is to an American, a portable meal. It's a molded ball of rice sprinkled with sesame seeds and salt and inside there's usually a red pickled plum. My mother once told me it was the mainstay of Japan's Imperial Army, the symbol of her flag: the red plum with the white rice around it. Nami poured tea for my father into a small white cup decorated with a few gold and black strokes—a tiny boat, a lone fisherman, a mountain, and a full moon.

The sound of my hungry father devouring his lunch, gulping his pale yellow tea, the rustle of wind in the cottonwood leaves, far away the put-put of a car, these lonely sounds depressed me so, I wanted to cry. Across the acres of flat land, the view to the house was unbroken. It stood bleached in the sun in awful isolation. A car turned up the road that led to our house.

I shaded my eyes. Far in the distance I saw the small form of my mother waving frantically and running toward us. I pulled my father's sleeve and the three of us ran to her. I could see she clutched our baby. His arms dangled limply. My heart fell.

The front of my mother's dress was stained with water. "I left him only a minute and I found him facedown in the water! I've killed him, forgive me, I've killed him!" she wept.

My father tore the baby from her, turned him upside down and thumped his back. He put him on the ground and blew into his mouth. The car that had entered our yard lurched over the fields. Reverend Umino, perspiring in his black suit and tie, jumped out. "Come, we will take him to a doctor," he said. My father looked dazed. He carried the naked baby into the car. He brushed away the dirt and grass from our baby's face. The Reverend drove off, steering the car like a madman, reeling and bouncing through the fields toward the road to town.

When Tets came home, we all cried together. Waiting for my father that afternoon, my mother roamed from room to room, holding Kenji's kimono to her cheek. We followed her everywhere. Outside, the tub sat with its four inches of water and the toy boat listing, impervious to our tears. From habit, I said my prayers, "Six, six, six."

The Reverend brought my father home in the evening. With his handkerchief he wiped the dust from our family shrine and lit the incense. He chanted his sutras while we sat behind him, our heads bowed. He stayed a while longer murmuring to my mother, assuring my father that everything would be attended to. My mother gave him clothes for the baby: diapers, pins, a small white gown he had never worn, a slip. The Reverend received them with both hands. He put a palm to my mother's shoulder and left.

There was only a small group in the chapel, those few who were unable to go away for the summer. The flowers that surrounded Kenji's white coffin were the last of summer and their scent along with the smoking incense filled the room. Candles flickered in tall brass stands and the light glittered on the gold of the shrine. In the first pew reserved for the family, we five sat huddled, at last together in our grief. Reverend Umino's sutras hung in the air with the oppressive smoke. He began the requiem: "An infant, child without sin, pure as the day he was born, is swept from the loving arms of parents and siblings. Who can say why? Trust. Believe. There is an Ultimate Plan. Amida Buddha is Infinite Mercy. The child is now in His sleeve, now one with Him…" In his elegant vestments, the tunic of gold brocade, the scarlet tassels, with his broad chest heaving, Reverend Umino swabbed at his swarthy neck. "Providence sent me to this family at this extreme hour. I was there to witness the hand of Amida…" And the Reverend went on to tell the story of that dreadful day in flawless detail. The assembly wept.

Pictures of that day, the days of the gestation, the summer before, the terrible days following the birth, reeled before me. Someone in the congregation sobbed. Scratching at his eczema, the Reverend motioned the group to rise. My father helped my mother to her feet. She gripped my hand. It seemed to me if my tears would stop, I would never cry again. And it was a long time before I could believe in God again.

MARGARITA LUNA ROBLES

Margarita Luna Robles is a poet, prose writer, performance artist, mentor, and former teacher of Chicano–Latin American culture at California State University, Fresno. She is a cofounder of Manikrudo, a program for aspiring and emerging writers. She lives in Redlands, California.

The following poem, "Self-Portrait on the Border Between Mexico and the United States (1932)," is taken from her poetry collection Tryptich: Dreams, Lust and Other Performances, *published in 1993.*

Self-Portrait on the Border Between Mexico and the United States (1932)

I can paint myself beautiful in pink
on the border
one foot resting on the past
the other on the future

Mexican sun spills fire
through an angry mouth in a white cloud
the steps of the pyramids wait
a pile of stones formed with wisdom
a forgotten clay Indian god
desert flowers bloom
the dead skull of the past now brittle
the dark cloud embracing a tilted crescent moon
crashing into the sun's sphere, a bolt of lightning
splits the sky

An American dream waves a U.S. flag
over the future
progress
coal mines shooting into the air
the oil will spill
I will slip
lose my balance
fall

And the pink dress will tear and get
dirty.

JEANNE WAKATSUKI HOUSTON and JAMES D. HOUSTON

Jeanne Wakatsuki Houston, born in Los Angeles in 1934, and James D. Houston, born in San Francisco in 1933, are the coauthors of the best-selling Farewell to Manzanar, *first published in 1973, since reprinted many times, and still read worldwide. The book tells the story of how Wakatsuki Houston's Japanese American family was forced from their Los Angeles home in 1942, when she was seven years old, and interned until 1945 at the Manzanar War Relocation Center, in the northernmost region of the Mojave Desert's remote Owens Valley. The husband and wife team also cowrote the teleplay for the NBC drama based on* Farewell to Manzanar, *winner of a Humanitas Prize for film and television writing in 1976. James D. Houston died in 2009 at the couple's home in Santa Cruz.*

The following chapter from Farewell to Manzanar *vividly describes the conditions at the camp where Wakatsuki Houston, her family, and ten thousand other detainees were forced to live during World War II.*

from *Farewell to Manzanar*

Manzanar, U.S.A.

In Spanish, Manzanar means "apple orchard." Great stretches of Owens Valley were once green with orchards and alfalfa fields. It has been a desert ever since its water started flowing south into Los Angeles, sometime during the twenties. But a few rows of untended pear and apple trees were still growing there when the camp opened, where a shallow water table had kept them alive. In the spring of 1943 we moved to Block 28, right up next to one of the old pear orchards. That's where we stayed until the end of the war, and those trees stand in my memory for the turning of our life in camp, from the outrageous to the tolerable.

Papa pruned and cared for the nearest trees. Late that summer we picked the fruit green and stored it in a root cellar he had dug under our new barracks. At night the wind through the leaves would sound like the surf had sounded in Ocean Park, and while drifting off to sleep I could almost imagine we were still living by the beach.

Mama had set up this move. Block 28 was also close to the camp hospital. For the most part, people lived there who had to have easy access to it. Mama's connection was her job as dietician. A whole half of one barracks had fallen empty when another family relocated. Mama hustled us in there almost before they'd snapped their suitcases shut.

For all the pain it caused, the loyalty oath finally did speed up the relocation program. One result was a gradual easing of the congestion in the barracks. A shrewd house hunter like Mama could set things up fairly comfortably—by Manzanar standards—if she kept her eyes open. But you had to move fast. As soon as the word got around that so-and-so had been cleared to leave, there would be a kind of tribal restlessness, a nervous rise in the level of neighborhood gossip as wives jockeyed for position to see who would get the empty cubicles.

In Block 28 we doubled our living space—four rooms for the twelve of us. Ray and Woody walled them with sheetrock. We had ceilings this time, and linoleum floors of solid maroon. You had three colors to choose from— maroon, black, and forest green—and there was plenty of it around by this time. Some families would vie with one another for the most elegant floor designs, obtaining a roll of each color from the supply shed, cutting it into diamonds, squares, or triangles, shining it with heating oil, then leaving their doors open so that passers-by could admire the handiwork.

Papa brought his still with him when we moved. He set it up behind the door, where he continued to brew his own sake and brandy. He wasn't drinking as much now, though. He spent a lot of time outdoors. Like many of the older Issei men, he didn't take a regular job in camp. He puttered. He had been working hard for thirty years and, bad as it was for him in some ways, camp did allow him time to dabble with hobbies he would never have found time for otherwise.

Once the first year's turmoil cooled down, the authorities started letting us outside the wire for recreation. Papa used to hike along the creeks that channeled down from the base of the Sierras. He brought back chunks of driftwood, and he would pass long hours sitting on the steps carving myrtle limbs into benches, table legs, and lamps, filling our rooms with bits of gnarled, polished furniture.

He hauled stones in off the desert and built a small rock garden outside our doorway, with succulents and a patch of moss. Near it he laid flat steppingstones leading to the stairs.

He also painted watercolors. Until this time I had not known he could paint. He loved to sketch the mountains. If anything made that country habitable it was the mountains themselves, purple when the sun dropped and so sharply etched in the morning light the granite dazzled almost more than the bright snow lacing it. The nearest peaks rose ten thousand feet higher than the valley floor, with Whitney, the highest, just off to the south. They were important for all of us, but especially for the Issei. Whitney reminded Papa of Fujiyama, that is, it gave him the same kind of spiritual sustenance. The tremendous beauty of those peaks was inspirational, as so many natural forms are to the Japanese (the rocks outside our doorway could be those mountains in miniature). They also represented those forces in nature, those powerful and inevitable forces that cannot be resisted, reminding a man that sometimes he must simply endure that which cannot be changed.

Subdued, resigned, Papa's life—all our lives—took on a pattern that would hold for the duration of the war. Public shows of resentment pretty much spent themselves over the loyalty oath crises. *Shikata ga nai* again became the motto, but under altered circumstances. What had to be endured was the climate, the confinement, the steady crumbling away of family life. But the camp itself had been made livable. The government provided for our physical needs. My parents and older brothers and sisters, like most of the internees, accepted their lot and did what they could to make the best of a bad situation. "We're here," Woody would say. "We're here, and there's no use moaning about it forever."

Gardens had sprung up everywhere, in the firebreaks, between the rows of barracks—rock gardens, vegetable gardens, cactus and flower gardens. People who lived in Owens Valley during the war still remember the flowers and lush greenery they could see from the highway as they drove past the main gate. The soil around Manzanar is alluvial and very rich. With water siphoned off from the Los Angeles-bound aqueduct, a large farm was under cultivation just outside the camp, providing the mess halls with lettuce, corn, tomatoes, eggplant, string beans, horseradish, and cucumbers. Near Block 28 some of the men who had been professional gardeners built a small park, with mossy nooks, ponds, waterfalls and curved wooden bridges. Sometimes in the evenings we could walk down the raked gravel paths. You could face away from the barracks, look past a tiny rapids toward the darkening mountains, and for a while not be a prisoner at all. You could hang suspended in some odd, almost lovely land you could not escape from yet almost didn't want to leave.

As the months at Manzanar turned to years, it became a world unto itself, with its own logic and familiar ways. In time, staying there seemed far simpler than moving once again to another, unknown place. It was as if the war were forgotten, our reason for being there forgotten. The present, the little bit of busywork you had right in front of you, became the most urgent thing. In such a narrowed world, in order to survive, you learn to contain your rage and your despair, and you try to re-create, as well as you can, your normality, some sense of things continuing. The fact that America had accused us, or excluded us, or imprisoned us, or whatever it might be called, did not change the kind of world we wanted. Most of us were born in this country; we had no other models. Those parks and gardens lent it an oriental character, but in most ways it was a totally equipped American small town, complete with schools, churches, Boy Scouts, beauty parlors, neighborhood gossip, fire and police departments, glee clubs, softball leagues, Abbott and Costello movies, tennis courts, and traveling shows. (I still remember an Indian who turned up one Saturday billing himself as a Sioux chief, wearing bear claws and head feathers. In the firebreak he sang songs and danced his tribal dances while hundreds of us watched.)

In our family, while Papa puttered, Mama made her daily rounds to the mess halls, helping young mothers with their feeding, planning diets for the various ailments people suffered from. She wore a bright yellow, longbilled sun hat she had made herself and always kept stiffly starched. Afternoons I would see her coming from blocks away, heading home, her tiny figure warped by heat waves and that bonnet a yellow flower wavering in the glare.

In their disagreement over serving the country, Woody and Papa had struck a kind of compromise. Papa talked him out of volunteering; Woody waited for the army to induct him. Meanwhile he clerked in the co-op general store. Kiyo, nearly thirteen by this time, looked forward to the heavy winds. They moved the sand around and uncovered obsidian arrowheads he could sell to old men in camp for fifty cents apiece. Ray, a few years older, played in the six-man touch football league, sometimes against Caucasian teams who would come in from Lone Pine or Independence. My sister Lillian was in high school and singing with a hillbilly band called The Sierra Stars—jeans, cowboy hats, two guitars, and a tub bass. And my oldest brother, Bill, led a dance band called The Jive Bombers—brass and rhythm, with cardboard fold-out music stands lettered J. B. Dances were held every weekend in one of the recreation halls. Bill played trumpet and took vocals on Glenn Miller arrangements of such tunes as *In the Mood, String of Pearls*, and *Don't Fence Me In*. He didn't sing *Don't Fence Me In* out of protest, as if trying quietly to mock the authorities. It just happened to be a hit song one year, and they all wanted to be an up-to-date American swing band. They would blast it out into recreation barracks full of bobby-soxed, jitterbugging couples:

Oh, give me land, lots of land
Under starry skies above,
Don't fence me in.
Let me ride through the wide
Open country that I love...

Pictures of the band, in their bow ties and jackets, appeared in the high school yearbook for 1943–1944, along with pictures of just about everything else in camp that year. It was called *Our World*. In its pages you see school kids with armloads of books, wearing cardigan sweaters and walking past rows of tarpapered shacks. You see chubby girl yell leaders, pompons flying as they leap with glee. You read about the school play, called *Growing Pains* "...the story of a typical American home, in this case that of the McIntyres. They see their boy and girl tossed into the normal awkward growing up stage, but can offer little assistance or direction in their turbulent course..." with Shoji Katayama as George McIntyre, Takudo Ando as Terry McIntyre, and Mrs. McIntyre played by Kazuko Nagai.

All the class pictures are in there, from the seventh grade through twelfth, with individual head shots of seniors, their names followed by the names of the high schools they would have graduated from on the outside: Theodore Roosevelt, Thomas Jefferson, Herbert Hoover, Sacred Heart. You see pretty girls on bicycles, chicken yards full of fat pullets, patients back-tilted in dental chairs, lines of laundry, and finally, two large blowups, the first of a high tower with a searchlight, against a Sierra backdrop, the next a two-age endsheet showing a wide path that curves among rows of elm trees. White stones border the path. Two dogs are following an old woman in gardening clothes as she strolls along. She is in the middle distance, small beneath the trees, beneath the snowy peaks. It is winter. All the elms are bare. The scene is both stark and comforting. This path leads toward one edge of camp, but the wire is out of sight, or out of focus. The tiny woman seems very much at ease. She and her tiny dogs seem almost swallowed by the landscape, or floating in it.

JOAN DIDION

Joan Didion was born in 1934 in Sacramento and is well known for her work as a journalist, playwright, essayist, and novelist. Much of her work focuses on the people and places of California. Her books include Slouching Towards Bethlehem *(1968),* The White Album *(1979),* Where I Was From *(2003), and* The Year of Magical Thinking, *which won the National Book Award in 2005. Didion lives in New York City.*

Didion's novel Play It As It Lays *(1970) depicts the emotionally fragile life of a woman married to a Hollywood film producer. The following selection takes place in the isolated desert town of Baker, California.*

from *Play It As It Lays*

The town was on a dry river bed between Death Valley and the Nevada line. Carter and BZ and Helene and Susannah Wood and Harrison Porter and most of the crew did not think of it as a town at all, but Maria did: it was larger than Silver Wells. Besides the motel, which was built of cinder block and operated by the wife of the sheriff's deputy who patrolled the several hundred empty square miles around the town, there were two gas stations, a store which sold fresh meat and vegetables one day a week, a coffee shop, a Pentecostal church, and the bar, which served only beer. The bar was called The Rattler Room.

There was a bathhouse in the town, an aluminum lean-to with a hot spring piped into a shallow concrete pool, and because of the hot baths the town attracted old people, believers in cures and the restorative power of desolation, eighty- and ninety-year-old couples who moved around the desert in campers. There were a few dozen cinder-block houses in the town, two trailer courts, and, on the dirt road that was the main street, the office for an abandoned talc mine called the Queen of Sheba. The office was boarded up. Fifty miles north there was supposed to be a school, but Maria saw no children.

"You can't call this a bad place," the woman who ran the coffee shop told Maria. The fan was broken and the door open and the woman swatted listlessly at flies. "I've lived in worse."

"So have I," Maria said. The woman shrugged.

By late day the thermometer outside the motel office would register between 120° and 130°. The old people put aluminum foil in their trailer windows to reflect the heat. There were two trees in the town, two cottonwoods in the dry river bed, but one of them was dead.

* * *

"You're with the movie," the boy at the gate to the bathhouse said. He was about eighteen and he had fair pimpled skin and he wore a straw field hat to ward off the sun. "I guessed it yesterday."

"My husband is."

"You want to know how I guessed?"

"How," Maria said.

"Because I—" The boy studied his grimy fingernails, as if no longer confident that the story illustrated a special acumen. "Because I personally know everybody from around here," he said then, his eyes on his fingernails. "I mean I guessed right away you weren't somebody I already knew."

"Actually I come from around here." Maria had spoken to no one else all day and she did not want to go into the bathhouse. She did not even know why she had come to the bathhouse. The bathhouse was full of old people, their loose skin pink from the water, sitting immobile on the edge of the pool nursing terminal cancers and wens and fear. "Actually I grew up in Silver Wells."

The boy looked at her impassively.

"It's across the line. I mean it's on the test range."

"How about that," the boy said, and then he leaned forward. "Your husband couldn't be Harrison Porter, could he?"

"No," Maria said, and then there seemed nothing more to say.

"My room, my game." Susannah Wood was sitting on her bed rolling cigarettes. "So turn up the sound."

Carter walked over to the bank of amplifiers and speakers and tape reels that Susannah had brought with her to the desert.

"Somebody's going to complain," Maria repeated.

"So what," Susannah Wood said, and then she laughed. "Maria thinks we're going to get arrested for possession. Maria thinks she's already *done* that number in Nevada."

BZ looked up. "Turn it down, Carter."

Susannah Wood looked first at BZ and then at Maria. "Turn it up, Carter."

Maria stood up. It was midnight and she was wearing only an old bikini bathing suit and her hair clung damply to the back of her neck. "I don't like any of you," she said. "You are all making me sick."

Susannah Wood laughed.

"That's not funny, Maria," Helene said.

"I mean sick. Physically sick."

Helene picked up a jar from the clutter on Susannah Wood's dressing table and began smoothing cream into Maria's shoulders. "If it's not funny don't say it, Maria."

* * *

"What about Susannah," Maria asked Carter. She was standing in the sun by the window brushing her hair.

"What about her."

Maria brushed her hair another twenty strokes and went into the bathroom. "I mean did you really like fucking her."

"Not particularly."

"I wonder why not," Maria said, and closed the bathroom door.

"Where's Carter," Maria said when she came into BZ's room.

"We had some trouble with Harrison, Carter stayed out there to block a scene with him. You want a drink?"

"I guess so. They coming back here?"

"I said we'd meet them in Vegas. Helene's there already."

"Let's not have dinner at the Riviera again."

"Harrison likes the Riviera."

Maria leaned back on Helene's bed. "I'm tired of Harrison." She licked the inside of her glass and let the bourbon coat her tongue. "Some ice might help."

"The refrigerator's broken. Roll a number."

Maria closed her eyes. "And I'm also tired of Susannah."

"What else are you tired of."

"I don't know."

"You're getting there," BZ said.

"Getting where."

"Where I am."

* * *

They had been on the desert three weeks when Susannah Wood got beaten up in a hotel room in Las Vegas. The unit publicity man got over there right away and Harrison Porter did a surprise Telethon for Southern Nevada Cystic Fibrosis and there was no mention of the incident. When Maria asked Carter what had happened he shrugged.

"What difference does it make," he said.

Susannah Wood was not badly hurt but her face was bruised and she could not be photographed. Carter tried to shoot around her until the bruise was down enough to be masked by makeup but by the end of the fourth week they were running ten days over schedule.

"Was it Harrison?"

"It's over, she's O.K., drop it." Carter was standing by the window watching for BZ's car. BZ had been in town for meetings at the studio. "Susannah doesn't take things quite as hard as you do. So just forget it."

"Was it you?"

Carter looked at her. "You think that way, get your ass out of here."

In silence Maria pulled out a suitcase and began taking her clothes from hangers. In silence Carter watched her. By the time BZ walked in, neither of them had spoken for ten minutes.

"They're on your back," BZ said. He dropped his keys on the bed and took an ice tray from the refrigerator.

"I thought they liked the dailies."

"Ralph likes them. Kramer says they're very interesting."

"What does that mean."

"It means he wants to know why he's not seeing a master, two, closeup and reaction on every shot."

"If I started covering myself on every shot we'd bring it in at about two-five."

"All right, then, it doesn't mean that. It means he wants Ralph to hang himself with your rope." BZ looked at Maria. "What's she doing?"

"Ask her," Carter said, and walked out.

"Harrison did it," BZ said. "What's the problem."

"Carter was there. Wasn't Carter there."

"It was just something that got a little out of hand."

Maria sat down on the bed beside her suitcase. "Carter was there."

BZ looked at her for a long while and then laughed. "Of course Carter was there. He was there with Helene."

Maria said nothing.

"If you're pretending that it makes some difference to you, who anybody fucks and where and when and why, you're faking yourself."

"It does make a difference to me."

"No," BZ said. "It doesn't."

Maria stared out the window into the dry wash behind the motel.

"You know it doesn't. If you thought things like that mattered you'd be gone already. You're not going anywhere."

"Why don't you get me a drink," Maria said finally.

"What's the matter," Carter would ask when he saw her sitting in the dark at two or three in the morning staring out at the dry wash. "What do you want. I can't help you if you don't tell me what you want."

"I don't want anything."

"Tell me."

"I just told you."

"Fuck it then. Fuck it and fuck you. I'm up to here with you. I've had it. I've had it with the circles under your eyes and the veins showing on your arms and the lines starting on your face and your fucking menopausal depression—"

"Don't say that word to me."

"*Menopause. Old.* You're going to get *old.*"

"You talk crazy any more and I'll leave."

"Leave. For Christ's sake *leave.*"

She would not take her eyes from the dry wash. "All right."

"Don't," he would say then. "Don't."

"Why do you say those things. Why do you fight."

He would sit on the bed and put his head in his hands. "To find out if you're alive."

In the heat some mornings she would wake with her eyes swollen and heavy and she would wonder if she had been crying.

* * *

They had ten days left on the desert.

"Come out and watch me shoot today," Carter said.

"Later," she said. "Maybe later."

Instead she sat in the motel office and studied the deputy sheriff's framed photographs of highway accidents, imagined the moment of impact, tasted blood in her own dry mouth and searched the grain of the photographs with a magnifying glass for details not immediately apparent, the false teeth she knew must be on the pavement, the rattlesnake she suspected on the embankment. The next day she borrowed a gun from a stunt man and drove out to the highway and shot at road signs.

"That was edifying," Carter said. "Why'd you do it."

"I just did it."

"I want you to give that gun back to Farris."

"I already did."

"I don't want any guns around here."

Maria looked at him. "Neither do I," she said.

"I can't take any more of that glazed expression," Carter said. "I want you to wake up. I want you to come out with us today."

"Later," Maria said.

Instead she sat in the coffee shop and talked to the woman who ran it.

"I close down now until four," the woman said at two o'clock. "You'll notice it says that on the door, hours 6 a.m. to 2 p.m., 4 p.m. to—"

"6:30 p.m.," Maria said.

"Well. You saw it."

"What do you do between two and four."

"I go home, I usually—" The woman looked at Maria. "Look. You want to come out and see my place?"

The house was on the edge of the town, a trailer set on a concrete foundation. In place of a lawn there was a neat expanse of concrete, bordered by a split-rail fence, and beyond the fence lay a hundred miles of drifting sand.

"I got the only fence around here. Lee built it before he took off."

"Lee." Maria tried to remember in which of the woman's stories Lee had figured. "Where'd he go."

"Found himself a girl down to Barstow. I told you. Doreen Baker."

The sand was blowing through the rail fence onto the concrete, drifting around the posts, coating a straight-backed chair with pale film. Maria began to cry.

"Honey," the woman said. "You pregnant or something?"

Maria shook her head and looked in her pocket for a Kleenex. The woman picked up a broom and began sweeping the sand into small piles, then edging the piles back to the fence. New sand blew in as she swept.

"You ever made a decision?" she said suddenly, letting the broom fall against the fence.

"About what."

"I made my decision in '61 at a meeting in Barstow and I never shed one tear since."

"No," Maria said. "I never did that."

SUSAN STRAIGHT

Susan Straight was born in Riverside, California, in 1960. She is a professor of creative writing at the University of California, Riverside, and among her many awards are a 2007 Lannan Literary Award for Fiction and a 2008 Edgar Allan Poe Award for her short story "The Golden Gopher." Straight is the author of six novels, including The Gettin Place; I Been in Sorrow's Kitchen and Licked Out All the Pots; Highwire Moon *(a finalist for the National Book Award); and* Blacker Than a Thousand Midnights. *She has also published one collection of short stories,* Aquaboogie *(winner of the Milkweed National Fiction Prize); one novel for young readers; and one children's book,* The Friskative Dog.

The following short story, "Cellophane and Feathers," first appeared in Aquaboogie *(1990). The story's protagonist, Roscoe, is a prisoner who has been assigned freeway trash pickup duty in the desert along Interstate 10.*

from *Aquaboogie*

Cellophane and Feathers

February

Mockingbirds must have been fighting over the territory at the edge of the prison, because Roscoe had heard them singing frantically, at their best, some time after three that morning. He lay there until daylight and couldn't go back to sleep, listening to the birds twitter, call, warble, each song more anxious and perfect than the last.

He knew what they were doing only because his son Louis had come home years ago from the junior-high library to tell him that the mockingbird who sat on the telephone wire above their fig tree wasn't singing for joy all day and night that spring, but fighting to keep other birds away. By the time the road crew got on the bus, Roscoe's head banged against the glass when he fell asleep; he sat next to the window, trying not to think about Louis, and shook his cheeks to the left and right to wake himself up. But Jesus Trevino's hair was black and iridescent, feathered as a crow's wing, in front of him. Damn, he thought, birds all night, birds all I'm likely to see today. Crow time right now, when at home on the Westside the only people who would be up were Red Man, Floyd and him, the ones who worked outside and started early. Roscoe let himself wonder what Louis was doing this early, where he was sleeping. Louis was the one who called it crow time, had his father thinking of it that way ever since. The huge black birds thought they *owned* the streets this time of morning, stalking up and down the asphalt and into the yards, diving and calling to each other. Louis used to come into Roscoe's bedroom, wake him up and ask, "Daddy, what are they saying to each other? They talking about something, huh? But I don't know what it is."

Roscoe saw desert outside the windows, not the landscaping that meant they were headed to work the city freeways, and he rubbed his eyes, the skin in front of his ears. If he'd been one of the old-timey, suspicious women from Mississippi, the ones who always sat talking in squinty whispers when he was small, he would think of magic, that someone was working a strange power on him. Yesterday had been exactly a month done of his three-month time, and the road crew had worked in Rio Seco, along the main freeway that ran through the city. He'd found a matchbook with an elegant, every-colored parrot glossy on the cover. One match left. The bird sat inside a fancy letter P, and on the back "Palms Resort Hotel and Country Club" curled across the shiny white. He stared at the parrot, saw his son's face rapt at the pet store, the vivid reds and yellows, blues and greens of the perfect feathers, and he'd put the matchbook in his pocket. Only two hours later, during the afternoon when the traffic from L.A. back to Rio Seco got heavy and slow, he'd seen Louis.

The Sixth Street exit, main offramp to the Westside, the last place he wanted to be in his bright orange county uniform, matching plastic bag in his fist, the outhouse bobbing along on the truck next to him and Jesus. He'd just dropped another nasty disposable diaper into the bag, and then he heard a thundering drum from a car stereo, jumping against his skin it was so deep and loud. When he turned, Louis' face was just sliding back around, away from him. Louis said something to Jimmy, who was driving the Suzuki Samurai. What all the boys called the dope man's vehicle of choice. Roscoe knew Louis wasn't dealing, but he hung around with Jimmy, partied with him and Lester. The Suzuki was caught on the offramp just ahead of him, and the bass thump went through Roscoe's ribs to his heart, it felt like; Louis didn't turn around, kept his shoulders tight until the music began to fade up the ramp to the green light.

Hell, what had he wanted the boy to do? Wave? Pass him a swig of soda? Holler out, "Thanks for going to jail for me, Pops?" Roscoe listened to the grumpy voices scattered through the bus. Damn, he thought, I'm too old for this shit. He pulled out the matchbook from his pocket, looked at the parrot. Birdman—that was Louis' nickname.

Shit, and whoever was working that evil on him had stirred something into something for the last two days, because when he looked up the desert was butting up against Palm Springs; he flipped the matchbook over to read the script again, shook his head. He'd been born in Palm Springs, hadn't been out this way for years. *Damn*, he was too old for this. Jesus started up again about his wife, the child support he couldn't pay, which was why he got picked up. He was getting out in two weeks, he whined, how was he gonna pay it now that he'd done time for three months? "Man, shut the fuck up," Williams said from across the aisle. "I'm so tired a hearin about your wife I'm ready to kill you so the bitch can get Social Security." The men laughed, and the bus tires crunched on the gritty sand at the edge of the highway just outside Palm Springs, pulling off to let the first ones out.

They didn't double up out here. The trash was scattered by the wind that swept through the desert every day, the wind he remembered from his childhood when it tossed and rubbed sand against the walls of their tiny house on Indian land. Not as much garbage stayed beside the road in the desert, not like the city freeways that they usually worked. All the Rio Seco County system, that stretched to the edge of Los Angeles County and came the other way almost to San Diego. Near the cities, the clothes and cans, shoes and icechests were thick in the oleander and ice plant along the banks, and he was always teamed with Jesus. He'd thought it would be better to work alone, instead of hearing Jesus complain all day, but now he thought that Jesus and his laments about wife and money were almost songlike. Yeah, a boring, repetitive song, but background music, where you could listen if you didn't want to let your brain work. And Jesus was a couple of miles ahead, probably talking to himself.

The wind blew the bag against Roscoe's legs and he stood, looking down from the roadside at the shifting grains of dust. The sun was winter-bright, and he felt his cheeks rise close to his eyes. He still looked young, not many lines around his eyes because he'd always worn a hat when he worked, he thought; no lines around his mouth because he didn't smile. All those years of trimming trees, hauling brush and branches to the dump with Red Man, but his skin was still smooth. Yeah, and so? So the judge hadn't believed he was almost forty-two when he sentenced him to road camp. Usually the younger guys got camp.

Red Man was only a few years older than Roscoe, but he looked sixty from all his squinting and frowning and laughing in the sun. If my brain could have wrinkles, though, Roscoe thought, from too much frowning and squinting and shit, it would damn sure be cracked and lined as the mud down there. The ditch that cut between the railroad track and the highway was dry patchwork. Like a quilt, he thought, mud in little patches, the edges curled up because no one had sewn them down. He stood very still, didn't feel the twisting bag slick against his arm, and saw the image, all colors and the rims of the mud squares, and then a car passed, tore a seam through the air. How poetic. He bent to the ground. Yeah, put another Coors can in the bag. You're not a poet anymore. Pick up the trash and leave the metaphors in the ditch.

But they kept coming, made everything look like what it wasn't. The brittle bush with stiff gray branches, bare clumps in winter, caught Kleenex and baggies and wrappers; some of the plants had so much shredded tissue hanging from the stems that they looked like cotton, ready to pick.

"Niggers and Mexicans picking cotton," he said into the wind. A Cadillac rushed past, going at least eighty, and he shouted, "Your tax dollars hard at work." Only February and hot as hell, sweat already between my shoulders, but don't worry about us in the heat. That's why we were imported from Africa in the first place, remember? We could stand it better than those paleface settler types. And Mexicans, they *beg* to work in the fields, right?

He pulled at the tissue. Where had he seen cotton? He walked, saw the long stretches of fields, the huge bales of picked cotton and the snowlike traces in the ditches by the road. The Golden State Freeway, Interstate 5, from L.A. all the way to up near Oakland, and he'd been riding with Claude Collins, keeping him company on his trucking route. Claude had stopped the truck for Roscoe, smiled when Roscoe got out and said, "Man, I've only seen this stuff in pictures! Look at it. King Cotton!"

"I seen enough of that in Mi-sippi to last me till I die, man. Get back in here, we got time to make," Claude said, shaking his head.

Everyone on the Westside—Red Man, Claude, Lanier, Floyd, their wives— they were all from Mississippi, Louisiana, Oklahoma. They laughed at Roscoe when he talked about being born in Palm Springs. He watched the cars flying past. "With me you got the best of both worlds today! How would you know that?" he shouted, asking them. "A nigger born in the desert, got some Mexican blood, some Agua Caliente Indian somewhere in here." He saw the hills purple behind the cars. "No, you say, you didn't know niggers *lived* in Palm Springs! Don't let that stop you—you gotta get to Palm Canyon Drive, to the shopping." He thought of bumper stickers he'd seen—Born to Shop, and Shop Til You Drop—and laughed.

"Those Indians owned all that land you heading to," he said to a dark blue Mercedes, taking out his matchbook again. Section 14, the only place colored people could settle when his father came from Mississippi, and since the Indians owned it, the few white people that he saw when he was small were selling something or collecting money. But it was all underneath hotels now. When he was eighteen, in what—'66?—the resorts wanted Section 14, so all the houses were condemned. No one would sell them land, except on the Northside of the city, another square, and Roscoe told his father he was leaving. "Ain't nobody telling me where I can live," he said, and his father shouted, "All them hotels goin up, you know you gon get a good job. Where the hell you think you goin?"

"To L.A. I'm a poet," Roscoe said, and his father laughed, threw balled-up sheets of notebook paper at him.

"You gon write poems to the bank?" he yelled. "You fixin to eat them goddamn poems?"

Roscoe touched the brittle bushes now, stroked the spiny branches, the spires. Blowing air out through his nose, he shrugged his shoulders. The first poem he'd ever written was for a bush. He walked through the desert, going home from high school, and close to the road he saw a big, pointed bush; someone must have emptied out an ashtray there, because butterscotch wrappers, deep clear gold, and foil from chocolate kisses studded the branches. In the desert sunlight, it was a Christmas tree, glittered and reflecting against his arms, and he wrote a poem for the bush. His English teacher said it was very good, but not what she'd expected from him. "Aren't there things you want to write about from your experience, your race?" she said.

The flatbed with the guards came toward him slowly, and he saw the outhouse rocking in the wind. He turned away. Nobody ever used it—there were plenty of bushes, gullies, where you could pee. And you held the other till you got back to "the facility."

Frayed steel belts, black and flattened like dead snakes, lay along the asphalt. He kept walking. They didn't keep an eye on you as close out here; where you gonna run? Long way to anything. And you definitely weren't getting a ride from somebody passing by—nothing but Cadillacs, BMWs, Mercedes filled with one kind of people, and the four-by-four trucks carrying off-road vehicles and the other kind that came out here. Less than likely to pick up a brother.

They were driving like maniacs, too, and the traffic was getting heavier. Why were there so many cars today? He paused to pull a piece of white plastic from a stretch of barbed wire; the wind puffed it tight as a pillow between the strands. OK, it was Friday, and Monday was a holiday. Presidents' Day, uh-huh, people starting their patriotism early.

The trucks weaved around the older people who cruised stately in their big tanks. A few trucks even passed on the shoulder, raising swirls of dust down the road from Roscoe. Shit, if he did that once he'd get a ticket so fast the dirt wouldn't have time to settle. He wondered whether the Rio Seco cops would be giving him tickets for the rest of his life, to pay him back, or if the three months he was doing would be enough to turn him into a good nigger. Three months time for tickets.

After the two white policemen had been shot on the Westside in 1973, he'd been taken down to the station along with every other black male wearing white shoes. That was what the suspects wore. How old had he been? He tied the plastic bag closed and left it on the edge of the road. He'd come to Rio

Seco from L.A.; he was twenty-two. Spent a weekend in jail with all the other Westsiders, getting yelled at, threatened. "Who shot them?" over and over.

Then last summer, an addict named Ricky Ronrico shot two cops coming for him on a parole violation. Shot them on the other end of the city, but of course they came looking for Ronrico on the Westside. And they'd been about to take Louis by the second night, because he and Lester and Jimmy were fool enough to stand on the corner and talk shit on that long weekend when cops were everywhere and angry. Rio Seco's finest cruised past again and again, and when Louis came in the house, Roscoe said, "go in the back room and shut up," even though he hadn't seen Louis for a couple of weeks. The two officers were in the process of tearing up the living room, talking about evidence and getting ready to work their way to the back, and Roscoe stood there and hollered that he'd sue till he was blue. The call came over their radio that Ricky Ronrico was found, in a house on the south end next door to where the shooting took place, but the next week, Roscoe started to get tickets.

The officers did it practiced and perfect as he and Red Man clearing out the trees and brush from a vacant lot. Red Man with the chainsaw, then Roscoe with the stumper, chewing the wood to chips. The motorcycle cop, stationed around the corner from Roscoe's house, then his buddy, pulling Roscoe over about a mile away to tell him he'd been weaving or making illegal lane changes. Fix-it ticket for the broken windshield on the truck. No seat belt, no insurance—that was a hundred bucks. Another seat-belter, and then no car seat for his baby granddaughter, Louis' girl. "And how you gon tell me what I have to do by law to save my life, or hers?" he said to the motorcycle cop, and the man's sunglasses slid closer to his moustache. Warrants were out by then, and when Roscoe's arms wouldn't meet behind his back because of their shortness and muscle, it was obvious that he was resisting arrest.

The Westside was cleaned out—some went for receiving stolen property just after two white guys came by selling tools and gardening equipment, and some went for old warrants. Make us all good niggers for another ten years.

And Louis was still in the streets, Red Man wrote in his letters. Nobody was at Roscoe's house; his granddaughter, whose mother had dropped her off with Louis, a three-month-old baby, and never come back, was with Big Ma. Around the corner from Red Man's house. Maybe they'd come to visit this weekend.

He heard droning from up ahead, a constant whining like giant wasps or mosquitoes, and he walked faster until he reached the bridge over a dry wash. A long line of four-by-four trucks was parked in the dusty streambed, in the shade of the bridge, and women sat in the truckbeds sipping drinks and watching the men and boys ride up and down the hills in their three-wheelers.

The tracks crisscrossed the hills like ribbons, like the land was gift-wrapped. Roscoe stared at the precise trails, but then he felt the riders begin to watch him, and heads turned to look. He walked off the bridge, away from the high buzzing and the smiles.

A forest of wind machines stood still and idle past the bridge. The wind scoured his ears, hot and insistent, but the blades didn't turn, and he remembered that some doctor Red Man cut yards for told him that the machines were tax shelters. It didn't matter if they worked. To Roscoe, they looked like trees, trees a maniac had pruned.

When he first began trimming trees, after he'd come to Rio Seco and gotten a gardening job at the university, he'd been afraid to cut too many branches. The supervisor yelled at him until he got ruthless, pruned them right. When Roscoe met Red Man, while they took their breaks and watched the students, he'd shown Red Man the poems about palm trees, about cars and women. But he'd never gotten any of them published; people he sent them to wrote back to say that a black poet had an obligation to write about his race, or that they didn't understand how his imagery fit with power and revolution.

He and Red Man quit the university to start their own gardening and tree-trimming business. Louis was three, Roscoe's wife Joyce was mad, and she left, went back to L.A., saying any fool who quit a job with benefits for a beat-up truck and climbing trees was too foolish to support *her* right. Roscoe told her to go to Palm Springs, visit his father's grave, and tell his father all about it. When she was killed in a car accident, her sister brought Louis back to Rio Seco, to live with Roscoe. Louis was six then, tall and thin, with fingers long as these brittle stems. He was quiet for months, walking around the yard, looking at the trees and birds, at Roscoe; he'd seemed soft, too quiet, but then he began to play with Red Man's sons, shooting baskets in their driveway.

Roscoe looked back at the wind machines, then ahead to the long stretch until the hill where he was supposed to stop for lunch. Nothing now but barbed-wire fence and telephone poles straddling the gully. He went partway down into the ditch and peed in a long crack that rainwater had worn into the bank. A red hawk bore down suddenly and landed on one of the telephone poles near him, settling to look out over the desert for rabbits. Roscoe stayed still for a moment after he'd zipped up his pants, but the hawk didn't even turn.

The red hawks still hunted the fields near the Westside; he wondered if Louis ever watched them now, hurt his neck looking up for so long, the way he had when he was small.

His son had always been in a trance when the flocks of crows passed over the house in winter, at five-thirty every night, Louis told Roscoe. It took them an hour to cross the sky, crowds of them and then a few stragglers, more pecking and fighting in a stream of black. They nested near the river-bottom, Louis said. He had books with pictures of every bird in North America, and he waited for robins, scrub jays, any of the birds with color. Once he came running inside to tell Roscoe he'd seen a cedar waxwing in the fig tree. Roscoe had said, "Okay, and now he's probably gone. And?"

The kids called Louis "Birdman" one summer when colored feathers were all he looked for in the streets; they laughed at him, pulled him away to the gym to play basketball. And the name got confused when he was in high school, when he was six-five and his hands as big as fig leaves, when the coach pressed him to play harder, to jump, and everyone thought he'd been called Birdman because he could leap. He had a vertical of forty-two inches, but the coaches said he was lazy, that his mind was always somewhere else. Roscoe hollered at him, told him if he didn't start playing better, he'd never get to college, and Louis began then to keep his face as distant and new as when he'd been six.

Roscoe came up out of the gully; he couldn't be disappearing like that for too long. The hawk swept down from the pole, but not because of him, he saw; it curved toward a spot far into the desert, and snatched something off the ground.

Roscoe faced the traffic coming toward him, and now the cars were moving so slowly faces were visible in the windows. A winter holiday in Palm Springs. Snowbirds—what the older tourists from Canada and the colder states were called, the ones who crowded the desert every winter. He saw them stare at him, saw small dogs in the windows sometimes. The kind that were either wrinkled or shaved, pressing their noses to the glass to see him. Some of the cars had kids in the back seats, and he watched all the mouths in those cars moving. "I'm on the chain gang," he said into the wind they made, and he knew all the windows were rolled up anyway. The parents were telling their kids, "Those men in orange are bad men. They committed crimes, and now they have to pick up trash."

Yeah, I *am* a bad man, he thought. I did lots of bad things. He couldn't stop seeing Louis' face kept turned away from him, the Suzuki pumping out music, and these faces all turned to watch him in silence, the purr of expensive motors lost in the racing air.

He'd done bad things. He'd taken Louis up that same highway where cotton laced the barbed-wire fences, up I-5. A small state college in Northern California had offered Louis a basketball scholarship, a free ride, and Roscoe drove him there in the truck, motor rattling loud in the cab. Louis had never been past L.A., and Roscoe hadn't gone anywhere since riding with Claude years before. Hawks sat on fenceposts along the road, and Louis' head swiveled to look at each one. His voice grew higher, excited, while he ignored Roscoe's talk about basketball and the team. "Look," Louis said, "the hawks guarding those fields. Like each one is saying, okay, this is *my* thirty-nine posts, and don't none a you come near them. I wonder if it's thirty-nine…"

He chanted words to Roscoe, to the window, all the way north. "Egrets," he shouted, and Roscoe hadn't heard him so loud, sitting so close to him, ever before. "I can't believe it, a crane, look, standing in that water by the field!"

Mud nests were mounded on the cement overpasses, made by swallows, Louis said, and once they saw a shimmering tornado of blackbirds rise from a plowed section of land, bodies turning and twisting in perfect unison, swirling and riding to fall and curve back against the sky. Louis was silent for a long time. And when they came closer to the college, Roscoe told him to forget about the bird foolishness and start thinking about practice. It would be harder than high school, and he didn't want anyone telling him his son was lazy and absentminded.

Like his own father had told him, all day, every day, in Palm Springs. Roscoe couldn't stop himself, though, his voice deep and harsh, saying, "You want to see some birds, buy a canary. Put him in your dorm room, and he can sing to you after you win."

On the way back down, after he'd seen the coach, shaken hands a couple hundred times and left Louis in the gym office, thick fog had settled near the college, and all the way down the interstate. Tule fog, he remembered Claude calling it. He could barely see the fences, the hawks hunched down into their thick necks, feathers puffed ragged against the cold wind. They were too close together on the posts, he found himself thinking, watching, and quickly he was angry. Who gave a damn about how close the birds were to each other? But when he passed the exits for Bakersfield, the sun came back, and he stopped to stretch. One of the big white birds that had made Louis' voice jump high in the cab stalked near a pool of irrigation runoff. Was that an egret? It was clean, stark white as the new socks he'd bought for Louis, and with a long, curved neck held high, the bird looked tall as a child.

Roscoe dropped another orange bag. People were still staring at him, he knew, but then he saw a circle of clear, blue-green water ahead, sparkling, wavering in the heat. He almost ran to be closer, and came to a mound of broken windshield glass in a small turnout, aqua chunks that were so greenish moss could be under their surfaces. He didn't even look at the strips of tire nearby, the burned car seats. He laid the new plastic bag down and sat on a rock near the glass, imagined it water. Not much water in Rio Seco; dry river, that was what the city had been named for. But Louis went to the man-made lake with the other boys, and they told Roscoe that while they fished and caught crawdads to boil for dinner, Louis watched the ducks, the geese. They laughed at him, said to Roscoe, "We told him he can't eat no duck, but homeboy wanna *feed* them ducks. Don't come home with nothin, either."

Louis lasted only a year at the college. He came back to the Westside loud and telling everyone he didn't get enough playing time, in a hick town anyway. No soul, nothing to do. Shit, Roscoe thought, listening to the plastic bag ripple near him, what would he have done learning about birds? Be a professor of birds? What was that word—he'd known when he worked at the university. It came to him after a few minutes—ornithology. Yeah, and what was Louis going to do with that? Look through some binoculars?

He could have been a dope dealer anyway, taught trained pigeons to deliver bags of rock cocaine that Lester and Jimmy sold. Hell, I don't know.

Roscoe heard a low noise, knew the flatbed, the outhouse, the guards were coming up the shoulder of the road, passing the stalled traffic. I don't give a shit, I ain't getting up. Crow time was long over, it was almost lunchtime. But cat time would be the hardest—when the afternoon turned purple and he was back at the facility, when the evening was coming fast and Louis used to say, "Why do the cats come out now? How come they all sit on the porches at the same time? How do they know what time it is?"

Roscoe pulled at his eyebrows, listened to the flatbed with his eyes trained on the glass. The mockingbirds would have a songfight again tonight. He closed his eyes to keep the heat inside them. He'd never known what to answer when Louis asked, when he sat at the edge of his father's bed. "Why's the mockingbird singing *now*? It's three in the morning."

"He's drunk."

"No, Daddy, he ain't singing like he's drunk. Like he's sending a message."

"Message is go to sleep."

When he woke again, Louis' eyes were shining, reflected in the streetlight. An hour had passed, and he was still listening, trying to decipher the song.

BECKIAN FRITZ GOLDBERG

*Beckian Fritz Goldberg was born in 1954 in Hartford, Wisconsin, and holds an MFA from Vermont College. She is the author of six poetry collections—*Body Betrayer, In the Badlands of Desire, Twentieth Century Children, Never Be the Horse, The Book of Accident, *and* Lie Awake Lake*—and she has won many awards for her work. A professor of English in the creative writing program at Arizona State University, she lives in Phoenix.*

The poem reprinted here, "A Dog Turns Back When You Turn Back," originally appeared in the collection Lie Awake Lake, *published in 2005 by Oberlin College Press.*

A Dog Turns Back When You Turn Back

The desert won't get off my back
not even here in the green heart of Italy.
All night I lie awake and the cuckoos
repeat from the lindens repeat
the hours reliving themselves.
Then I lie on my back with

the desert beneath me.
My very gravity I feel depends
on this—cool wind, a flock of pansies
lighting in the window box, a flock of inks,
a flock of backward looks
you won't forget. I know

you can live in two places at once:
Whenever I go to bed the desert's there.
I say, "I thought I left you, I thought
we had it out." The desert
hardly stirs. My leg in the sheets—
the rustle of a lizard
and no breeze: No matter which way

I turn, the desert stretches out,
content. I say, "Goodbye."
I say, "I don't love you."
And the whole pillow exhales its

creosote, its turpentine bush,
and even moonlight full of
strange silhouettes and valleys.
I blame my eyes and roll onto
my side and stick out
my elbow just for spite. But

you know the desert doesn't budge,
lies like a drunk who hasn't shaved,
who can't remember where he went
or how, and drinks to not
remembering—a little birthday
eternally when the sun comes up.
When the sun comes up I open one eye
and, since it's still there, flat
and littered with arms of chainfruit cholla,
"I hate you, I hate you,"

I say and nothing listens. Of course,
nothing. Though the quail
go strolling the dry arroyo down
to the bedpost. I blame
my breastbone. I blame my ear.
I blame the globe mallow
and the city of rock squirrels.
Cursed, I begin to sing.

The song goes, "Here, here
where the sea got up and left
I lie down." I blame my skin,
dry and withered and,
at that moment when the desert
drifts off into dream, I put
my hand into the blank space
beside me: out from it go
rings of terrible childhood, Sundays
in the shite glare of gravel and
dusts, a brand of endlessness
I both fear and crave in large

doses—the result of this lullaby,
this thing near death
beside me, clatter of mesquite pods
falling on one another
and the sound of paper,
the sound of shore, the song of
what was the song...

that lovers and ex-lovers ask
when they try to go back—
Nothing follows *them* around, no
scorpion waiting in their shoes,
no mean little needles pricking
through their clothes. "Leave me
alone, leave me alone," I cry
as if I were not already

thirsty and desolate. Blaming
my feet. Blaming my tongue.
Love is never, no matter how many times
we say it will be, any different.
I come from a brutal, unbearable place
and every time, even in the trees
and grapevines terraced
in the valley, the desert sits on my chest
and begins to beat.

KENT NELSON

Kent Nelson has published four novels, including Land That Moves, Land That
Stands Still *(2003), and five collections of short fiction, the most recent of which
is* The Touching That Lasts *(2006). He and his daughter edited* Birds in the
Hand: Fiction and Poetry about Birds *(2004). Many of his stories are based on
observations and interactions with the natural world. He has a JD from Harvard in
environmental law but has worked at various unusual jobs such as city judge, squash
coach, and hired hand on an alfalfa ranch. He lives in Ouray, Colorado.*

The following short story, "Irregular Flight," takes place at the Salton Sea.

Irregular Flight

Claire heard about the vagrant Cook's Petrel at the Salton Sea on a Thursday
morning in October on the rare-bird tape in Los Angeles, and she called me right
away in Tucson. She had a mandatory lunch meeting at her lab in Pomona, but
offered to meet me at five at the post office in Indio. Indio was six hours from
Tucson, but how could I refuse? Cook's Petrel was a *Pterodroma* that nested on the
islands off New Zealand—thirteen inches from head to tail, black M pattern across
its gray wings and back. Except for its breeding period, the Cook's, like other
gadfly petrels, spent its life wandering erratically around the Pacific. It fed on the
wing and rarely alighted on water. Its status off the California coast was unclear.
Over the years, several individuals had been photographed on pelagic trips, but
none had ever been seen inland, none ever before at the Salton Sea.

When I arrived at the post office, Claire was sitting on the ground in the
parking lot in the shade of her Land Rover, dressed in shorts and a loose khaki
shirt and hiking boots, beautiful as ever. Since I'd last seen her months before,
she'd cut her hair and had lost weight, but what struck me more than her
appearance was that she was older, if thirty could be called old. Something
had changed in her life. She tilted her head and shielded her eyes from the
sun, then stood up, uncoiling her body from the earth. Her movements were
patient, even restrained, but they belied the intensity of her personality: She
was too smart to be casual or calm.

"Let's take my car," she said. "The light's going fast. We should hurry."

"Is it all right to leave my car here?"

"Park it on the street," Claire said. "No one will notice."

I drove my Corolla to a side road, fetched my scope and overnight bag from
the trunk, and loaded what I had into the back of her Land Rover, beside the
cooler and campstove and a sack of groceries from Von's. My binoculars were
already around my neck.

"Who found the bird?" I asked when I got into the passenger side of the Rover.

"Strachen Donelly."

"You think the sighting is reliable?"

"One hundred percent. You know Strachen. He gave excellent directions on the tape."

"Is it far?"

"North shore, mouth of the White River. There are details. It's maybe twenty miles from here. He saw it late yesterday afternoon."

"So it might be gone already."

"A petrel could be anywhere," Claire said, handing me a piece of paper. "Here, you navigate."

In Tucson Claire was married to a man I never met. She was a biologist, dark hair to her shoulders, a little overweight. We shared a ride to the research institute on Tanque Verde, thirty-five minutes each way, and spoke mostly about our ongoing projects monitoring the effects of radiation on flora and fauna in the Southwest. Our relationship was professional: she was a biologist, I a chemist. She never asked me a personal question and, though I knew she was married, never confessed anything about herself. Still, I was attuned to her moods. She was cheerful early in the mornings, moody and quieter in the afternoons. I couldn't know whether these differences were derived from what she had left at home or what she was going toward in the evening or whether they were related to what happened during the day at work. Always, though, morning or evening, I was surprised by her observations. She noticed wind directions and cloud formations, whether a street vendor had changed his prices, that a woman's baby carriage was missing a wheel. She paid attention to everything around her, except me.

She knew nothing about me. I lived alone, had never been married, and was at ease in my solitariness. I'd never needed to be with people and had few expectations or desires. My father had died when I was a child, and my mother, sensitive and repressed, lived a recluse's life in Boston. I'd done well at Haverford and had a PhD in biochemistry from Princeton, after which I signed on with the government research program. The work was challenging, but I had little ambition. I spent my free time looking for birds.

I was an insomniac, though, and to appease the darkness, I watched the news on television at all hours of the night. Something was always happening somewhere, in another time zone where it was day—I might see a bomb fall into a building in the Gaza Strip or feel an earthquake shake a town in Indonesia or watch the unearthing of a mass grave in Rwanda. I witnessed not only the aftermath of events, but actual occurrence. Even if what I saw was videoed by a hand-held camera or filmed by a reporter, it was real. I felt as if I knew everything, everywhere all at once.

One day on a fact-finding excursion to a missile range west of Tucson—there were six of us; I was driving—a bird flew low through a tangle of palo verde and mesquite. It was long-tailed, grayish, bigger than a robin. A thrasher, I thought.

"Crissal Thrasher," Claire said.

I reconfigured what I'd seen: yes, larger than a Sage Thrasher, darker, too, not a Curve-billed Thrasher because it had a rusty undertail patch. Its bill was curved more like a California Thrasher's, but that bird wasn't in Arizona.

"Do you know birds?" I asked.

"A little," she said. "Do you?"

"A little."

"I like thrashers because they're secretive," Claire said.

"So is the government," someone else said.

And we talked about the high contaminant levels we had found on our core sampling area, the radiation in the ground-water, the two dead lizards. But I was still thinking of how Claire had identified the Crissal Thrasher. How had she observed so much in so brief a moment? And why had she kept such a secret as bird-love from me for so long?

After that on our daily commute Claire and I talked birds. She knew courtship rituals, food sources, habitat, range overlaps. She had pursued birds with a passion in quest of rarities, to Alaska, South Texas, Maine, Point Pelee and Key West. She had done boat trips off Hatteras and Monterey. What she knew was far beyond my ken. She had spent weeks once on a rocky island in the Bering Sea studying the behavioral relationship between Arctic Foxes and cliffside bird populations. Foxes had been introduced to kill rodents, but once the rodents were hard to find, the foxes preyed on the nests of kittiwakes and puffins. But the birds adapted, too, especially the kittiwakes. When a fox appeared, they flew away in alarm, leaving their nests exposed, and the fox was lured onto the cliff by the cries of abandoned fledglings. When he'd gone too far, the birds swooped back and knocked him from the cliff into the water two hundred feet below.

Compared to Claire, my skills were amateurish. But in all her travels and hours in the field and on the water, she had never seen a Cook's Petrel.

* * *

We drove from Indio to Mecca, passed dozens of migrant workers hoeing lettuce in red and yellow and blue shirts and dresses, slid by the groves of date palms and oranges. To the west were the treeless, sun-weary Santa Rosas, and southeastward, the Orocopia Mountains and the Chocolates that on either side framed the valley with a receding jagged horizon.

"How is work?" Claire asked.

"I can't complain."

"You never do. No new discoveries?"

"We found radiation at Cabeza Prieta," I said, "but we knew we would."

"The government denies it?"

"You know the story. They pay for the research and then hide the results. What about you? Do you like the new job?"

"Oh, you know." She laughed and didn't elaborate.

I remembered last April when we were driving home—it was a Friday, and Claire was at the wheel. She'd slowed for traffic at a red light. It was hot. Neither of us liked air-conditioning so the windows were open. A Cactus Wren was singing in a gully nearby. My arm was slung out on the window ledge. Her hand was on the stick shift between us. "My husband got promoted," she said. "We're moving to Los Angeles."

The Cactus Wren sang again. The light changed, and the cars ahead of us moved forward. I smelled exhaust. I remember the colors around me faded—green palm trees, red neon, the blue sign of a bar. It was the first mention of her husband. I conjured up questions to ask—What about your job? How can you leave the desert? Is this what you want? But I asked nothing. We passed through the green light and stopped again in the dry heat.

"Where from here?" she asked.

I roused myself from the pointless memory and looked at the directions. "Proceed to Mecca," I said. "Then we turn right."

Mecca was a few rundown pastel buildings—a general store, a gas station, and a café. The houses and trailers had trellises covered with brilliant red bougainvillea, and in the dirt yards were empty fruit crates stacked high.

"Turn on Road 42," I said. "Go a half-mile to an adobe house. There'll be a bridge. Turn left toward the sea."

"I told you the directions were clear," Claire said.

"They don't show us the bird," I said. "Where is the water?"

Through the windshield the mountains were visible, sunlit on one side of the valley, shadowed on the other. There was no sea, only the absence of trees where the water must have been, rows of date palms in the fading light, and the enormous, wide sky.

* * *

After Claire left Tucson, I was lonely at the institute, and I focused on my projects with a vengeance. I proved chromatic genes had been altered in rattlesnakes, that the petals of cactus flowers were contaminated, that Verdins were sick, but so what if genetic predispositions had been altered for generations? The government was silent. No accounting would be made, no guilt assigned to responsible parties, nothing would change. I knew my work suffered.

Weekends I went to canyons in the mountains and camped—Cave Creek in the Chiricahuas, Madera in the Santa Ritas, French Joe in the Whetstones. In the mornings I found Sulphur-bellied Flycatcher, Grace's Warbler, and Hepatic Tanager, and in the afternoons I read in the sun and let the heat erode my body. At night I listened to the soft calls of poorwills and the Elf Owl's chattering song.

Always there were other events in the world: a train crash in Rhodesia, the assassination of the prime minister in Norway, a wildfire in the Carpathian Mountains.

And Claire was gone.

The Land Rover raised dust behind us on the dirt track. Ahead was the adobe house and to our left the steel bridge appeared. We clanked across it and ran a hundred yards or so along a broken fence line. The smell of orange blossoms was thick on the breeze. A narrow lane opened up. Claire threaded the Land Rover between a tangle of mesquite and a broken-down Ford parked in the weeds by a canal.

"Is this right?" I asked.

"You have the directions."

"They don't say anything about after the bridge."

We proceeded to a nearly dry alkaline pond where four Black-necked Stilts dipped their long bills into the scum. A few peeps flew up and swerved over the canal. We climbed a low rise, maybe ten feet, and there before us was the sea, a great blue-gray sheen without wind or sunlight. High, white cirrus clouds were reflected in the water, and the mountains in the distance on both sides were shrouded in a blue haze.

The sea lay in a sink created eons ago by the uplifting and erosion of the mountains. The original river, now called the Colorado, had silted in and changed its course eastward, leaving the Salton Basin without water. For thousands of years nothing happened. But history proceeded. The West was settled and grew; California had a good climate and became the most populous state. It was also a food producer, and at the turn of the last century, the U.S. Army Corps of Engineers decided to construct an irrigation canal to the Imperial

Valley. Typically, the government miscalculated, and during a high spring run-off, the river cut a new channel before the canal was ready. For eighteen months the Colorado River flooded into the Salton Basin, and a new sea was born.

In the winnowing light, this was the place we'd come to, and Claire drove ahead to the collar of gray sand bordered on either side by thick saltillos.

"Strachen saw the bird here," Claire said. "From the beach."

Ducks and grebes floated on the near sheen, and a few gulls whirled in the air above us and out over the water. We got out and scanned the sea with binoculars. The near birds were familiar—wigeons, coots, shovelers, cormorants. A small flock of sandpipers flew eastwards through the circle of my glasses.

"Do you see anything farther out?" Claire asked.

"Terns and gulls," I said. "And a Black Skimmer. More gulls. Cormorants."

"Let's set up the scopes."

We set up our tripods and scopes on the rise behind the beach. My Questar gave good resolution to forty power, and through it the indistinguishable birds far out became Common Terns, two Black Skimmers, a flock of Cinnamon Teal. Three White-faced Ibises flew as silhouettes against the paling hills.

"Nothing unusual," Claire said. "Do you agree?"

"What was farther out is closer," I said, "but now there are more birds farther still."

"Like stars, when you look through the scope at the night sky," Claire said. "But the petrel has an irregular, bounding flight. We might be able to see that."

"If it's there."

"Yes, if it were there."

The sun vanished from the high clouds, and without the refracted light, the sea diminished to gray. I saw no bounding arcs, no birds I could identify, and while Claire kept looking, I abandoned my scope and walked down to the shore. The smell of acrid, brackish brine rose to my nostrils, but I took off my shoes and socks—I already had shorts on—and waded out around the saltillos at the edge of the beach. The water was as warm as the air. On the other side of the brush, the irrigation canal fed in, and there was a cove filled with drowned trees. Before the sea was filled, before the land had become orange groves, before chemicals had leached in from irrigation, this must have been a riparian woodland fed by springs. Now it was a wasteland, the bare branches of the trees spidering into the air, with herons and cormorants perched on them like huge, grotesque, faintly colored leaves. I felt as if I had stepped into a world already destroyed.

And why did I perceive it so? Was this dark interpretation the creation of my lonely soul? Had I let myself be affected by tenuous conversations with Claire, reading into her words the meanings I wished them to have and yearning for what was not possible? Or was it purely the moment *now*, being with Claire again, that conjured up these feelings? Her desire to know and see birds, which I had once thought was a sharing bond between us, was now a division.

I imagined an earthquake happening at that moment in Malaysia, or a train wreck in France, children starving in Ethiopia—yes, these may have happened. Something like these events happened all the time. But all I knew, really, was the world before my eyes—the cove, the reflected orange of the clouds, the rim of blue mountains. I saw a few date palms with their spiky fronds silhouetted against the barren hills, the dead trees in front of me with a few herons sitting on the branches, the darkening shore beyond.

"Slater?" Claire's voice came to me from the other side of the saltillos.

"Right here," I said.

"Look west, flying low."

I raised my glasses. "How far out?"

"Three o'clock, against the hills."

I waded out clear of the brush and lifted my glasses. Dark birds, backlit, winged over the water. I made out gulls and terns and dark birds on the water.

"Do you see the bird I see?"

"The light isn't very good. Are you seeing the petrel?"

I circled the saltillos to where she was, and she raised up from her scope. "I saw a possible possible," she said. "A different pattern of flight…"

"I didn't see what you saw."

"You're right, though. The light isn't good enough to be positive." She collapsed the legs of the tripod and leaned the scope against the fender of the Land Rover. "Are you hungry?" she asked. "I bought ham-and-cheese sandwiches, potato salad, and beer."

It had not occurred to me to be hungry, and I was surprised by her casual tone and the ease with which she let the petrel go. "I could drink a beer," I said.

She raised the back of the Land Rover and got two Coronas from the cooler. "So there's tomorrow morning," she said, "but I have to leave by nine to be back at the lab."

"I have accumulated leave," I said, "so I took the whole day."

She opened the beers and handed one to me. "But you're going back?"

"Yes."

"I mean, you're staying at the institute. You're frustrated, but you won't quit."

I drank more beer. "It's what I know. What choice do I have?"

"You have infinite choices."

There was a silence, and we stood looking out, sipping our beers. Lights were coming on the shore across from us, tiny points drifting over the water.

"My husband was never promoted," Claire said. "I wanted to quit, so I did."

"Why did you say he was, then?"

She turned toward me, but her expression was unclear.

"I have to pee," she said.

She set her beer on the bumper and walked past the Land Rover and up over the rise.

I climbed onto the hood of the Rover and rested my back against the windshield. The beer was warming, and I drank it quickly and stared out over the paling sea. There was a wisp of orange on the surface, the soft lapping of the waves, insects chirring far away and near. Ducks and coots squawked in the shallows. On the east shore, a train was barely visible, coming toward Mecca. A star or two emerged from the blue that hovered over me.

I waited and closed my eyes, and I must have dozed for a few minutes because when I opened my eyes it was night. Insects were chirring in the saltillos, but the birds were silent. I watched the red blinking light of an airplane cross among the stars. Claire was leaning into the fender of the Land Rover, but in the darkness I couldn't see her well.

"Do you love me?" she asked.

I sat up from the windshield. "You're married."

"Is that an answer?"

"Isn't it?"

"All that time driving together, you never called me."

"You never called me," I said.

"I did today. When I heard about this bird, I thought you might meet me."

"Today there was an earthquake in China," I said. I looked out over the sea to the faraway lights of towns along the black shore.

Claire walked around to the grille and climbed up on her hands and knees onto the hood. "What do you feel, Slater? Do you feel anything?"

I didn't know how to answer.

"Don't turn away. Look at me."

She raised up so she was on her knees and unbuttoned her shirt. Beneath it she was naked, and her skin was pale in the warm air. Her breasts were luminous. I don't know what gave her the courage to risk herself in this way, but I had to look at her.

She took her shirt off and scrabbled toward me. "What do you see?"

Again I didn't answer.

She unsnapped my shorts and pulled them down. "Is this all right?" she asked.

I didn't resist. I felt the warm air slide over my skin.

"Let me," she said.

She touched me. I was afraid of falling and wanted to fall. I was afraid to move. The insects, the lights of towns, the stars dissipated into one sensation, and for the first time in my life I felt the eerie helplessness of desire.

We spent the night on the hood of the Land Rover, and at dawn we dressed and heated water on the campstove and made coffee in a sieve. The air was heavy with dew. The sun threw down its wash of pink into the mountains, and across the water the pale, faraway colors of houses appeared. The sea was flat and windless and reflected the sky.

We took up our vigil on the beach and scanned through our scopes back and forth, back and forth across clouds of gulls and terns already feeding. We were looking for the irregular, bounding flight of the petrel that may never have been there.

JUAN FELIPE HERRERA

*Juan Felipe Herrera, the son of migrant workers, was born in 1948 in Fowler,
California. The author of more than two dozen books of poetry and prose, he has
received many awards and fellowships, including the National Endowment for
the Arts' Creative Writing Fellowship and the 2008 PEN USA Literary Award
for Poetry. He teaches creative writing at the University of California, Riverside,
where he holds the Tomás Rivera Endowed Chair. Also a film and stage actor, he
appeared in 2004 in* The Upside Down Boy, *a musical for young audiences based
on his book of the same title. Herrera's latest book,* 187 Reasons Mexicanos Can't
Cross the Border: Undocuments 1971–2007, *was published in 2007. He lives in
Redlands, California.*

*The poem that appears here, "Loss, Revival, and Retributions (Neon Desert
Collage)," is taken from his poetry collection* Night Train to Tuxtla, *published by
the University of Arizona Press in 1994.*

Loss, Revival, and Retributions (Neon Desert Collage)

I've been bad-mouthing love too much
—chipped white in sadness; the solar collar

above me; an onyx flame. I am running,
smashing my Dodge over relief maps.

There's a peering scarf ahead, maybe Andalusian—
a Ziegfeld dancer with a tender belly? I sing: Rise.

I command her: Rise! Too many roulettes haunt me;
there is a catch in the black and red velvet air.

Nothing left,

but to reach for the pinball-soul inside. Swagger,
stay there—(please stay there) for once. A chrome

pearl speaks in southern Spanish accents. A Moorish boot
swings out: my new master—in oxblood. In chalk dust.

Infinity burns on my tongue.
If only I could spit again, bruise the wastelands;

a polychromatic insignia from this gut vest—to please.
I am not like that anymore. O square-shouldered guitars,

pop me another drink. An off-beat hooligan flamenco
slaps my back—the wind. Bring me fortune, I say—

fortune again—a shredded red star—on the inside
of my knee; my mother's tiny prophecies.

I believe in the Goddess of Oasis and Velocity,
her ragged sky skirt to pull me. My last destination.

There is elegance—a glassy palm tree sparkles ahead.
A starry fruit, bluish hands, waiting. Desire. Or is it electricity?

I am coming up fast without water. The number 27, someone
stands there, with orchids. I can see it all, now. Above
the wild lonely palms.

LURE OF THE DESERT

MARGARET ERWIN SCHEVILL

Margaret Erwin Schevill, a poet and artist, was born in Jersey City, New Jersey, in 1887, and graduated in 1909 from Wellesley College. She later moved to Arizona, where the desert landscape inspired her work as a poet and painter, and where she also taught English at Tucson High School and developed an interest in Native American culture. She lived with her husband for many years in Berkeley, California, and died in Tucson in 1962.

The following poem, "Desert Center," was first published in 1952.

Desert Center

There's a magnet in the desert earth,
Or a meteor buried deep,
An old Indian said.
It fell there long ago,
A black star with a tail,
A long, lizard tail,
And it fell
To show
The earth people
Where the center was.
Deep down in the earth
It is there,
For it draws us
And it draws us,
And it draws you, too,
And you will always return.

JOHN C. VAN DYKE

The writer John C. Van Dyke, who was also an art history professor at Rutgers University, was born in 1856 in Brunswick, New Jersey. His landmark book, The Desert, *was first published in 1901, widely acknowledged as a classic, and was one of the first works of modern literature to characterize the desert as a place of beauty and allure. The book is based on observations and experiences Van Dyke gathered during the nearly three years, starting in 1898, he spent traveling across the deserts of the Southwest. As Van Dyke put it, the desert "never had a sacred poet; it has in me only a lover." He died in 1932.*

The following excerpt is taken from Chapter 2 of The Desert, *"The Make of the Desert."*

from *The Desert*

The first going-down into the desert is always something of a surprise. The fancy has pictured one thing; the reality shows quite another thing. Where and how did we gain the idea that the desert was merely a sea of sand? Did it come from that geography of our youth with the illustration of the sand-storm, the flying camel, and the over-excited Bedouin? Or have we been reading strange tales told by travellers of perfervid imagination—the Marco Polos of to-day? There is, to be sure, some modicum of truth even in the statement that misleads. There are "seas" or lakes or ponds of sand on every desert; but they are not so vast, not so oceanic, that you ever lose sight of the land.

What land? Why, the mountains. The desert is traversed by many mountain ranges, some of them long, some short, some low, and some rising upward ten thousand feet. They are always circling you with a ragged horizon, dark-hued, bare-faced, barren—just as truly desert as the sands which were washed down from them. Between the ranges there are wide-expanding plains or valleys. The most arid portions of the desert lie in the basins of these great valleys—flat spaces that were once the beds of lakes, but are now dried out and left perhaps with an alkaline deposit that prevents vegetation. Through these valleys run arroyos or dry stream-beds—shallow channels where gravel and rocks are rolled during cloud-bursts and where sands drift with every wind. At times the valleys are more diversified, that is, broken by benches of land called mesas, dotted with small groups of hills called lomas, crossed by long stratified faces of rock called escarpments.

With these large features of landscape common to all countries, how does the desert differ from any other land? Only in the matter of water—the lack of it. If Southern France should receive no more than two inches of rain a year for twenty years it would, at the end of that time, look very like the Sahara, and the flashing Rhone would resemble the sluggish yellow Nile. If the Adirondack region in New York were comparatively rainless for the same length of time we should have something like the Mojave Desert, with the Hudson changed into the red Colorado. The conformations of the lands are not widely different, but their surface appearances are as unlike as it is possible to imagine.

For the whole face of a land is changed by the rains. With them come meadow-grasses and flowers, hillside vines and bushes, fields of yellow grain, orchards of pink-white blossoms. Along the mountain sides they grow the forests of blue-green pine, on the peaks they put white caps of snow; and in the valleys they gather their waste waters into shining rivers and flashing lakes. This is the very sheen and sparkle—the witchery—of landscape which lend allurement to such countries as New England, France, or Austria, and make them livable and lovable lands.

But the desert has none of these charms. Nor is it a livable place. There is not a thing about it that is "pretty," and not a spot upon it that is " picturesque" in any Berkshire-Valley sense. The shadows of foliage, the drift of clouds, the fall of rain upon leaves, the sound of running waters—all the gentler qualities of nature that minor poets love to juggle with—are missing on the desert. It is stern, harsh, and at first repellent. But what tongue shall tell the majesty of it, the eternal strength of it, the poetry of its wide-spread chaos, the sublimity of its lonely desolation! And who shall paint the splendor of its light; and from the rising up of the sun to the going down of the moon over the iron mountains, the glory of its wondrous coloring! It is a gaunt land of splintered peaks, torn valleys, and hot skies. And at every step there is the suggestion of the fierce, the defiant, the defensive. Everything within its borders seems fighting to maintain itself against destroying forces. There is a war of elements and a struggle for existence going on here that for ferocity is unparalleled elsewhere in nature.

The feeling of fierceness grows upon you as you come to know the desert better. The sun-shafts are falling in a burning shower upon rock and dune, the winds blowing with the breath of far-off fires are withering the bushes and the grasses, the sands drifting higher and higher are burying the trees and reaching up as though they would overwhelm the mountains, the cloud-bursts are rushing down the mountain's side and through the torn arroyos as though they would wash the earth into the sea. The life, too, on the desert is peculiarly savage. It is a show of teeth in bush and beast and reptile. At every turn one feels the presence of the barb and thorn, the jaw and paw, the beak and talon, the sting and the poison thereof. Even the harmless Gila monster flattens his body on a rock and hisses a "Don't step on me." There is no living in concord or brotherhood here. Everything is at war with its neighbor, and the conflict is unceasing.

Yet this conflict is not so obvious on the face of things. You hear no clash or crash or snarl. The desert is overwhelmingly silent. There is not a sound to be heard; and not a thing moves save the wind and the sands. But you look up at the worn peaks and the jagged barrancas, you look down at the wash-outs and piled bowlders, you look about at the wind-tossed, half-starved bushes; and, for all the silence, you know that there is a struggle for life, a war for place, going on day by day.

How is it possible under such conditions for much vegetation to flourish? The grasses are scanty, the grease wood and cactus grow in patches, the mesquite crops out only along the dry river-beds. All told there is hardly enough covering to hide the anatomy of the earth. And the winds are always blowing it aside. You have noticed how bare and bony the hills of New England are in winter when the trees are leafless and the grasses are dead? You have seen the rocks loom up harsh and sharp, the ledges assume angles, and the backbone and ribs of the open field crop out of the soil? The desert is not unlike that all the year round. To be sure there are snow-like driftings of sand that muffle certain edges. Valleys, hills, and even mountains are turned into rounded lines by it at times. But the drift rolled high in one place was cut out from some other place; and always there are vertebrae showing—elbows and shoulders protruding through the yellow byssus of sand.

The shifting sands! Slowly they move, wave upon wave, drift upon drift; but by day and by night they gather, gather, gather. They overwhelm, they bury, they destroy, and then a spirit of restlessness seizes them and they move off elsewhere, swirl upon swirl, line upon line, in serpentine windings that enfold some new growth or fill in some new valley in the waste. So it happens that the surface of the desert is far from being a permanent affair. There is hardly enough vegetation to hold the sands in place. With little or no restraint upon them they are transported hither and yon at the mercy of the winds.

W. STORRS LEE

Born in 1907 in Connecticut, W. Storrs Lee was for many years an English instructor and a dean at Middlebury College in Vermont. He was an editor, historian, biographer, and scholar, and although he lived most of his life on the East Coast, he wrote several books about California, including The Great California Deserts, *published in 1963. He also edited* California: A Literary Chronicle, *published in 1968, and compiled similar anthologies about the literature of Colorado, Hawaii, Maine, and Washington. He died in Maine in 2004.*

In this excerpt from the chapter "Goodness Gracious, All That Scenery for Cinema!" from The Great California Deserts, *Lee chronicles the starring role the California desert has played in cinema, starting in the 1920s.*

from *The Great California Deserts*

Very early, the role of the desert itself as a major antagonist was recognized, and picture after picture employed the dramatic theme of struggle against thirst, solitude and trackless waste, in one way or another. Cecil de Mille used it in *Heir to the Hoorah*, in which the tattered hero, lost in the desert, carrying an empty canteen and slowly dying of thirst, reaches a water hole only to find it dry. Half-crazed and delirious, he sees a magnificent mirage across the shimmering sands and struggles on until the lone figure disappears over an outcropping ledge. Slow fade-out. THE END.

Thomas Meighan took the title role, and de Mille motored his cast and truckloads of equipment down to Palm Springs. "This was in the days when Palm Springs had not yet been adopted by the movie colony as a desert retreat," commented the director. "The town was only a cluster of weather-beaten dwellings, served by a typical country store—a desert settlement in an Indian reservation. If you walked a few hundred yards from the little oasis of the village, you were in the trackless waste of the desert itself, with all the mysterious sense of loneliness, its mirages, its deceiving distances and its cruel grandeur."

De Mille quickly discovered the handicaps of location work in the desert. At the foot of Palm Canyon, where cameras were set up, thermometers registered 130°. There was no ice within fifty miles. To prevent film from melting in the sun, assistants had to keep wet compresses over both cameras and cases. "The heat was so killing," de Mille reckoned, "that I didn't let Tom play up in rehearsal—just walked him through it to get camera composition and position, and also to be sure that the route he traversed was fairly clear of rattlesnakes, scorpions, centipedes and other items which might take an actor out of the proper mood."

Around a dried water hole the property boys planted the collection of skulls and bones brought all the way from Hollywood, and at last the touching sequence of Meighan's demented wanderings and disappearance was started. "It was really quite an effective scene," gloated de Mille, "with miles of desolation as a background and no sign of human habitation visible."

All went off exactly as the script specified. Tom's figure got smaller and smaller in the distance, to represent "the insignificance of man opposed to the desert's silent power." The director was delighted. "Cut!" he called.

But the order was scarcely issued when a yelp of agony came from the distance into which the hero had faded. Apprehensive over the fate of the irreplaceable star, fearing that he had broken a limb, stepped on a rattler or too closely identified himself with the crazed man he was portraying, de Mille led the rescue rush across the sands.

They found Meighan immobile, sprawled on the edge of a rock ledge, his face contorted with the misery he was suffering. To make a retreat from camera vision at the end of his scene he had vaulted to another shelf, without first surveying the terrain, and had landed bottomside down on a massive cactus plant. Meighan was all but impaled on a cushion of two-inch spikes.

With quantities of prickly pear still adhering to his rear, Tom was cautiously lifted from his perch and laid face down on the sand. The rest of the afternoon was spent out of camera range and out of visual range of starlet admirers as his tattered trousers were clipped away and the needles plucked from his hindquarters with mechanic's pliers. Meighan recovered from the surgery, but never from his embarrassment.

In 1925 the public was given the ultimate in the man-versus-desert theme, straight from Death Valley. The story came packaged in the famed film called *Greed*, and to this day it can provoke such an overpowering thirst in an audience that viewers invariably stampede from the theatre to the nearest bar.

Erich von Stroheim directed it—130 reels of mining, money grabbing, misery and murder from Frank Norris' *McTeague*. But the theme ran away with the director and he did not know when to stop. Eventually he cut the film to twenty-six reels, insisting that they be shown in two installments. Metro-Goldwyn-Mayer ruled otherwise, trimmed the film down to ten reels, and in the process eliminated all but the bare essence of a story.

But left in with every grim detail was the final episode in which the implacable avenger catches up with murderer McTeague on the broad salt flats of Death Valley. In all the heat of the desert the two fight it out, and just before the pursuer is killed he succeeds in snapping handcuffs on Mac. There, a hundred miles from water, against a horizon of endless salt and sand, Mac finds himself manacled to a dead man. In the fade-out he submissively lies down beside the corpse to await his own fate.

Greed was an artistic triumph and a magnificent box-office flop—one of the greatest flops in Hollywood history. Moving pictures of the day, agreed the critics, had to serve up either amusement or entertainment. Von Stroheim's masterpiece offered neither. There was too much avarice, too much aridity and too much Death Valley. For a year or two the picture even helped to scare tourists away from Scotty's Castle.

Though the desert failed to provide a palatable setting for *Greed*, with its complex moral preachment, it proved thoroughly adequate as a location for the string of Biblical extravaganzas that began pouring from the studios in the twenties. Finding just the right spot to look like the environs of Jerusalem, Mount Sinai or the Sea of Galilee, took some doing on the part of location scouts, but in general the Mojave was a fair double for the Holy Land.

The first *Ten Commandments* set the standard for picture testaments that followed. Cecil de Mille traipsed the Israelites halfway across Southern California, with Pharaoh and the Egyptians in hot pursuit. But the filming chronology had nothing to do with Biblical chronology. Moses picked up the Decalogue in Red Rock Canyon; the great scenes of the Exodus and the crossing of the Red Sea were shot on the sand dunes at Guadalupe, near Santa Maria; while Muroc Dry Lake—now Edwards Air Force Base—bore witness to some of the more chilling plight of the wandering Children.

At Guadalupe a huge tent city was set up, with a compound for 2,500 people and 3,000 animals. And it was there that a major crisis of the screening developed. De Mille was sure that he could get the best performance out of a cast of orthodox Jews, and he scoured Los Angeles for them. Eager to re-enact the critical events in the annals of their fathers, they assembled at Santa Maria in hundreds, trusting that they would be properly provisioned during their acting debuts.

De Mille had the right intentions, but there was some oversight in the commissary department, and the Israelites from Los Angeles walked between the parted billows of the Red Sea as famished as their forebears. For lunch they had been offered nothing but ham. A few made an unfilmed exodus before the cameras started grinding.

For the biggest scene of all, Muroc Dry Lake was chosen. Down the rugged hills to the north, Pharaoh's horses and chariots were supposed to charge and chase the Children into the Red Sea.

The Muroc sequence was so stupendous that there were not enough horses and horsemen in all Los Angeles to fill the bill. A hurry call was sent to the Presidio in San Francisco. Responding cheerfully to the emergency, the commanding general dispatched an entire cavalry brigade. And that sparked another crisis.

The Army had nothing but contempt for the Hollywood cow-punchers who had previously been engaged, and behind scenes they hatched a plot that was not in the script. In the precipitous descent to Muroc Dry Lake, the San Francisco campaigners intended to show up the screen professionals and ride them down. The plot leaked to the opposing camp, and half the cowboys refused to get into the melee at all. Nevertheless, that charge into Muroc Dry Lake was one of the all-time spectacles of the screen.

On *Ten Commandments* de Mille sunk almost $1,500,000, to the dismay of his employers in New York. It was an unprecedented expenditure for a film of the early twenties, but as an indication of the market value of California desert scenery in terms of celluloid, the picture eventually earned over $4,000,000.

To eliminate some of these expensive filming safaris the big cinema companies soon had their location "ranches" spotted in the canyons northeast of Hollywood or on the edge of the desert. There the same settings could be used over and over in turning out adventure epics or cowboy quickies.

When talkies came in, everyone in Hollywood assumed for a few years that Westerns and magnificent outdoor spectaculars were washed out. The day of the soundproofed studio had arrived, and technical problems of reproducing authentic sound in the wide-open spaces were insurmountable.

Because the microphone was considered to be immobile, the free-flowing action of the silents was abruptly displaced by static, staged presentations. Pictures were studio bound and the desert less frequently visited. But it did not take long for producers to catch on to the technique of intersplicing close-up recordings with long-range scenes and dubbed-in sound effects.

Talkies such as *In Old Arizona* and *Cimarron* paved the way for bigger and better desert epics. Over the years they came in a steady flow: *Stagecoach, Jesse James, Treasure of Sierra Madre, Broken Arrow, High Noon*, and desert-made Westerns—long regarded by the first-run theatres of the big cities as box-office poison—proved as popular there as they had everywhere else.

Then came television, with a demand for mass production of Western serials. Places like Red Rock Canyon and Fairview Valley were busier than ever with the influx of Hollywood idols. Although the movie ranch at Corriganville, only 30 miles from the film capital, would fall short of complying precisely with either the tourist's or the geographer's conception of desert, its 2,000 acres of canyons, bold rocks and parched highland easily satisfied the screen criterion.

Corriganville catered to the rubberneck crowds and the kids, as well as to studios. Open to them were such familiar sets as Fort Apache, the Burma Road, a Corsican Village, a Mexican border town, and the typical Wild West street. And its proprietors could boast that over 3,000 pictures had been made there in 20 years.

That movie ranch was headquarters for TV series like *Bat Masterson, Wyatt Earp, Gene Autry, Cisco Kid, Kit Carson, Gunsmoke, Have Gun, Will Travel, Tales of Wells Fargo, Wagon Train, Rin Tin Tin* and *Robin Hood*. For flatter, sandier scenes to fill out the continuity, the Mojave and Colorado deserts were just over the hills.

Over the years, of course, many a good desert location of the past had been lost to the real-estate developers. Palm Springs was sold out, Victorville had turned into a modern metropolis, and scouts had to go farther and farther afield to find sand dunes where a golf green or the roof lines of a smart ranch house did not spoil the site for a panoramic shot of old Araby.

More and more the moviemakers were depending on public lands and even military reservations in the search for desert locations. For the opening scenes of *Spartacus* the cast traveled all the way to Death Valley's Zabriskie Point, and X-15 was filmed at Edwards Air Force Base. Still, the desert's most profitable industry was far from doomed. There were enough wasteland settings left so that young fry would not need to worry for awhile about discontinuance of their favorite bang-bang telecasts, or their elders about future Foreign Legion, Wild West and Holy Land screen entertainment.

COLIN FLETCHER

Colin Fletcher was born in 1922 in Cardiff, Wales, and moved to the United States in 1956. He is the author of many books about the deserts of the American Southwest. The Thousand-Mile Summer, *published in 1964, is an account of his 1958 walk up and down California's eastern border, which includes the Colorado and Mojave Deserts. He died in 2007.*

The following excerpt is taken from The Man Who Walked Through Time *(1968), which chronicles Fletcher's walk through the entire length of the Grand Canyon.*

from *The Man Who Walked Through Time*

About lunchtime the weather began to back up the Mojave's bid to delay me. Thin clouds drifted down from the north. They circled and coalesced. The breeze died. The overcast thickened, and humidity magnified the heat. The sand grew softer. I plodded on, telling myself again and again that this was not going to be one of those days on which the will to walk just seeps away. At a mid-afternoon halt I leaned against my pack—and woke up fifteen minutes later. At teatime, while I was wondering how two furry black caterpillars managed to cross the glowing embers of my fire without turning a hair, I fell asleep again.

And so it went on as I trudged down the slope of a stage that was as rich in sideshows as a circus—and as unlike the Mojave of evil reputation as the windswept plateau had been.

Afternoon eased into evening. The sand grew progressively softer. I forced up my pace—and kept finding it had slackened again. At last, about six o'clock, I came to a familiar lava flow. Six weeks earlier I had parked the car beside it and carried a huge bottle of water over the lava flow, down a gully on the far side, and into a sandy wash.

Now, the trail on which the car had stood no longer existed. Heavy rains had scoured deep into the gravel. If rain had also scoured the wash on the far side...

I climbed up onto the lava flow. Six weeks earlier it had been a gloomy, brooding place. Now, even in fading light, it blazed with flowers. I walked across the lava through air heavy with perfume. I reached the edge of the lava. At my feet ran the gully. I peered down. The sand had not been scoured or built up; and beside a bush lay the black marker stone. I was almost sure it had not been moved. Tiredness gone, I bounced down the gully.

Then I stopped dead.

I heard myself say "Oh!" The extravaganza had kept for its finale the most breathtaking spectacle of all. At the foot of the gully, two ethereal yellow orbs floated above the sand. Their petals glowed with a luminous magic.

As I reached for my camera it occurred to me that no sane man who had walked across three hundred miles of flower-filled desert, who had been crushing flowers underfoot all day long, who was tired and hungry and still not certain his cache was intact, no sane man would...I opened the camera.

By the time I had taken one photograph the light had lost its luminosity and the two yellow visions their ethereal glow. They had become flowers. Very beautiful still, but mere earthbound flowers.

Sad that the curtain had fallen at last, I went over to the bush, moved the stone, and dug out the huge bottle. The water was clear as the day I buried it. The note was smudged by rain but still legible.

I drank a cupful of the cool water, then walked out into the wash to look for a better campsite.

As I walked, the gray desert began to grow lighter. I stopped and looked up. Directly overhead, an arc of sunlight streamed out of a break in the clouds and plunged like a flamethrower onto a range of black lava hills.

But the lava hills were no longer black. They were not even fiery red. They had passed beyond mere heat, beyond incandescence, to something purer. They glowed with a radiant magenta that was never one single and definable color but bloomed and swelled and expanded into a thousand transplendent hues until the whole line of hills was a pulsating mosaic held fast between black lava and gray sky.

The end was a slow diminishing. Finally, a time came when the purple of infinite royalty was no longer purple but black.

And standing there in the gray wash, a little breathless, with the wind blowing suddenly cold in my face, I knew that in the extravaganza's finale I had at last seen the climax the desert had withheld that evening beside the Colorado when I looked out beyond the tules and watched the Arizona hills catch fire. Now, remembering the pulsations that had just died, I understood for the first time something of the magic that binds people to the desert; and I seemed to hear as well not only another of the hints the jumping fish had given—those hints of a real reason for The Walk—but also a final assurance that I need not try to force the secret. In fact, that I must not try. If I left the reasons to look after themselves it would all be clear in the end.

I walked slowly back to the water cache and went to sleep among the flowers.

JOHN W. HILTON

John W. Hilton was born in 1904 in China to missionary parents and grew up in Minnesota. After living in Los Angeles, he moved to Oasis, California, where he owned a gem store adjacent to the historic Valerie Jean market. In 1950 he moved to Twentynine Palms, where he lived and painted for many years, eventually gaining fame as a desert artist. Hilton wrote and illustrated several books, including Sonora Sketch Book *(1947) and* Hilton Paints the Desert *(1964). He was also a poet, musician, geologist, miner, and iconic desert personality. Also appearing in this anthology is an excerpt from Hilton's biography, in which he discusses his friendship with General George Patton in and around Indio during World War II. Hilton died in 1983.*

The following essay, "How To Be a Desert Rat and Like It," was first published in 1948 in "The Desert Rat Scrap Book," a Southwest desert newsletter edited by another colorful Coachella Valley resident, Harry Oliver, from 1945 to 1967.

How To Be a Desert Rat and Like It

There are as many deserts as there are people in them, just as there are as many worlds. Being a "Desert Rat" is like being a lover, for the desert is like a woman with a million faces.

Her flowery make-up in the spring is like a young chorus girl. There's real beauty there if you can get past the sand verbenas and cloying perfume. Most visitors see only the surface charm. She's in a flirting mood then and is anybody's girl. If you want to be a "stage door Johnny," you can pay a high price for a cheap thrill and feel a bit silly afterward, carrying home a mass-produced flashy painting of the desert in bloom like a Paris postcard.

The desert possesses a rich dowry and some woo her for that alone, but the money leaves a biter taste with a secret hatred for its source. For business reasons or to cover up their frustration, such persons may try to act and talk like real desert rats, but it won't work! Even if they ride silver-saddled Palominos, wear embroidered shirts, teach their butler a western drawl and serve champagne in tincups.

Anyone can be a desert rat who can see and love the beauty of the desert in all of her moods. There's beauty and wild music in a desert sandstorm. The lightning and thunder of a summer cloudburst are the flashing eyes—the emoting and tears of a high-spirited, beautiful actress "putting on a scene." They are soon over. There's beauty in crisp cool winter mornings and hot sultry summer afternoons, but most of all there's the intimate beauty of being alone with her on long walks or lying on her warm breast on balmy summer nights counting the stars in her hair—listening in the silence to your own heartbeat as it matches hers.

Being a desert rat and liking it is like being in love—you just can't help it.

ENID SHOMER

Enid Shomer is a prolific fiction writer, poet, and lecturer who lives in Tampa, Florida. Her publications include the poetry collection Stars at Noon: Poems from the Life of Jacqueline Cochran *(1992) and the short story collection* Imaginary Men *(1993), which won the Iowa Short Fiction Award. She is the recipient of many other awards and fellowships, and her work has been published in a number of magazines, literary journals, and anthologies.*

The following poem, "In the Desert," is from Stars at Noon, *a collection based on the life of aviation pioneer and mid-twentieth-century Coachella Valley resident Jacqueline Cochran. Cochran ran for Congress in 1956 and lost to Dalip Saund, an Imperial Valley resident who became the first Indian American congressman. Cochran lived at her home in Indio, California, with her husband, Floyd Odlum, until her death in 1980. This poem shows her enjoying her desert ranch home shortly after her unsuccessful bid for Congress.*

In the Desert

There is a woman
in the moon tonight, hazy
as the last breath
I once saw fogging a mirror
when a mill worker died.
She dips her toothless jaw
into a thin gruel of clouds
and is not nourished....

God rested here in the desert
after Creation. He wanted
to empty himself of all
those designs for fin,
thorn, and claw. He was sick
of the gap-toothed smile
of mountains, the insistent wishing
of streams.

Whenever I find the skull
of an animal, I tell myself
it's a bead He rubbed
Himself to sleep with.

I want to empty myself like that.
I want to fall all the way
to ordinary.

DICK BARNES

Dick Barnes was born in San Bernardino, California, in 1932 and grew up in the Mojave Desert. His desert-based poetry collections include Few and Far Between *(1994) and* A Word Like Fire *(2005). He taught creative writing and medieval literature at Pomona College, and he specialized in translation of the poetry of Jorge Luis Borges. He died in 2000.*

The poems included here, "Bagdad Chase Road in July" and "Song: Mojave Narrows," are taken from A Word Like Fire.

Bagdad Chase Road in July

Within the immense circle of the horizon
only the two of us on two legs
that don't have feathers on. Hello,
horned lark. Hello, loggerhead shrike.
Hello, dove-size bird with black fan-tail
fluttering along the ground, a jackrabbit
would jump as high. And for the vast
absence of our own species,
thanks, thanks, thanks. Not that you
didn't dig the mines and make this road
we're on; but it's your absence
today that earns my gratitude. Thanks too
for the monument and bronze tablet
to mark where Ragtown was, and the railroad
going down to Ludlow, so I can rejoice
they've already all disappeared
with hardly a trace. Thank you sky
for speaking only after lightning. Hello, jackrabbit,
hello groundsquirrel, good luck raven,
I never saw you hover like that.
Thank you, rain, for flavoring our jaunt
with a hint of danger, and for the splashy mist
when you lashed the desert hills to show
what you can do when you mean business.
Thank you, other twolegged bare featherless creature,

for sharing the jagged horizon of my life.
Thank you rainbow over the East Mojave
low to the ground so early in the afternoon:
thank you for being here with us.

Song: Mojave Narrows

To say that death is a river
and my love for you a star reflected in the river

that the river has worn a channel through boulders, and flows
north and east from The Forks a hundred miles without tributaries

to say a dry lake is a mirror, where the river
gives itself up to the sky

JUDY KRONENFELD

Judy Kronenfeld was born in New York City in 1943. She has published two collections of poetry—Ghost Nurseries *(2005) and* Light Lowering in Diminished Sevenths *(2008)—and has taught creative writing at the University of California, Riverside, since 1984. She lives in Riverside.*

The following poem, "Minding Desert Places/Winter 4 P.M.," was first published in 2007 on the online Innisfree Poetry Journal.

Minding Desert Places
Winter—4 P.M.

Shadows lay themselves down
on the bare hills, darkly
soft, breast to breast.

Every tree and bush
in the wash—mesquite,
creosote, tamarisk—
is articulate
in its loneliness.

Cholla blink here,
there, guttering out.

Light slides from the warm
rock's upturned face.

You still see nothing
that is not there,
but now you sense,
everything that is.

JON KRAKAUER

Jon Krakauer was born in 1954 in Brookline, Massachusetts. An outdoor adventurer and mountaineering expert, he is the author of Into Thin Air *(1999), a firsthand account of a Mount Everest climbing tragedy, and* Into the Wild *(1997), the alluring yet tragic story of Christopher McCandless, who wandered the western United States, deliberately living as minimally as possible, until he met his untimely death in Alaska at age twenty-four.* Into the Wild *was made into a major motion picture in 2007.*

In the following excerpt from Into the Wild, *McCandless spends time near the Salton Sea at a winter campground known as Slab City and in nearby Salton City before leaving for Alaska.*

from *Into the Wild*

After McCandless bid farewell to Jan Burres at the Salton City Post Office, he hiked into the desert and set up camp in a brake of creosote at the edge of Anza-Borrego Desert State Park. Hard to the east is the Salton Sea, a placid ocean in miniature, its surface more than two hundred feet below sea level, created in 1905 by a monumental engineering snafu: Not long after a canal was dug from the Colorado River to irrigate rich farmland in the Imperial Valley, the river breached its banks during a series of major floods, carved a new channel, and began to gush unabated into the Imperial Valley Canal. For more than two years the canal inadvertently diverted virtually all of the river's prodigious flow into the Salton sink. Water surged across the once-dry floor of the sink, inundating farms and settlements, eventually drowning four hundred square miles of desert and giving birth to a land-locked ocean.

Only fifty miles from the limousines and exclusive tennis clubs and lush green fairways of Palm Springs, the west shore of the Salton Sea had once been the site of intense real estate speculation. Lavish resorts were planned, grand subdivisions platted. But little of the promised development ever came to pass. These days most of the lots remain vacant and are gradually being reclaimed by the desert. Tumbleweeds scuttle down Salton City's broad, desolate boulevards. Sun-bleached FOR SALE signs line the curbs, and paint peels from uninhabited buildings. A placard in the window of the Salton Sea Realty and Development Company declares CLOSED/CERRADO. Only the rattle of the wind interrupts the spectral quiet.

Away from the lakeshore the land rises gently and then abruptly to form the desiccated, phantasmal badlands of Anza-Borrego. The *bajada* beneath the badlands is open country cut by steep-walled arroyos. Here, on a low, sun-scorched rise dotted with chollas and indigobushes and twelve-foot ocotillo stems, McCandless slept on the sand under a tarp hung from a creosote branch.

When he needed provisions, he would hitch or walk the four miles into town, where he bought rice and filled his plastic water jug at the market–liquor store–post office, a beige stucco building that serves as the cultural nexus of greater Salton City. One Thursday in mid-January, McCandless was hitching back out to the *bajada* after filling his jug when an old man, name of Ron Franz, stopped to give him a ride.

"Where's your camp?" Franz inquired.

"Out past Oh-My-God Hot Springs," McCandless replied.

"I've lived in these parts six years now, and I've never heard of any place goes by that name. Show me how to get there."

They drove for a few minutes down the Borrego-Salton Seaway, and then McCandless told him to turn left into the desert, where a rough 4-x-4 track twisted down a narrow wash. After a mile or so they arrived at a bizarre encampment, where some two hundred people had gathered to spend the winter living out of their vehicles. The community was beyond the fringe, a vision of post-apocalypse America. There were families sheltered in cheap tent trailers, aging hippies in Day-Glo vans, Charles Manson look-alikes sleeping in rusted-out Studebakers that hadn't turned over since Eisenhower was in the White House. A substantial number of those present were walking around buck naked. At the center of the camp, water from a geothermal well had been piped into a pair of shallow, steaming pools lined with rocks and shaded by palm trees: Oh-My-God Hot Springs.

McCandless, however, wasn't living right at the springs; he was camped by himself another half mile out on the *bajada*. Franz drove Alex the rest of the way, chatted with him there for a while, and then returned to town, where he lived alone, rent free, in return for managing a ramshackle apartment building.

Franz, a devout Christian, had spent most of his adult life in the army, stationed in Shanghai and Okinawa. On New Year's Eve 1957, while he was overseas, his wife and only child were killed by a drunk driver in an automobile accident. Franz's son had been due to graduate from medical school the following June. Franz started hitting the whiskey, hard.

Six months later he managed to pull himself together and quit drinking, cold turkey, but he never really got over the loss. To salve his loneliness in the years after the accident, he started unofficially "adopting" indigent Okinawan boys and girls, eventually taking fourteen of them under his wing, paying for the oldest to attend medical school in Philadelphia and another to study medicine in Japan.

When Franz met McCandless, his long-dormant paternal impulses were kindled anew. He couldn't get the young man out of his mind. The boy had said his name was Alex—he'd declined to give a surname—and that he came from West Virginia. He was polite, friendly, well-groomed.

"He seemed extremely intelligent," Franz states in an exotic brogue that sounds like a blend of Scottish, Pennsylvania Dutch, and Carolina drawl. "I thought he was too nice a kid to be living by that hot springs with those nudists and drunks and dope smokers." After attending church that Sunday, Franz decided to talk to Alex "about how he was living. Somebody needed to convince him to get an education and a job and make something of his life."

When he returned to McCandless's camp and launched into the self-improvement pitch, though, McCandless cut him off abruptly. "Look, Mr. Franz," he declared, "you don't need to worry about me. I have a college education. I'm not destitute. I'm living like this by choice." And then, despite his initial prickliness, the young man warmed to the old-timer, and the two engaged in a long conversation. Before the day was out, they had driven into Palm Springs in Franz's truck, had a meal at a nice restaurant, and taken a ride on the tramway to the top of San Jacinto Peak, at the bottom of which McCandless stopped to unearth a Mexican serape and some other possessions he'd buried for safekeeping a year earlier.

Over the next few weeks McCandless and Franz spent a lot of time together. The younger man would regularly hitch into Salton City to do his laundry and barbecue steaks at Franz's apartment. He confided that he was biding his time until spring, when he intended to go to Alaska and embark on an "ultimate adventure." He also turned the tables and started lecturing the grandfatherly figure about the shortcomings of his sedentary existence, urging the eighty-year-old to sell most of his belongings, move out of the apartment, and live on the road. Franz took these harangues in stride and in fact delighted in the boy's company.

An accomplished leatherworker, Franz taught Alex the secrets of his craft; for his first project McCandless produced a tooled leather belt, on which he created an artful pictorial record of his wanderings. *ALEX* is inscribed at the belt's left end; then the initials *C.J.M.* (for Christopher Johnson McCandless) frame a skull and crossbones. Across the strip of cowhide one sees a rendering of a two-lane blacktop, a NO U-TURN sign, a thunderstorm producing a flash flood that engulfs a car, a hitchhiker's thumb, an eagle, the Sierra Nevada, salmon cavorting in the Pacific Ocean, the Pacific Coast Highway from Oregon to Washington, the Rocky Mountains, Montana wheat fields, a South Dakota rattlesnake, Westerberg's house in Carthage, the Colorado River, a gale in the Gulf of California, a canoe beached beside a tent, Las Vegas, the initials *T.C.D.,* Morro Bay, Astoria, and at the buckle end, finally, the letter N (presumably representing north). Executed with remarkable skill and creativity, this belt is as astonishing as any artifact Chris McCandless left behind.

Franz grew increasingly fond of McCandless. "God, he was a smart kid," the old man rasps in a barely audible voice. He directs his gaze at a patch of sand between his feet as he makes this declaration; then he stops talking. Bending stiffly from the waist, he wipes some imaginary dirt from his pant leg. His ancient joints crack loudly in the awkward silence.

More than a minute passes before Franz speaks again; squinting at the sky, he begins to reminisce further about the time he spent in the youngster's company. Not infrequently during their visits, Franz recalls, McCandless's face would darken with anger and he'd fulminate about his parents or politicians or the endemic idiocy of mainstream American life. Worried about alienating the boy, Franz said little during such outbursts and let him rant.

One day in early February, McCandless announced that he was splitting for San Diego to earn more money for his Alaska trip.

"You don't need to go to San Diego," Franz protested. "I'll give you money if you need some."

"No. You don't get it. I'm *going* to San Diego. And I'm leaving on Monday."

"OK. I'll drive you there."

"Don't be ridiculous," McCandless scoffed.

"I need to go anyway," Franz lied, "to pick up some leather supplies."

McCandless relented. He struck his camp, stored most of his belongings in Franz's apartment—the boy didn't want to schlepp his sleeping bag or backpack around the city—and then rode with the old man across the mountains to the coast. It was raining when Franz dropped McCandless at the San Diego waterfront. "It was a very hard thing for me to do," Franz says. "I was sad to be leaving him."

CLARA JEFFERY

Born in Baltimore, Maryland, in 1967, Clara Jeffery is the editor-in-chief of
Mother Jones *magazine and a former editor at* Harper's Magazine. *She has been*
nominated for a National Magazine Award seven times, in multiple categories, for
pieces she has edited. She lives in San Francisco.

    The following excerpt is taken from her article "Go West, Old Man,"
first published in Harper's *in 2002. It is a colorful look at the people of and*
environmental issues surrounding California's southernmost desert lands, along the
periphery of the Salton Sea.

from "Go West, Old Man"

Slab City

Entering Niland from the south, I pass the "Chamber of Commerce and RV
dump" (a fenced-in dirt lot that hosts a weekly swap meet) and a downtown
consisting largely of vacant storefronts. Three miles east of town a small butte
marks the old shoreline of Lake Cahuilla, but geology cannot compete with
what Leonard Knight has done to it. Using donated paint and mud adobe,
Leonard has created an enormous Technicolor landscape—part Pennsylvania
Dutch, part Pentecostal—called Salvation Mountain. The work's centerpiece is
a giant heart containing the message: JESUS I'M A SINNER PLEASE COME UPON
MY BODY AND INTO MY HEART.

    Slab City itself is an odd combination of mobility and decay. A gold-
rush town without the gold. A pioneer town with satellite TV. Residents
turn junk (and junk is everywhere) into shelter, into art. But this is a place
where people come not to make something of themselves but to unmake
themselves, or at least to leave their pasts so far back in the rearview mirror
that it doesn't hurt anymore. Most have fled life's ordinary tragedies, but
there are honest-to-God outlaws here, "people whose radar you don't want
to be on," one resident warns.

    To the casual visitor, the Slabs is a curiosity, one more stop on the see-
America-in-your-RV tour. Next will come Yuma, or Quartzsite, or wherever
the yen for mobility and a good swap meet takes you. But a lack of resources
is the main reason people come to Slab City. A huge gaggle of elderly flock
here because their VA and Social Security checks can't cover the cost of a $10-
a-night RV park. They come because free is all they can afford, because it's
warm, and a short drive to Mexico, where maybe they can buy the medications
they need. And they come because although they treasure their independence

they're scared of being alone.

During the winter, the population swells to a few thousand residents, the vast majority of whom pack up and leave before summer, when the temperature frequently reaches 110 degrees. Each year a few folks stay on—usually because they're dead broke—and if they survive the heat (I was told that in 2000 nine people did not) they might never leave again. For them, full-time RVing or biking or drifting has become full-time slabbing.

February is the peak season, and all the best spots—known by function, description, or resident—are taken. Driving down Main Street, I pass Poverty Flats and School Bus slab, where the kids who go to school wait for the bus. At the north corner of a dusty intersection Solar Mike sells and repairs the solar panels and batteries that serious RVers depend on. I turn right toward a blue trailer emblazoned with large white wooden letters, WELCOME SLAB CITY SINGLES ("and auxiliary," reads a small caveat).

The Slab City Singles clubhouse is composed of three trailers—a pantry, a library, and a game room—encircling a covered kitchen area; tarps provide a limited wind break. Out front a bunch of raggedy furniture serves as a living room; a fire pit sits out back.

Once inside the dining area, I am approached by a man named Chuck. With a bright blue alpine hat woven from plastic Wal-Mart bags, a cheerful, white-bearded face, a yellow bandanna tied jauntily around his neck, and an Arkansas drawl, Chuck is reminiscent of the benevolent figures found in Disney forests. As is Art, with his elfin ears, stooped shoulders, and a sort of utility belt, cinched not far below his armpits, from which various objects, including a cup, dangle.

"Are you single?" Chuck asks. He's forty years my senior, but I assent.

"Can you cook?"

Before I can answer, a rooster crows from the center of his chest. I jump. He taps a device on his sternum and looks up bashfully. "That's to remind me to take my glaucoma medicine."

Gary wanders in, wearing a poncho. "I'm the token Indian," he says. A former cop in his early fifties, Gary is one of the club's youngest members, as is Wilson, who has an English accent, wears a kepi, and lives in a tent. We all sit out front, shielding our eyes from the sun as David approaches.

"Who've we got here," he says, taking a chair that promptly collapses. After a moment's embarrassment he begins grilling me, cocking a cauliflowered ear to hear the answers to his questions: what am I doing here and where am I from and how old am I and can I cook.

The last question becomes a familiar refrain. The club has about seventy members, mostly unattached men (Chuck and his wife, Peggy Sue, form the "auxiliary"), but new women seem to be valued less as fresh meat than as fresh meat preparers. "Most of these old guys aren't looking for a girlfriend," Wilson later says, "but a mother, I swear to God." There are two other singles clubs in town. The primogenitor is Loners on Wheels, which has chapters all over North America, as does its offspring and rival, the Loners of America (LOAs). Despite the slogan "where singles mingle," the LOWs in Slab City, mostly women, rigorously enforce a no-cohabitation rule to the point of celibacy. They do host the occasional square dance, Wilson says, but "I wound up with some eighty-year-old guy getting all upset because I'm dancing with some old bird who he thinks is his girlfriend. And I need that like a bloody hole in my head." The point of Slab City is to escape that kind of bullshit, those kinds of feuds and rules and conventions, he says, they all say, before detailing the various feuds and rules and conventions currently causing them strife.

In addition to the singles clubs there are annual migrations of Canadians who cluster down by the canal, fortified bunkers scattered here and there, the Apple Dumpling Gang dune-buggy enthusiasts—the endlessly recombining and sometimes squabbling constituencies of Slab City. A central tenet of RV living is: if you don't like your neighbor, just move. But if you have a good spot, with a slab and a gopher hole and maybe even a mesquite tree, and your neighbor is a jerk, why should you move? It's easier, as I would learn, to burn him out.

For the time being I am mercifully ignorant of such territorial remedies, though I know enough to ask where to park my rented RV. After some consideration, they pick a central spot where they can "look out" for me. And for the next two and a half weeks, they do, particularly Wilson, who develops a habit of magically appearing at my side.

Once established, I head to the [local] bird festival. The participants have gathered at a country club where Orvis dealers, wildlife photographers, and environmental groups hawk their wares. In a conference room off to the side, Steve Horvitz, superintendent of the Salton Sea State Park (he has since become chairman of an NGO called Save Our Sea II), regales us with the area's ecological problems. Horvitz talks of the sea's heyday in the early sixties, when the state park received more visitors than Yosemite; when 600,000 boats were launched; when the Salton City 500, called "power boat racing's richest event," was televised on CBS; and residents complained of traffic. Yet at a recent picnic of park employees, he says, everyone paused, forks in midair, to watch a boat go by. "One boat on California's biggest lake and it stopped conversation." He adds that a 1989 survey found that more than half of the people who once visited the sea are now afraid of even being near it.

And who can blame them? Come to the sea and chances are as good as not that you'll be confronted by rafts of dead fish that can stretch on for acres and a stench that seems to combine all the world's worst smells—urine-soaked hallways, skunk, manure, vomit, sulfur dioxide—into one hideous potpourri so malodorous that a Palm Springs attorney who lives forty miles away once demanded Horvitz put an end to it or face a lawsuit.

The smell and the dead fish can be traced back to the problem of nutrient loading. Treated but still phosphate-rich sewage combines with nitrogen-rich runoff to form a kind of MiracleGro that causes algae to bloom. This has several effects. Brown algae is the main staple of the far too abundant tilapia, in turn the main food source of the corvina. Thus, too much brown algae equals too many fish. Worse is the green algae. Its blooms can cover huge portions of the sea, and when they die off the algae decomposes, taking oxygen out of the water and causing a horrible stench. Fish in oxygen-depleted waters surface for air and, failing to get it, die. The death throes of the fish, some of which carry avian botulism, attract birds. In 1997, the disease killed 10,000 endangered brown pelicans. The irony is that the sea has too much life—too many nutrients, too much algae, too many fish.

Fierce debates rage over which is the more dire threat to the sea: the salinity or the nutrient loading. Either one would cost tens of millions, perhaps billions, to address, a tab no one is rushing to pick up, and certainly one that Imperial County, California's poorest, cannot pay alone. This quandary plays right into the hands of the sea's true nemeses: the water managers of Los Angeles, San Diego, and Palm Springs. One water manager told Horvitz that the 1.3 million acre-feet (one acre-foot equaling roughly 326,000 gallons) that flow into the sea each year were "worth a great deal of money to us...We want it. We're going to get it. And there's nothing that you can do to stop us."

After Horvitz concludes his talk, I take him aside. His rusty hair and beard are neatly cropped and flecked with gray, and he possesses that solemnity about the planet common to geology majors. Horvitz believes the sea could be saved if the state and federal governments would commit to doing so, but they're "really only good at moving water around. Destroying where it comes from and destroying where it goes." "The sea itself has no rights to the water," he says, so "there's a lot of angling for it," and the metropolitan water authorities would gladly see the ecosystem collapse, because then nothing would stand in their way. California's water rush has lasted for a hundred years, but the more sources that tap out the more valuable are those that remain.

He ticks off a list of examples: In 1993 the billionaire Bass brothers of Texas (and Biosphere 2) bought 42,000 acres of Imperial Valley farmland to sell their allotment of water to San Diego. In 1989 the Imperial Irrigation District cut a deal with Los Angeles whereby the IID conserves 107,000 acre-feet of water a year, and L.A. gets that much more Colorado River water. Now a hotly contested deal is under way in which the IID would "conserve" another 300,000 acre-feet so that it can "transfer" (sell) that "saved" water to San Diego and Coachella Valley (the northern part of the Salton Sink, containing Palm Springs). Although this deal is by no means done, hanging in the balance is a seven-state water treaty, and if California cannot decide the fate of the Salton Sea by December 31 [2002], the Interior Department has threatened to withhold Colorado River water, which would almost certainly lead to rationing in southern California.

The forces aligned against the sea are powerful, which is perhaps why, historically, most environmentalists haven't rushed to its defense. Although California's sprawl problem is at least as attributable to its post-Proposition 13 reliance on commercial development for revenue as to immigration, environmentalists fear that to fight for the sea is to risk accusations of racism, says Horvitz. And Washington is sick of California's disproportionate share of budgetary attention; the state created its natural-resource crises, why should Congress continually bail it out?

I am late for the communal dinner back at Slab City Singles. Four o'clock did not mean, as I had assumed, that the cooking would commence at 4:00 but that the dishes would be done by 4:30. "Some of these guys are really old," explains Wilson. "Look there at Art"—who was shuffling off toward his trailer—"he's going to bed. You'll never see him out past dark." Art, who'd celebrate his eightieth birthday while I was in camp, isn't the oldest. The oldest is Irv, who is eighty-eight and has a great-great-grandson. "He shouldn't be here," someone explained, "but his family doesn't want him, he's too damn mean."

Slab City lacks any kind of medical facility, but in some ways life here isn't that different than at your average retirement community. There are a bewildering array of clubs and activities for those who wish to join. Banter, coffee, tall tales, and card games are also plentiful; only they take place at the fire pit instead of in a cafeteria.

The dean of the fire pit is Frank, a rail-thin World War II veteran who seems cast as the ol' prospector. He pulls off black boots and reveals long thin feet covered in what seem to be ladies' trouser stockings: sheer, navy blue, with vertical piping. He sits back and theorizes about the hidden levers of power, linking King Fahd to the Hudson's Bay Company to Pamela Harriman. Frank is a vegetarian. He practices yoga. He is also referred to by Slabbers as "the local Shylock." "I saw a number of unsavory characters stop at his rig, go in for a moment, and come back out," says a club member. "You didn't hear that from me." And perhaps because of his profession, though others claimed that a feud within the club was the cause, his rig was burned down last year.

"Accidentally?" asks Richard, a community-theater director who resembles the cowardly lion.

"Accidentally on purpose," says Wilson.

Social hour ends a good six hours before my normal bedtime. Back in my rig, hyper-vigilance to potential arsonists makes sleep impossible, so I open *Salt Dreams* and read about another misbegotten utopia found across the sea, Salton City.

> Think about the picture you have in mind of the perfect place, and the ideal setting. Wouldn't it be much like this? A place ringed by snow-capped mountains and bathed in warm sunshine winter and summer, and cooled by sea breezes....I have never been able to stand on that rise of land above the Salton Sea without seeing a grand resort city. Now our dream is coming to life.
>
> —M. Penn Phillips, First Developer of Salton City

It is easy, with hindsight, to make a distinction between a dreamer and a liar, a seer and a sucker. But a steady diet of lies can sustain a human being for a long time, especially when the liar believes as well. It's impossible to say whether M. Penn Phillips knew he was running a pyramid scheme, or really believed he was building the pyramids. It doesn't matter now. Salton City functions best as metaphor, the endgame of manifest destiny.

In 1958, Phillips spent $150,000 on 19,600 acres of barren land on the sea's western shore, where the temperature exceeds 100 degrees one day in three. He bulldozed a maze of roads, laid sewer and water lines. He planted 9,000 fan palms along grand boulevards with improbable names like Avon and Acapulco. For as little as $250 down and $29 a month, so went his pitch, ordinary people could enjoy a piece of the "Salton Riviera," which would be "bigger than Capri and Monaco and Palm Beach combined...the most valuable piece of resort property on earth." He took prospective buyers on aerial tours, so they could scout property from the clouds. That many buyers would have no equity until the property was entirely paid off did not deter them. On opening day May 21, 1958, Phillips made $4.25 million.

Two years and another $20 million in sales later, Phillips pulled out. He'd drawn back his curtain, but people were too invested to admit that they'd bought into an illusion. The Holly Corporation promptly pursued making Phillips's dream a reality. They built golf courses and marinas and yacht clubs and hotels; they staged their Salton City 500; they lured the Beach Boys, Frank Sinatra, Dean Martin, and President Eisenhower to Salton City, though it is unclear whether Frank and Deano sang or only played golf, like Ike, on the fairways sloping toward the sea. The celebrities were part of the mirage. Around 1971, the Holly developers took their money and ran to Lake Havasu, where they would soon relocate the London Bridge.

Strangely, when the plug was pulled, the waters rose, and soon all that was left of Salton City's ring-a-ding days was the space-age roof of the Yacht Club looking forlornly out over vanished marinas and half-submerged telephone poles. The golf courses have reverted to sand, the palm trees are mostly dead. Sick of being sued for property damage, the IID bought as much waterfront property as it could and razed the Yacht Club and the entire archipelago of sodden aspirations. Today the area boasts one state dump zone for hazardous waste (buried atop ten fault lines), and soon another will receive up to 20,000 tons of L.A.'s trash per day. "Essentially what they are doing to our county is throwing us bones while they're trying to destroy the Salton Sea," says Norm Niver, a community activist and gadfly who wears a Hawaiian shirt and has played music at Merv Griffin's house. We are in the Salton City Chamber of Commerce, which is a step up from that of Niland, but a small step.

Norm's dream for Salton City isn't that different from what Phillips planned, but he has a better idea of what it's up against, starting with a Coolidge Administration law that mandates how much Colorado River water each state in its watershed receives annually. California was allotted 4.4 million acre-feet but has been using 5.2, which nobody minded until Arizona and New Mexico caught the development virus. In 1998, the Clinton Administration brokered a "Quantification Settlement Agreement" whereby California would gradually reduce its take and various water districts would settle their disputes. All provisions have been agreed to, except for the matter of diverting 300,000 acre-feet that now go to the Salton Sea to supply San Diego and Coachella Valley. It's the sea or sprawl. And sprawl has all the clout.

"They're growing a cancer over there," says Norm. "They've got their golf courses going in, their country clubs going in, and all that is 'reasonable, beneficial use' of the water. Everything that's going on in Coachella Valley for the richest 2 percent of people on Earth, maybe 1 percent—101 golf courses, two-mile-long water-skiing lakes where homes start at $1 million...the water shopping mall going in at Rancho Mirage because those people 'like to shop around water'—that's reasonable and beneficial. Why is this not reasonable and beneficial? We have all these birds and fish, we lose 50 percent of our birds if we keep screwing around with the sea. People say, well, if they don't come here, they'll just go someplace else, but that's not true, because the scientists tell us that they are taxed. They are broken down. They don't have the energy to go find another place."

Norm can rattle off any number of reasons why destroying the sea would be morally wrong and even economically stupid ("it'll cost them a heck of a lot more to do nothing; the dust bowls, the lawsuits, just like Owens Valley"), but he realizes that the plight of property owners doesn't stoke public outrage, and environmental issues do. As a friend put it to him, "It's the birds, not your asses." Unfortunately, the fates of all concerned are in the hands of entities that, seen in the best possible light, are slow to act. Since none of the usual cavalries are coming to the rescue, Norm has, in the best Salton City tradition, pinned his hopes on a quick fix, in this case Terra Organics, a bioremediation company that sketchily claims it can treat the water with microorganisms that will consume the nutrients that feed the algae. "These 49 microscopic Pacmans have given a lot of people a lot of hope out here," says Norm.

"Hey, kiddo," Wilson says the next morning. "You want to go visit the Rhinos?" He's been hoarding government issued food to bring to a Slab family. The kids are always hungry; the father—Rhino—is a wild man. There's trouble there, drugs, mental illness, it's hard to say exactly what, but it'll "break your heart." On the way, he and Ron, a big biker who once chaperoned Secretariat around the world, will show me some local attractions.

We load sacks of potatoes and cans of generic food into the back of Ron's minute Datsun pickup and squeeze into the front seat, my feet propped on the dash so that Ron can shift; he grunts an apology with every gear change. First stop is Slab City's homemade desert golf course, eighteen holes, Ron says, "but all you need is a sand wedge." The biggest hazard is an active bombing range that borders the course. Scavengers have been killed or maimed trying to gather artillery casings. Most hail from the Badlands, an outlying area of Slab City full of dealers and crazies who live in compounds fortified by fences, dogs, KEEP OUT signs, and the occasional exchange of gunfire. Wilson points out one such bunker. "Don't ever think you can go over there, luv. I don't go over there."

At the Rhinos' the warning is more profound. Cans and bottles and soggy toys are strung to their chain-link fence, which surrounds rotting piles of clothing and refuse and a few dilapidated vehicles. "Wait here," says Wilson as we climb out. "They don't know your truck." What looks to be an eighteen-year-old girl approaches. She's wearing a sort of outback hat over coppery skin and dark eyes. Her hands are twitching. "There's Mrs. Rhino." She looks at him foggily. "It's Wilson," he says, but the kids—three boys and a girl, the oldest maybe eight—have already recognized him and grab at his jeans. "What'd you bring, what'd you bring?" they ask. We open the back of the truck and they set upon us like refugees. "What's this?" asks the oldest boy, holding up a can that clearly says corn. Another little boy named Harley gives me a shy smile. His teeth are made of metal.

We carry the food through the compound to the main trailer, also filled with garbage. The kids paw through the offerings and look as if they might eat the potatoes raw. "Thank you, thank you," they cry. Mrs. Rhino thanks us, too, and tells us that Rhino is out trying to sell their pickup. Wilson and Ron express dismay. Loss of mobility, though no one says it, can only spell more disaster for this family.

Back near the Singles Club, Wilson spots Rhino leaning over the bed of his truck talking with a few guys. As we climb out, Wilson warns me to keep my mouth shut. Rhino seems at least six four, a big but not fat frame packed into overalls, an outback hat topping a huge black beard and maniacal eyes. His gestures are simultaneously menacing and Falstaffian. He's talking to a white-haired, pockmarked guy in head-to-toe camouflage that matches the netting strung over his camp in the background. The other two guys are filthy, mostly toothless cooter types sunburned a deep pink; their blue eyes glitter from tweaking or poor wiring.

"We've just been out to see Mrs. Rhino. Brought you some food," says Wilson. Rhino thanks him. "Hear you're selling your truck?" "Got to, man," says Rhino. "I gotta file a lawsuit to get my kid back." He tells the tale of one or perhaps two kids being taken by social services. The caseworkers won't visit his place anymore. "They say they're scared. Look at me, am I going to hurt anybody? Government says that my children are in the lower fifth, a 'failure to thrive.' I tell them all my kids start out small, but look at me!" he says, drawing himself up to his full height. He's so animated that he seems to occupy more than three dimensions. "DEA, FBI, HHS, ATF—I got four government agencies with initials that know everything about me. I've been examined, plucked, and prodded. They know how many public hairs I got and how many on my ass."

"You know where this is going," says the vet guy. "Weapons." The cooters nod.

"That's right," says Rhino. He briefly lowers his voice. "I've got it on good authority that in fifty-one days thirty-seven different militias will join forces. Something big will go DOWN." Not that he's in a militia mind you. "They'd either kill me or make me their king!" he shouts.

There should be a certain rhythm to my days, learning about the sea while it's light, returning to the Slabs in the evening, but something's always a little off. Like the way I am awakened abruptly. Yesterday Gary pounded on my door and shouted, "Get your raggedy ass out of bed!" and left a cup of coffee on my steps. Today it is David who knocks. He silently hands me a poem that begins, "Our deepest fear is not that we are inadequate. Our deepest fear is that we are powerful beyond measure."

It's been raining off and on for the last few days, at a rate that elsewhere would be entirely unremarkable but here—where the average annual rainfall is

2.92 inches—is a catastrophe, filling the papers with gloomy harvest prognoses and the ditches with tractor trailers. The weather doesn't bode well for me either. I had arranged for a tour of the Imperial Irrigation District's facilities, but nothing's being pumped now, because the farmers want less water, not more. Instead I meet with Dave Bradshaw, supervisor of the irrigation-management unit. Bradshaw appears drawn by Charles M. Schulz: a bigheaded, genial man who sketches diagram after multicolored diagram of irrigation systems on a giant whiteboard so fast that I feel like I'm witnessing an astrophysics lecture on fast-forward.

Bradshaw creates a flowchart of the water's path from river to field. Of the 3.1 million acre feet the IID transports, 98 percent is used for agriculture. Farmers tell the IID how much water they want a day ahead of time. Using past averages, the IID has already told the U.S. Bureau of Reclamation how much water it will need for the week. Based on the requests of the IID and other water districts in years past, the USBR has sluiced the appropriate amount through the Hoover Dam. Thus a farmer's water order has been en route for seven days before he requests it. Such a prognostic system depends on uniform, reliable weather—as faith in such, the IID lacks large holding reservoirs—but all bets are off "if it rains like this," says Bradshaw.

He whips an eraser across the board and draws a cross section of an imaginary field. The Imperial Valley lacks natural drainage, so six feet under their crops farmers have installed corrugated pipes to collect excess drainage. In the 1989 transfer deal, Los Angeles paid for tailwater-return systems that recycle runoff, known as tailwater, back through cropland. In essence, the sea now gets 100,000 acre-feet less a year, an amount L.A. takes "right out of the Hoover order," says Bradshaw. Revenue was the overwhelming incentive, but the IID also embraced the transfer because, prodded by thirsty constituencies, the state ordered it to "conserve" tailwater. "We had people over here with video cameras taking pictures of the water running off the fields," says Bradshaw.

With a transfer of three times as much water in the works, "now the state's saying [we've got to] keep all that water going to the Salton Sea because if you lower it one inch, there's going to be an environmental problem, so it's a Catch-22 situation," says Bradshaw. "Some were suing when it was going up, some when it was going down...At one time there were eighty-six studies [on how to save the sea] going on, so depending on which study you want to believe and which one the state's pushing hardest..."

And then there are physical barriers to the agreement, which calls for lining the eighty-two-mile All American Canal, and some of the 16,000 miles of smaller canals, with cement. "It's hard to line something with water in it," Bradshaw notes dryly. The only way to do so is through diversion, which is how we got the Salton Sea to begin with. "Catch-22," he repeats.

That night I take Wilson to Niland for dinner. Over cheeseburgers and beers, we discuss Slab City justice, vigilantism that ranges from relatively mild mischief making—tacks under tires—to homicide. Some years ago, Wilson says, a Slab kid who despite multiple warnings continued to rob other residents was found floating in an IID canal, decapitated. There had been other violent deaths at the Slabs, and suspicious fires like Frank's were fairly common. Theft is epidemic, largely attributable to drifters and, in the case of food and water, illegal aliens. Still, given that it is a basically anarchic community, the place seemed pretty peaceable; most people I met seemed generous and neighborly in a way that's utterly incompatible with the suburban isolation most Americans experience. When a homeowner on the other side of the sea had commented that "you have a group like that and nobody trusts each other because nobody's trustworthy," I had bristled.

Wilson interrupts my train of thought. "I'm on the run, you know," he says. He's scrupulously avoided being photographed, so I'd figured, but I flush anyway. "Don't worry, luv," he says. "I didn't murder anyone."

If people come to Slab City running from anything, it's precisely what I am finding myself immersed in: stories of depression and marital unhappiness, of roads taken and not taken, of regrets and obligation. It is exhausting. In search of a break from two weeks of darkness, I decide to visit Salvation Mountain. Leonard Knight, now seventy-one, arrived here sixteen years ago. He tried to launch a hotair balloon bearing the slogan GOD IS LovE, but it "rotted out" on him. So he thought he'd take a week, build "an eight-foot mountain, but God had another plan."

At about 5:00 nearly every morning, Leonard gets to work, lifting bales of hay—"as I get older, they get heavier"—which, like the paint, his trucks, and his tractor, are donated. The mountain needs constant maintenance, which he does "as often as the paint comes in. Like if a pretty gallon of orange comes in, wow, I paint all the flowers orange. I just play it by ear." In 1990, Salvation Mountain collapsed under the weight of all that paint—Leonard estimates he's applied more than 100,000 gallons—and he rebuilt.

In 1994 county supervisors declared the mountain a "toxic nightmare," to be buried in a hazardous-waste site in Nevada. "The *Los Angeles Times* said that the laws of God and the laws of man are going to collide in Niland, California," recalls Leonard. "And we collided with every museum on my side, because every old painting in the world has lead in it." That, and Leonard is considered an important outsider artist, compared with Grandma Moses. "Gee that thrills me," he says. "I can't believe it. I must be the biggest counterfeit in the whole world."

Later that night, back at the fire pit, Leonard pulls up in his fantastically painted truck and steps out with a battered guitar to serenade us for the evening. As the fire starts to die, he compares his battle with the county to the periodic threats of various developers to turn Slab City into a paid campground. Who, he asks, would pay to be next to a bombing range and to smell the Salton Sea? "I'd almost give a dollar a day not to smell the Salton Sea," he cackles.

A hundred and fifty years ago, Horace Greeley made famous the expression "Go West, young man, and grow up with the country." The trouble is that the country grew faster than anyone could have anticipated, and the flow west has become backwash. We pushed from sea to shining sea, and now our aspirations—sadly diminished as they are to the manicured-lawn variety—have grown too big for the land. We are not living within our means, and at the Salton Sea you can see the bills mount. Here is where the Cadillac desert blows a gasket, where the hucksters go to die, where salvation comes hand in hand with lead poisoning, where the last of the pioneers squat on the last sorry piece of free land, where suburban expansionism and a tidal wave of immigration conspire and collide. That coming bubble of uninsured elderly we keep hearing about? That's here. The speed, crank, and meth that are sweeping through blue collar America? Those drugs are cooked and shipped from here.

The sea itself will probably not survive the seeming necessity of Pizza Huts, of three-car garages and grass in the desert, of our God-given right to golf. Since I left, the urban editorial pages have become increasingly shrill: How dare a bunch of bird lovers and dirt farmers stop progress? When Senator Dianne Feinstein warned that fields could be fallowed voluntarily with compensation, or by force and without, an IID board member called her a "bureaucratic gasbag, pig-eyed sack of crap." In September the state suspended a law that barred the killing of thirty-seven endangered species. The Interior Department is barreling ahead with the transfer, even though it won't release a congressionally mandated environmental-impact statement—a fait accompli that prompted the Sierra Club, the Cabazon Indian tribe, and others to seek a federal injunction in September. The IID hesitantly agreed to limited fallowing, but farmers fear being sued if the sea perishes and the area becomes a dust bowl. Even if a solution to the sea is found, sprawl is like crime: push it out of one neighborhood and it pops up elsewhere.

Yet for all the grinding shortsightedness the valley represents, it is also full of people adept at change, those who shed their skins and start anew. "It's nice to know," someone back in the real world said to me, "that there's a place to go if your life goes to hell."

Slab City is that respite, but it is more. At first glance its residents appear as faded as the fifties landscape they inhabit. But that undervalues their weedy tenacity, their tolerance of eccentricity. Anyway, they don't care what you think. Ideas of conventional success, of anything conventional, they're past that now, if they ever cared to begin with. Whatever combination of adventure and avoidance caused them to hit the road, once a year they gather among the like-minded and together they improvise a meal, a community, a family. One of Leonard Knight's songs puts it best:

The road treated me so poor, so cold, wet, and damp.
I came to Slab City, just looking like a tramp.
Then a family gave me a smile,
So I think I'll linger here awhile.
Here in Slab City,
California, U-S-A.

DAVID N. MEYER

David N. Meyer, a film and music writer, teaches cinema studies at the New School in New York City, where he lives. He is the author of a biography of Gram Parsons, Twenty Thousand Roads: The Ballad of Gram Parsons and His Cosmic American Music, *published in 2007.*

Gram Parsons, born in Florida in 1946, was a singer-songwriter who rose to popularity in the 1960s and 1970s folk rock scene as a member of the Byrds and the Flying Burrito Brothers. He found inspiration for his music in Joshua Tree National Park, a two-hour drive from Los Angeles, and helped spark the high-desert music scene that endures to this day. The following selection from Meyer's Twenty Thousand Roads *describes the nighttime trips Parsons and members of the Rolling Stones made into the park to look for UFOs. Parsons died of a drug overdose at the age of twenty-six at the Joshua Tree Inn in 1973. Infamously, his remains were partially cremated, illegally, in a remote section of the park.*

from *Twenty Thousand Roads*

Joshua Tree National Monument, about 140 miles east of Los Angeles, is a desert like no other. Joshua trees are large, slow-growing yuccas that are remarkably humanoid in shape and evocation. Some in the park are over thirty feet high and almost two hundred years old. With their thick, bulbous stalks and multiple branches that reach into the sky like upraised arms, the trees have an air of timeless suffering. Dotting the park are enormous piles of softly rounded boulders that, like the Joshua trees, seem somehow animated. They form phantasmagorical shapes against the park's infinite blue sky and deep silences. Joshua Tree feels like the end of the world, but a benign one. There is no escaping that the park is a magical place. As guitarist Bernie Leadon puts it, "Joshua Tree is everybody's power spot." The consumption of LSD or other psychedelics could only intensify this feeling. Gram developed a profound attachment to the landscape.

Somewhat less magical was the town of Joshua Tree: a short row of cheap motels, bars, and restaurants along Highway 10, an east-west truckers' artery. The summer heat in Joshua Tree is paralyzing; the winters are pleasant. In the late sixties the town was home to a small community of hippies and desert rats. When Gram visited, he usually stayed at a roadside motel called the Joshua Tree Inn, an L-shaped, one-story, flat-roofed joint of no distinction. The Inn offered the advantage that its pool and courtyard were invisible from the road, so that guests could do as they pleased. Perhaps for that reason, the motel had a reputation as a haven for fifties Hollywood libertines.

Gram discovered its charms via Ted Markland, whom he'd met hanging out in Topanga Canyon. Markland is a character actor who'd had roles in Burt Lancaster westerns, any number of TV series, Peter Fonda's acid western, *The Hired Hand*, and Dennis Hopper's psychotic (though compelling) western deconstruction, *The Last Movie*. Markland discovered Joshua Tree in the early fifties when he came out to the desert to attend a Spacecraft Convention at nearby Giant Rock. He returned frequently and eventually hauled a swivel chair to the top of one of the park's mountains so that he could enjoy panoramic views in comfort.

The area has a long history of attracting seekers. Native Americans regarded Giant Rock—a huge freestanding boulder—as a sacred spot. In 1953, an aircraft mechanic named George Van Tassel was meditating under the rock when he received a telepathic message from denizens of Venus, who later visited him and showed him techniques for reinvigorating human cells. On their instructions Van Tassel built the Integratron, a three-story-tall dome he claimed was able to fight gravity and "recharge the cell structure." During the fifties and sixties, Spacecraft Conventions at the dome drew thousands of people, including UFO contactees and mad scientists of all stripes.

The acceptance of crackpot ideas and the pursuit of seemingly forbidden or concealed knowledge were key aspects of the counterculture—and of living in Los Angeles—that Gram embraced. Joshua Tree enabled him to marvel at an apparently spiritual natural beauty and to imbue it with outer-space qualities. Blending the two seamlessly provides a portrait of a certain early-hippie mind-set. As part of that mind-set, Gram showed little self-consciousness about espousing his thoughts on UFOs. Despite his social sophistication, Gram was only twenty-one years old. Instead of the education he and his Joshua Tree regulars skipped in order to rock, Gram had enthusiasms. He and his fellow UFO seekers made a belief system out of being willfully naïve and wacky. Soon hundreds of thousands of Americans would follow their example.

"I showed Gram this whole area, and he went a couple of times up on the mountain," Markland told the *L.A. Weekly* decades later. "Then, through Phil Kaufman, I brought the Stones up there. Marianne Faithfull was with them. Mick kept comparing Joshua Tree to Stonehenge, or various Druid sites he'd been to."

Gram and his friends would go out into the desert with an array of stimulants and spend the night searching the sky for extraterrestrials and blowing their own minds. "They all seemed like one endless night," [Keith] Richards later recalled. "It took a thousand years, but it was over too quick."

"We had binoculars, loads of blankets, and a big stash of coke," [Anita] Pallenberg says. "That was our idea of looking for UFOs! Did we believe in UFOs? Well, it was all part of that period. We were looking for something."

The group also experimented with mescaline and peyote "and tried to talk with the local Indians," in Keith's words. In emulation of Markland's chair-topped mountain, Gram and Keith dragged an old barber chair to the top of another peak they claimed as their own.

"It was wonderful," Faithfull says. "Staying up all night, driving out to Joshua Tree and walking along as the dawn came up. We would leave the cars somewhere and go off. We didn't bring anything—food, water, nothing. In that state we could have gone off in the wrong direction and gone around in circles forever, but somehow we didn't."

One night as the moon came out, Faithfull heard "this unearthly sound, a sound I'd never heard before. It was so thrilling, like being in India with the wolves howling." What is that? she asked Gram. He answered, "Why, Marianne, don'ja know that's just a little old coyote?"

GAYLE BRANDEIS

Gayle Brandeis was born in Chicago, Illinois, in 1968. She is the author of several books, including the novels The Book of Dead Birds *(2003), which won Barbara Kingsolver's Bellwether Prize for Fiction in Support of a Literature of Social Change;* Self Storage *(2008); and the forthcoming* Pears and My Life with the Lincolns. *She teaches creative writing at the University of California, Riverside, and at Antioch University.*

In the following poem, "Climbing at Joshua Tree," Brandeis writes about the celebrated granite rocks in Joshua Tree National Park, which attract rock climbers year-round from all over the world.

Climbing at Joshua Tree

The rock will hold your hand,
but you have to work for it,
fingers hunting for that gritty
ridge, that sharp ledge that will
accept your weight, help you
pull your sticky soled shoes
up to the next subtle foothold.
The world disappears as you
climb, becomes nothing more
than the sun on your back, the rock
before you, its face impassive,
unimpressed, as your muscles
seize and burn. But that's what
the desert is, isn't it? Life boiled
down to its most basic parts:
sun and stone and the twitch
of survival. And when you hoist
yourself up to the top of the rock,
each limb trembling, throat
a dry ache, the desert offers up
its lunar landscape, its shaggy
prehistoric trees, a sky so big
it jars your skull, so big

you have to lie flat
on your back and let the heat
of the rock seep through
your shirt, let its steady presence
ground you before you
make the trip back down.

DEANNE STILLMAN

Deanne Stillman was born in Cleveland, Ohio. She is the author of Twentynine
Palms: A True Story of Murder, Marines, and the Mojave, *a Los Angeles*
Times *Best Book in 2001; and* Mustang: The Saga of the Wild Horse in the
American West, *published in 2008. Her writing has also appeared in* Rolling
Stone, Slate, Orion, *and other magazines. She lives in Los Angeles.*

*The following essay, "Rocks in the Shape of Billy Martin," which originally
appeared in* The Village Voice, *is a playful meditation based on one of the author's
frequent visits to Joshua Tree National Park.*

Rocks in the Shape of Billy Martin

I know a place in the Mojave Desert where there are rocks in the shape of Billy
Martin. I visit the rocks every year to commemorate the return of spring. It
makes perfect sense to me that the rocks are in the desert and not a mountain
range or forest because the gone-but-not-forgotten Yankee manager was a kind
of dug-out djinn, an electrical force who materialized to kick funny dust in the
other guy's face and then vanished until he had to do it again.

Where did he go since we last saw him? Where all legends go—back into
the desert, that big sandbox that holds America's deepest secrets. Significantly,
the baseball diamond—which began on a sandlot and invokes forever—is
America's most appealing attempt at taming the desert. Yet perhaps not for
much longer: With consistently low television ratings for the national pastime,
who knows whether it will soon be overtaken by the shifting sands?

I grew up far away from these sands, under the gray skies of Cleveland, Ohio,
the place that tells you it's okay to dream, but not really. I guess that's why I always
preferred the New York Yankees to the Cleveland Indians (although felt like a
traitor for rooting for them until I moved to New York) and why I used to send
away for cactuses—I know you're supposed to say "cacti" but I don't like the sound
of it—that you could get from places with names like Kaktus Jack's and Desert
Botanicals and keep them on a window ledge near my bed. I don't know if my
window ledge faced the west or not, but seeing the outlines of my little cactuses
against those cloudy skies fueled my fantasies of the never-neverland where the
turnpike went, the land where the misunderstood found understanding, the land
where Zorro and Bat Masterson and Wyatt Earp wouldn't let anyone hurt you, the
land where a girl named Jane lives forever as a Calamity, the land where the only
thing anyone or anything really wants is a drink of water.

Much later I moved to Los Angeles, at the edge of the sands, and have lived here for the past eighteen years. In the beginning, I toiled in the television mines of Hollywood—a task not unlike hauling borax out of nearby Death Valley with a twenty-mule team—and found myself making frequent trips to the desert. Week after week I would flee Hollywood, the Xerox machine of America's dreams, and head for the Mojave, where they all started. I felt at home in this vast space where, if you happened to be near the right dune at the right time, you might stumble across a cosmic joke in the form of a shamanic workshop at the corner of Highway 111 and Bob Hope Drive, a culinary epiphany in the form of the best Hungarian restaurant this side of the Danube, a cultural oasis in the form of a biker with a used bookstore and an espresso machine, or endless miracles of nature such as the desert frogs that leap out of the sands after a rainstorm. In the Mojave, I came to understand that Los Angeles was, like my feelings for it, fleeting, a momentary metropolis, and I came to appreciate it as the punch line to a desert joke. Like every enclave of castles in the sand, it's overrun with fakirs. Deal-proffering bedouins named Steve wander the dunes, searching for temporary oasis. Dreams rise and fall with the caprice of studio wizards. The real thing, the elusive connection for which all who have attempted to decipher Los Angeles have yearned and failed to take into account, is the Mojave Desert, where the glitter is refracted not in the sheen of a limousine, but in flecks of obsidian and pyrite and quartz; the Mojave, where the silence is not the thunder of an unreturned phone call, but the flap of a butterfly's wings in the springtime.

The faint, ever-present L.A. pall begins to dissipate as soon as I plan to head for the desert, for the very word "Mojave" itself is comforting to say; the harmonic tones with the beginning sound of "M" or "Mo" with a soft "o" suggests mother, a safe haven, a grounding, and, in fact, the desert is female, a wide open space that is always there, waiting. And so in the time when the days begin to get longer, and there is talk of baseball, of trades and possibilities, it is to the Mojave I return. It's not difficult for me to get to the Mojave, just a one-hour drive to the north, up the 405 and over the San Gabriel Mountains, or to the east just twice as far, across the 10 (formerly route 66), through San Bernardino, and turning off at one of my favorite signs, the one marked "Other Desert Cities," just before you get to Palm Springs.

I know I am close to the Mojave when the L.A. radio stations fade from Grammy Award winners to Christian advice shows and I start receiving transmissions of other bearded evangelicals, primarily Z.Z. Top. The sun is out, my top is down, and the traffic thins. The native urge to drive fast naturally assumes command. This is fun for motorists and highway patrolmen, but not for that other Mojave denizen, the endangered desert tortoise for which I have occasionally swerved to avoid crushing as it lumbers across the pavement.

Who says California has no history? I wonder, while a baby version of one of the world's oldest reptiles clambers onto the freeway shoulder and makes for some tiny blue flowers.

I cruise on and then—Oh joy! Another scenic distraction—my first Joshua tree! Now this is the true Mojave! Hi, big guy! The Joshua tree grows in only one place in the world and that is the Mojave Desert, and only at an elevation of two to six thousand feet. This misunderstood plant has taken a backseat to the towering saguaro, the Charlton Heston of cacti, the one that appears in many Westerns, sometimes wearing a sombrero, and looks like a big, welcoming, goofy person. To me, the Joshua tree is more appealing, a misfit that is the very picture of beauty and terror, a forgiving although freaky mirror that doesn't care what your name is, what you do for a living, or what kind of addiction you do or do not have. Maybe the Mormons were on to something in 1851 when they named this weird manifestation in the middle of nowhere the Joshua tree. Its shape, believed the westering followers of Brigham Young, with its uplifted and multitudinous arms, mimicked the Biblical supplicant Joshua frenetically gesticulating toward the Promised Land. Of course, they were right. But to them, the Promised Land was the future site of Salt Lake City. As far as I'm concerned, the Joshua tree is not telling people to go someplace else; it's pointing the way to other Joshua trees, whose lily petals are unfurling now to catch the morning sun. It's pointing to the rest of the Mojave, and sometimes, if you look hard through the shifting bars of light, even a coffeehouse.

Inside, a cross-section of desert locals bellies up for cheap espresso—rock climbers, handymen, end-of-the-line types who are stranded here because of DWI busts and the ensuing revocation of their drivers' licenses. I hang for a little while, but spring has sprung and I don't want to miss the fragile wildflowers that have popped open in a frenzied response to the heavy winter rains. I order a double shot in honor of Minerva Hoyt, the Pasadena socialite who in the early twentieth century lobbied for preservation of the Joshua tree, which people from Europe were uprooting and trucking out of the Mojave by the dozens, replanting them in the old country for display in botanical parks due to a cactus craze that had resulted in yet another plundering of the desert West. "L'chaim, Minerva," I say and head for Joshua Tree National Park, heading east on Highway 62, a one-way in, one-way out high-tension wire that stretches from Interstate 10 through the desert hamlets of Morongo, Twentynine Palms, and beyond. According to *Outside* magazine, Joshua Tree National Park is home to "more weirdos per square mile" than any other national park. Read on, and be forewarned: You may count me among them, and you may be right. Once inside the park, I leave the visitor center in the dust (although not before checking the day's activity list and seeing that there are no scheduled interpretive talks, for which I always brake and swerve), anxious to see all the colors of the season and check in with my favorite cactus, which is really a member of the lily family and therefore biologically not really a cactus at all.

Deep inside this bizarre preserve, which is carpeted with the ecstatic vegetable, I park my ragtop, grab a bottle of water and hike up a trail. I pass more campers from Europe than from America, and think about this paradox: Inside the park, Joshua trees are now protected from desert-crazed Euros but outside the park, and all over the West, cacti—yes, here "cacti" sounds perfectly appropriate—are routinely blown away by gun-crazed Americans who go to the desert to shoot. But as I continue up and down a trail that is lined with paloverde and ocotillo and cholla and sage, the Mojave, as it always does, cleans my slate, and once again I am aware of only breath and blood moving through my body. The desert sand verbena is in full bloom and there is a creeping plant that looks like orange spaghetti strewn across the tops of the low-lying bushes that hug the path. In a little while, I reach my destination, a Joshua tree that is about two hundred years old and somehow makes me feel as if I were sitting in my maternal grandparents' rock garden where the daffodils and crocuses shot through the Midwestern thaw every spring, where if you got really quiet, you could hear big Rocky Colavito crunching across the sands of Lakefront Stadium as he stepped up to the plate and took the first swing of the season.

I sit down on a warm granite boulder and gaze up into the Joshua tree as the sun pulses behind. "Hey, you," it says, an alfresco support group minus the sob stories and cigarettes. "We knew you'd be back. We've been waiting. Calm down. Stop running. Tommy Hilfiger is not the heartbeat of America. I am. Bring me the arm of Fernando Valenzuela. Do you see how the gringos have stolen his stuff?" What will happen to A-Rod, I wonder, but the tree goes on. "Yes, this is what the old ballpark looked like before George Steinbrenner and Pete Rose, before cactus lamps, before all-night mini-marts, before twenty-four-hour Bible theme parks, before rivers were forced to flow backwards in order to build a showcase for Kenny G. So slather on the jojoba oil and step up to the plate. We've got a fastball with your name on it. And don't worry if the game goes into extra innings. You'll have plenty of time to get home because, well, this is home...which is why we don't count strikes here, we don't even keep score...By the way, how come they got rid of ten-cent beer night?"

As the sun sets behind this cactus that's not really a cactus that grows only in the Mojave, I realize that that's the best thing about the desert: Just when you think that it explains everything, it turns around and admits that it's clueless. It takes a big piece of geography to do that; I toast the Joshua tree with my canteen and hit the road.

On my way out of the kingdom of the Joshua tree, I make my customary stop at the rocks in the shape of Billy Martin. I'm a little concerned. Has the latest swarm of earthquakes disturbed them? Apparently not; like Yankee Stadium, they haven't moved. The petrified Billy Martin is still here, gazing across the sands at the dream team, forever signaling a game-winning hit-and-run, and, as always, waiting for a drink.

Now, if you are ever out in the Mojave, the once-and-future baseball diamond, and you don't immediately come across the rocks, don't worry. Although the desert is open twenty-four hours, it has some secrets it can reveal to you only in its own time. Sooner or later you'll find them, or the rocks will find you. And if you listen closely, you may hear a distant crack of the bat, or a faint cry—"Yankee franks! Springtime! Programs!" For it's always the first day of the season out here in the sands that generate the national pastime, it's always opening day.

MAKING THE DESERT HOME

CLIFFORD E. TRAFZER

*Clifford E. Trafzer, a professor of history, holds the Rupert Costo Chair in American Indian Affairs at the University of California, Riverside, where he is also director of graduate studies in history and former director of the California Center for Native Nations. He is of Wyandot ancestry and recently completed a manuscript with George Sioui on the life of Wyandot medicine woman and spiritual leader Eleonore Sioui. For the past twenty years, he has been a member of the California Native American Heritage Commission. Trafzer's work with the Chemehuevis of Twentynine Palms began in 1997 and resulted in a book—*Chemehuevi People of the Coachella Valley—*which he cowrote with Luke Madrigal, director of Indian Child and Family Services of Temecula, and Anthony Madrigal, counsel for the California Native American Heritage Commission. Both Madrigals have BAs in history from the University of California, Riverside, where Anthony also received his PhD.*

The following creation story of the Chemehuevi people reveals their close relationship with the Mojave and Colorado Deserts. To this day, they are actively involved in the preservation of their homeland through their support of the Native American Land Conservancy and Salt Song Project.

Chemehuevi Indian Creation

Chemehuevi Indians, the southernmost band of Paiute people, are part of a larger group of Native people that call themselves Nuwu. The Nuwu are desert people, and their sacred Salt Songs intimately tie them to the deserts and mountains of Nevada, Arizona, and California. Chemehuevi people say Charleston Peak, located northwest of Las Vegas, Nevada, is their sacred mountain. They call their central mountain Nivaganti, Snow Having or Snow Mountain. From this place the first Southern Paiutes emerged, but only during the second creation. Before the arrival of human beings, including the first Nuwu, a series of dramatic events brought the world into being.

During the winter months, Chemehuevi families gathered in their lodges located throughout the Mojave and Colorado Deserts to listen to their own creation stories. For them, these were not myths or fantastic tales that grew with the telling. They constituted the foundation of their history, literature, geology, and biology—metaphors provided in story form so that the people would repeat and remember the narratives in an oral tradition. The people used the stories to explain their world, especially their spiritual beliefs that closely linked human beings to the natural world. For Chemehuevis, the animate and inanimate worlds lived in a real and integral manner. The Nuwu enjoyed a rich tradition of stories that survive today in several versions. Thus, the stories are alive, living in the hearts and minds of Chemehuevi people.

Much of what we know of Chemehuevi creation stories come from the words of George Laird and Dick Fisher, two tribal elders who lived in the late nineteenth and early twentieth centuries. Carobeth Laird, George's wife, left two accounts of creation stories, *Mirror and Pattern* and *The Chemehuevis*. In addition, contemporary Salt Song singers Matthew Leivas of the Chemehuevi Reservation and Larry Eddy of the Colorado River Indian Reservation know some of the stories and help keep them alive through the oral tradition.

Chemehuevi accounts say that at one time, salt water covered the entire earth, which existed with a sky above. A worm fell from the sky into the ocean below and it changed form and became a female spiritual being that lived within the ocean. Southern Paiutes call her Hutsipamamaa or Ocean Woman. She became the mother of creation. She did not like the fact that only water surrounded the earth, so Ocean Woman had a vision of solid land. From that vision, solid land would come about. She began the process that led to the formation of the land.

Ocean Woman dove to the bottom of the sea to gather handfuls of mud, which she stored on top of her head while she used other elements of her body to make the land. She rubbed the inside of her thighs to gather sweat, dry skin, and oils that she spread upon the ocean, mixed with the mud. Around this time, she used the same elements to create Coyote, Wolf, and Mountain Lion—three major characters named in the early creation stories. They too helped in the creation but only after Ocean Woman began to shape the land. After spreading her magical mixture onto the sea, Ocean Woman placed her body onto the mixture, which grew as she began to shape the land. With her feet, legs, arms, hands, fingers, head, and torso, she shaped the earth. She kept her head to the west and stretched the earth while Wolf, Mountain Lion, and Coyote ran in many directions to report on the development of the Western Hemisphere. Coyote and Wolf walked upon the earth to report on the creation. They reported that the land had not been fashioned quite yet, until they finally concluded that Ocean Woman had finished her work.

Ocean Woman put the land into motion but the hard surface continued to change as mountains, rivers, lakes, valleys, canyons, hills, and other geographical features evolved and moved. During this first creation, no humans existed, but plants, animals, and places came into being, making the earth ready to receive the first Native Americans. The emergence of humans came after the first creation, and this sequential motif is common among Native Americans and other peoples throughout the world.

California emerged during the first creation, but the landforms changed over time, as elements that composed the earth came ashore from the west to the east. Although Ocean Woman had made Coyote first, Wolf emerged as first among the beings. Wolf personified a positive spirit and power. He became divine and superior, a being of wisdom and understanding. Coyote, however, personified future humans. Although he could be a creative and clever fellow, he could also be unreasonable, selfish, uncaring, impulsive, and carnal. Coyote thought about and envisioned sex, and when the opportunity arose, he chased a young female being.

Wolf and Coyote lived in a cave on the sacred Snow Mountain, a spirit mountain of central importance to all Nuwu. One day, Coyote became bored and left the cave to venture out into the world. First he found a track, and he knew the track belonged to a woman. This aroused his interest, so he followed her tracks, moving rapidly to overtake her. Tribal elders call her Louse, but some elders say she is actually Ocean Woman dressed in a small rabbit-skin apron that hid her genitals. But as she moved, her apron kept flipping up, exposing her and enticing Coyote. Coyote tried to convince Louse to go home with him so they could sleep together. But Louse insisted on traveling toward the setting sun to visit her home on an island in the Pacific Ocean. Coyote then proposed to make love to Louse, but she protested having sex in the open. Louse asked Coyote to go ahead and build a shelter for them so they could make love. Coyote raced forward, built the house, and laid down to rest. When Louse got to the house, she sent sleep medicine into Coyote and made him sleep deeply. When Coyote awoke, he realized Louse had passed by, so he raced to catch up with her and asked her to have sex. Again she insisted that he race forward to build a house for them, and so he did. And once again, she caused Coyote to sleep while she passed by.

Louse tricked Coyote four times without Coyote understanding Louse's actions or motives. At the time, she apparently had no interest in having sex with him. Finally, the couple reached the ocean, where Louse explained that she was going to swim across the ocean to an island. Coyote wanted to make love with her but Louse insisted that she was going home. At this point, Coyote decided to go with her but he could not swim. Louse invited Coyote to get onto her back so she could swim them both to the island. Coyote agreed and off they went into the Pacific. When Louse had taken Coyote far into the ocean, she dove into the water and dumped Coyote off her back. Coyote nearly drowned, but when he reached the surface of the water, he turned into a water spider. With his many long legs, he skirted quickly across the water. He arrived at the island before Louse and found an elderly woman sitting in front of a lodge. Coyote snooped about and then entered the lodge, where he went to sleep.

Some time after, Louse arrived and told the elderly woman about her adventures, traveling across the land and meeting no one but Coyote. When she began to tell her story about drowning Coyote, the elder woman told her Coyote had arrived before her. He waited for her inside the lodge. She joined him in the lodge, where he tried to make love to her, but all night long he cried out in pain. Making love to Louse proved impossible because she had teeth in her vagina. But Coyote still tried to make love with Louse four nights in a row. In the meantime, Coyote made arrows and went hunting, while the elder woman began making a woven basket. When Coyote went hunting, he met another elderly woman, who instructed him to kill a female bighorn sheep and butcher it, paying special attention to connected neck bones. He did as she said and cleaned the neck bones. That night, he penetrated Louse with these bones, which broke her vaginal teeth. Louse now became Coyote's wife. They slept together each night, making love.

For several days, the elder woman wove a beautiful basket with a funnel neck. When she had nearly finished it, she asked Louse to help her tie off the end of the basket. But before they closed the basket, Louse deposited eggs from her body into the basket. The elder woman then called Coyote and told him to take the basket home with him. She instructed him not to open it until he reached Wolf, who would know what to do with it. Coyote transformed himself into a water spider and took the basket to the mainland. But by the time he reached the shore, the basket had become quite heavy, so he opened it to see why the basket had become so weighted. As he did, people oozed out of the basket, the first Native Americans. Coyote could not put them back into the basket or stop the flow of people out of the container. When the flow of people ended, Coyote peered into the basket and saw a few people that had been crushed, and the waste of so many people. He closed the basket and continued his journey back to Snow Mountain.

Once home, Coyote showed Wolf the basket. Wolf admonished Coyote for not following the directions given by the elder woman. He took the crushed bodies of the Indians left in the basket and blew on them, bringing them to life. These few people became the first Nuwu, the hardiest of all human beings, the children of Wolf and Coyote. At first the people lived on the mountain with their relatives, Wolf, Coyote, and other animals. But eventually they came down from the mountain and began spreading out over the huge geographical area of present-day Nevada, Arizona, and California. The southern Nuwu became known as Chemehuevi, and they lived by hunting and gathering near places of water throughout the deserts. Some of them went into the deserts, but others went south to live along the Colorado River. They inhabited lands to the north, south, and west of Mojave Indians.

For generations the Nuwu and Chemehuevis have lived with and near diverse people, changing in many ways but forever maintaining their cultural beliefs in songs and stories. When contemporary people mention creation and the important characters in their past, the people say the stories come alive again, animating the deserts, mountains, and significant places. The ancient stories play important roles in the culture of Chemehuevi people, just as they have since the beginning of time.

L. FRANK and KIM HOGELAND

L. Frank, born in 1952 in Santa Monica, California, is of Ajachmem/Tongva ancestry. She is a cultural activist, artist, and one of the founding board members of the Advocates for Indigenous California Language Survival. Kim Hogeland, born in 1980 in Mt. Shasta, California, has a BA in history and Native American studies from the University of California, Berkeley, and an MA in history from the University of California, Davis. A freelance writer and historian, she lives in Oakland, California.

First Families: A Photographic History of California Indians (2007), coauthored by Frank and Hogeland, provides a comprehensive view of California's Indian peoples as well as a rich sampling of family photographs from throughout the state. The following excerpt from the chapter on Inland Southern California looks at the historical and present-day diversity of the many indigenous people who live in California's deserts.

from *First Families*

"The land was important. Mainly because it gave everything that you needed to live on. Well, we came from the land, from the earth itself. And it was created especially for you, and so you had a place within that, and you were going back to it."[1]

Many people see the Mojave and Colorado Deserts of California as inhospitable environments, vast expanses of seemingly barren flats and boulder-strewn mountains. Yet today in the interior of San Diego County there are a dozen reservations and rancherias, mostly Kumeyaay. Riverside County supports another nine reservations, mostly Cahuilla. In San Bernardino County there are four reservations: two Chemehuevi, one Serrano, and one Mojave. The Fort Yuma Indian Reservation in Imperial County is Quechan. And straddling the border with Arizona is the Colorado River Indian Tribes Reservation of native Mojave and Chemehuevi people, along with Navajo and Hopi who were relocated here in relatively modern times.

It is no surprise that when most of these reservations were created early in the twentieth century, those in power selected locales that they considered wastelands. Yet for some native people, these are lands of plenty. The desert served as the source of life, and when they were able to hunt and gather freely in those vast lands, a good and sustainable life was possible. To those who know this land intimately, it provides food and medicine. Says Katherine

[1] JoMay Modesto, in Deborah Dozier, *The Heart Is Fire.*

202

Saubel, Cahuilla elder of the Morongo Reservation near Banning (quoted in Deborah Dozier's *The Heart Is Fire*):

> The Cahuillas used plants for food, for medicine, for housing, for clothing—for everything they used the plants, especially the medicine. A lot of the medicine that grows now, grows on the high hills. The ones that come from the mountains are the best medicine that can be used. That is why I fight so much to preserve all this, so we can continue to use these things. I still get my medicine plants from the mountains where I was born. That is why I would like to have it protected.

While modern political boundaries have placed these tribes in a state now called California, culturally they have more in common with the people of the Colorado River, the Southwest, and northern Mexico than they do with, say, the people of the Central Valley or the coast. While the salmon-laden rivers of northwestern California and the oak-rich valleys of central California supported larger villages and denser communities, the carrying capacity of the desert called for smaller units. These were often extended families and clans with wide territories. Among their traditional foods were six varieties of acorn, mesquite beans, piñon nuts, and the fleshy parts of several cactuses. Desert peoples made yucca-fiber cordage for nets and used mescal fibers for sandals. Music from elderberry flutes and rattles made of turtle shells, deer hooves, and gourds, and most importantly, the human voice accompanied much of daily life. Houses were typically made of brush. Petroglyph sites—their meanings mysterious and certainly spiritual—abound in the region; in these sun-baked areas of the world, the spirit seems close. Testament to this, perhaps, is the way that the Mojave went to war not only with warriors and war chiefs, but also with soothsayers and dreamers.

Like native people everywhere in California, those who dwelled in the desert were skilled basketweavers, using a surprisingly wide range of native materials—deer grass, juncus, devil's claw, flicker quill, willow, porcupine quill, bulrush, and yucca root, for instance—to weave fine, useful, decorative baskets.

In addition to baskets, the people of the desert made pottery, often excised and painted, by coiling the clay and paddling the coils against a smooth stone. The beauty of this style of ceramics is in its functionality: construction methods are practical; the pots are often thin-walled but durable; and their forms are ideal for storing food and water. Potters knew where to get just the right kind of clay and how to fire it to make pots that are sturdy but lightweight. Because California pottery is not as decorative as the pottery of the Hopi, the Tewa, and others of the Southwest, it never attracted the attention of collectors, and the practice fell into disuse. Traditional techniques were lost for many years, until Cahuilla potter David Largo and others revived the art in relatively recent times.

While all Californian peoples cultivated and managed their land through conservation, burning, pruning, digging, and other techniques, only along the Colorado River can we find anything that looks like European-style agriculture. The Colorado River—which is similar to the Nile in its yearly floods—deposits thick, fertile silt as it shrinks from its high-water stage during the summer; and several tribes planted fields of corn, squash, beans, and melons here. So central was this agriculture that when the Spanish appropriated the best Quechan fields in the late eighteenth century, it led to a full-scale uprising. Mojave elder Elda Butler affectionately remembers the remnants of traditional farming and the presence of the as yet undammed river (which today is more like a lake) in her childhood:

> All they had to do was plant. And that was it, and then just wait for [the crops] to ripen. But that river was something else. Just small kids, just grammar school, you know, and we'd swim across the river, whole bunches of us. And we'd have to start way up here, and by the time we crossed we were way down there because the water was so swift...And then we'd get melons and then we'd swim back. All that work, and then we'd swim back with the melons, and then when we got to this side we'd sit and eat them...We were down at the river all the time.

For a while, the allegedly undesirable nature of this land paradoxically protected the desert tribes from some of the atrocities suffered elsewhere in the state by way of gold mining, grazing, logging, corporate farming, and runaway residential development. When Europeans and Americans finally gained a strong foothold in the region late in the nineteenth century, they claimed the best land. Native people were unable to move freely in their traditional hunting, fishing, and gathering places. Like many of the peoples from other parts of the state, they were forced to live on small, isolated, poverty-stricken reservations.

Life has gotten somewhat easier in recent times—at least for tribes with casino income—but land-use issues continue. Still embedded in the minds of the dominant culture is the sense that this is a "wasteland," and by the end of the twentieth century "wasteland" suddenly had value: a nuclear waste dump was proposed for Ward Valley, twenty-one miles west of Needles, a site long sacred to the Mojave and Chemehuevi. The dump—alarmingly close to an aquifer and the Colorado River—might well have had disastrous effects on both tribal and nontribal residents of California, Arizona, and Mexico. Following an occupation of the site by members of five tribes and various environmental activists, the Department of Interior officially cancelled their plans in 1999.

The Ward Valley fight was only one of many. For years the tribes have been defending their land rights, water rights, and sovereignty rights. The Quechan are currently battling a proposed open-pit gold mining operation, nearly nine hundred feet deep and a mile wide, which would destroy sacred sites and petroglyphs near Indian Pass and despoil land for miles around it.

The Colorado River tribes, unlike most others in California, have always been known for their warrior traditions. For the Mojave and Quechan, specialized weapons such as a "potato masher" club of strong mesquite wood and fire-hardened wooden spears were common. From a very young age, boys trained in dodging arrows and shooting at targets—even practicing for war by throwing mudballs at hornet nests, in formation. War leaders organized and trained men around hand-to-hand combat. From earliest days they successfully defended their homelands against the Spanish and early Anglo intruders. Generations later, though methods differ, they have not stopped.

Standing side by side outdoors, usually on a hot, dusty day, accompanying themselves with gourd rattles, the singers of bird songs sound faint at first. Despite their contemporary clothing—jeans, cowboy boots or running shoes, Southwest-style ribbon shirts, tribal t-shirts—with their faces lined by the southern California heat, they look as if they've been there forever. As one draws closer, the reverberant rumble of men's voices begins to sound strong and ancient.

For the Cahuilla people, the bird songs tell the story of their ancestors' migrations in the early days after the death of their Creator. In the Cahuilla creation story, the brothers Mukat and Temayawet had made the world together. They both built clay humans, but Temayawet worked in haste and did a poor job of it: his "people" had webbed digits and double-sided faces and stomachs. Mukat, on the other hand, worked slowly and carefully, and the people he made were beautiful and efficient. Mukat looked dubiously at Temayawet's faulty creations. With webbed fingers how could they pick things up? With faces and stomachs on both sides, how could they sling carrying baskets over their shoulders to transport heavy loads? In shame and anger, Temayawet withdrew underground, taking his creatures with him and leaving the world to Mukat and the humans he created.

Over time Mukat changed, and instead of helping people he turned against them, introducing death, giving poison to rattlesnake, inventing the bow and arrow and tricking people into shooting at each other. The people decided to kill Mukat. And, as Katherine Saubel (Cahuilla) explains,

> The people burn Mukat's body. From Mukat's ashes, all the food plants grow—acorns, squash, chia, sage, all the food of the Cahuilla. In his death, Mukat has given his people their way of life, their laws, customs and ceremonies, and their food.

The Cahuilla bird songs deal with the time immediately following Mukat's death. Having killed their Creator, the people were sorrowful, and they scattered in many different directions. Time passed, and the few Cahuillas who remained in the original homeland decided to bring the others back, and they set off on a great journey to find them. Some say they circled the continent, others say they journeyed only through southern California and northern Mexico. The people who were living away realized that their original homeland was the best place after all. The bird songs, in hundreds of short stanzas, recount incidents and stories of these travels as the people separated and then came together again, moving and migrating in the way that birds do.

Bird songs are also a feature of Kumeyaay, Mojave, and Quechan culture, as well as indigenous cultures in northern Mexico. In traditional times, the linked stanzas of the song were sung for several nights in a row from dusk to dawn—beginning songs started the event, at midnight the returning songs began, and everything was timed to end with mornings songs when the sun rose. The songs are usually in a language understood by the singers, but often embedded in the songs are fragments of other languages thought to be ancient and mysterious. Although they deal with spiritual issues, the bird songs are not sacred, nor are they restricted; there is no special requirement for being a bird singer other than an honest interest in the songs. Bird singers sometimes stay up all night, singing and dancing, stopping only occasionally for short breaks. At other times samples of song provide welcome entertainment for a fiesta or celebration.

By the last decades of the twentieth century, the singing of bird songs had declined, was no longer practiced, reduced to a memory borne by fewer and fewer. Alvino Siva, realizing what it would mean to lose the tradition, began singing again in 1975. He, Robert Levi of Torres-Martinez, Tony Andreas of Agua Caliente, and others began to revive the tradition, pulling out of their collective memories the songs, their sequence, and the stories relating to them. And they trained younger apprentices such as Mark Macarro, Luke Madrigal, and Leroy Miranda, who in turn trained others. Tony Andreas describes some of the story (in *News from Native California*, vol. 5, no. 4):

> As I got to be about thirty, I started to sing with Joe Patencio and also Bert Levi. They used to say, 'Come on. Sit down and sing with us,' so I'd start singing with them...I knew some of the songs, but I didn't know them right. And Joe Patencio and Bert Levi taught me how to sing them the right way. I still don't know, but at least they gave me a path to follow. It's all synchronized. The songs have their own pattern and their own order.
>
> I have nobody to go back and ask questions—to say, 'Am I doing it right, or is this the way?' They used to tell me, 'Well you'll figure it out.' Figure out what?! I didn't know what to figure out. But now I think I know what they meant...And I hope I've figured it out right.

In December 2005, the Agua Caliente Reservation in Palm Springs sponsored a day-long Bird Song Festival. Singers from all over came and sang for an audience of many hundreds. The performers, men mostly in their thirties and forties, were skilled and proud, their singing robust and confident. Here and throughout the area bird singing is once again connecting California's desert people, their land, and their past.

FRANCISCO PATENCIO

Francisco Patencio was born in Chino Canyon, near present-day Palm Springs, California, in 1856. He was a Kauisik Cahuilla of the Agua Caliente Band of Palm Springs, and an important religious leader, storyteller, and elder. In 1943 he published Stories and Legends of the Palm Springs Indians. *Lowell Bean, an anthropologist who has devoted his life's scholarship to researching and writing about the Cahuilla, is currently at work on an expanded edition of Patencio's book, to be published by the Agua Caliente Cultural Center. Patencio died in 1947.*

The following excerpt is taken from Desert Hours with Chief Patencio, *published in 1971 and edited by Kate Collins. The significance of birds to Cahuilla culture is evidenced both in Patencio's story and in the bird songs and dances that continue to this day among the Cahuilla in Palm Springs and neighboring desert regions.*

The Birds

As I [Kate Collins] started up the path to Chief Patencio's house, the birds were singing and twittering gayly in the Mesquite bushes. I could see a rose-breasted Linnet trying to catch a wary insect on a branch. A Mockingbird scolded me as I passed by his Pyracantha bush. It seemed as though a hundred sparrows were holding conversation in the Palo Verde. As I walked along with nature's music filling my ears a tiny ruby-throated Hummingbird darted alongside the trail. He followed me into the Chief's yard where he hovered in the mouth of a Chuparosa blossom. The Chief was observing him, too.

"Our desert brothers—they are good to watch," he said.

"I used to think there were no birds on the desert before I came here," I replied.

"Sit down," said the Chief. "Let me tell you about our birds. When the birds were first created they had no legs, no wings, only heads and bodies, so that they were like worms and could not walk or fly, but had to lie on the ground in one place all of the time. The birds suffered much. They were eaten by other creatures, and stamped on. They could not help themselves.

"Then they began to wonder why they could not get about like other creatures. Why couldn't they make noises like other creatures? So, they tried to make noises; they made a rattling noise and they found that they had a voice. So they began to practice. First they practiced on one note, then two and even three notes. Then they began to make songs to the Great Spirit to come and help them. They all sang—those with one note, those with two notes and those with three notes.

"As they sang they began to raise themselves up. Then after some time they found that something was happening to them. The Great Spirit had heard them and he was caring for them. First he gave them feathers, warm feathers to cover them. Now the birds never stopped their practicing. They began to learn more than three words.

"Next the Great Spirit gave them legs and wings. When the legs first began to appear the birds had to learn to walk. They would hold out their little wings and try to stand. Little by little, step by step, they learned to walk. And all this time they sang their songs of one word, two words and sometimes as many as five words.

"Then one wonderful day they saw the sun come up. They had had only the dim light of the stars and moon before, and could not really see each other. Now they looked at each other and saw that everyone was different—different in kind, size, color. So, they separated and went away in tribes.

"Then they started walking toward the sun. They went far toward the East and they were very happy at everything they saw. They saw the sun, the hills, the sand, the water, the grass and the seed. They were so happy that they sang all the songs that they had learned as they walked.

"Finally they turned toward the West. Then they saw that it was a long, long way that they had walked; they thought they had better return. They talked this over together and then turned around to start back.

"Now, everything was dry and dusty. They had a hard time crossing over the desert. They suffered from thirst and hunger. Then the Meadow Lark soared up into the sky and made the rain to come. Everything grew wet. The Owl and the Rats and the Cotton-tail Rabbits and the Wild Turkeys, they all found caves and hid themselves from the rain for many days.

"Then the rains ceased and the land became dry. There was no more water than there had been before. During the rain the Mockingbird had come to the water hole and built her nest. Now, because there was no water at any other place, all the birds and animals came to the Mockingbird's water hole to drink.

"Now, the Mockingbird did not like this. She was hatching her young. The talking and the singing of all the other birds bothered her and she bothered the other birds with her chattering and scolding. She sang all day herself and half the night.

"One morning, early, the Mockingbird was scolding again when a bird called *Hem-o-ic-kal*, meaning a hooked nose, a ground bird something like the Road Runner, came along. This bird lays very pretty black eggs covered with green spots—the earliest eggs hatched among the birds. This bird got tired of hearing the Mockingbird scolding all the time. So, she began mocking everything that the Mockingbird said, until the Mockingbird got tired of hearing it, and stopped her scolding.

"Then the birds left the water hole and went away to the San Jacinto Mountains. There they found the rain and there was snow. They had nothing to eat and they were very cold. They looked down on the other side and saw a valley. It was not raining there. It looked pretty and green. So, they decided to go down into it. From there they went to the coast and decided to settle there. In the winter, when it was cold, they came back to their old home in the desert.

"Now, the birds sing many songs but some birds have many more notes than others. But, whether they sing two or three words, or whether they sing a long song or a short one, they still sing, always their song of thankfulness to the Great Spirit who remembered them in their sorrow and brought them to be birds of the air as well as birds of the earth."

A little Cactus Wren whirred past the Chief's head and headed for her nest in the Cholla. The Chief turned to watch her.

"Come back tomorrow," he said. "There is much I would say to you of our desert."

I turned, and as I did, I saw a Quail, top-knot bouncing in the breeze, scampering across my path and into the shadow of the Creosote Bush.

EDMUND C. JAEGER

Edmund C. Jaeger was born in Nebraska in 1887 and grew to become a renowned botanist and naturalist closely associated with desert environments. He taught at Riverside Community College for many years and authored numerous books and articles about desert animals, plant life, and ecology, including Desert Wildlife, *published in 1950. After his 1946 discovery, near Desert Center, California, of a common poorwill in a state of torpor, he became the first person to document hibernation in birds. Jaeger died in 1983.*

The following passage from Desert Wildlife *describes the quirky and captivating nature of the quintessential desert bird, the roadrunner.*

from *Desert Wildlife*

California Road Runner

Of all the feathered dwellers in the desert there is none that has such an amazing stock of peculiarities as the California road runner. He is the desert's bird wag, as full of comical manners and as mischievous as the jay or the nutcracker; yet, unlike these birds, he is never obtrusive in his familiarity. Every morning he goes down on the trail below my shanty and saunters along, waiting for me to come with my pail for water, well knowing that I will chase him and give him the fun of beating me to the corner. Just as I am almost upon him, he leaps into the brush out of sight and is not seen for an hour or two. This born gamester has been found time and again pursuing the ends of surveyors' chains as they were dragged along by the lineman, or, on golf grounds, running down stray balls with the eagerness of a playful dog.

You will never mistake the road runner. The bristle-tipped topknot which he raises and lowers at will, the reptile-like face with its deep-slit mouth, and the long tail which so unmistakably registers his emotions, make him a bird of most singular appearance.

This strange cousin of the cuckoo has earned his name from his habit of sprinting along roadways, especially when pursued by horsemen or moderately slow-going vehicles. In the old days of California, when tourists were frequently driven over country roads in tallyho coaches, it was no uncommon sight to see this bird running a half-mile or so in front of the fast-trotting horses. Another common name, "chaparral cock," is given in allusion to his living in the chaparral or scrub forest of the semideserts; and he is called "ground cuckoo" because of his inability to leave the ground in long-sustained flight.

Formerly the range of the road runner included the grassy plains, chaparral-covered hills, and arid mesas from Kansas to the Pacific Ocean and from central California to Mexico. With the settlement of the land and the increase in the number of gunmen, this unique bird is rapidly becoming rare, and the familiar Maltese-cross footprints which he leaves along dusty roads are now all too seldom seen except in the wildest portions of his former range.

The road runner makes no regular migrations and is seldom seen except when alone. He prefers the protection of thorny, low-growing mesquite and saltbush thickets, and once he chooses a clump of mesquites, he seldom leaves the vicinity and may be found there year after year.

Like a policeman, the road runner apparently has his beats, and anyone who watches him day after day will note how punctual he is in passing certain points at definite times. An invalid on the Colorado Desert recently called my attention to the fact that a road runner passed her porch regularly at 12:25 P.M. every day for over a week, never varying by more than a minute or two. A gentleman who some months ago put up a new board fence tells me that a road runner almost daily jumps on the upper rail and runs at top speed the full length of one side of the fence. He invariably does this at the same time of day—just about noon.

I became acquainted with a young road runner as a pet in the home of O. H. Wickard, at Antelope, California. The bird stayed in the house at night but went out early in the day.

First on its program was generally a sun bath. Mounting one of the granite boulders about the house, he would puff out his feathers until his body looked round as a ball, spread his wings, and lower his tail. In this position, he would sit quietly for some 30 minutes. Sun bath over, he would go into a cactus patch. In a succession of dizzy leaps and jumps and hurried flights and runs, he would go round and round a cactus clump, perhaps as many as 20 times. He always seemed to make an effort to assume as many different clownish attitudes as possible. This over, the road runner might then go to the house and annoy the cat a while. Dogs he had no use for. Even a very small Pekingese would send the bird fleeing in what appeared to be terror.

Cat-teasing consisted of rushing toward the animal with spread wings, extended neck and head, and wide-open mouth, followed by a snapping of the beak and a strange sound of rattling in the throat. The cat often countered this noisy, showy charge with a quick bat of its paw. Then the bird would deftly retreat and rush at the cat again. Both seemed to enjoy the sport. It generally ended after the cat ran away, with the road runner giving chase for a yard or two.

This road runner was unusually fond of fuzzy objects such as loose wads of cotton and milkweed seeds. He would run round and round the room, holding the object in his beak. Crayons he found in a box on a shelf were picked up one by one and thrown to the floor; flowers in a vase fared likewise.

From time to time his amusements were varied by running from the open door, picking up a leaf in the garden, and then dashing inside with it. Once he picked up a sizable piece of glass, brought it to show to his mistress, and to her amazement, swallowed it without ill effects. As she sprinkled the family wash he hugely enjoyed having her sprinkle water on him. When the neighbor's barefooted children came in, he would rush at them, make a rattling noise in his throat, and then peck at their toes, often frightening them from the house. If they had shoes on he pulled at the buttons or the ends of the shoestrings.

At night the bird slept inside the house on a branch of a tree nailed above the door, or on top of the iron-cased wall clock, his tail pushed flat up against the wall. The striking of the clock he wholly ignored, and he went to sleep even with the lights on; to evening conversations he gave no heed.

One evening a stranger came in and sat down in a chair near the clock. Old Roady jumped from his sleeping perch onto the visitor's head, and from there to his lap, where he made repeated stabs with his beak at the visitor's fingers.

One summer morning one of the boys found the pet caught in a spring trap set out for the rabbits that had been eating the garden vegetables. One foot was so nearly severed from the leg that amputation was necessary. With only a stump left, it was no longer able to leap or run, either in sport or while hunting for food.

But Old Roady was resourceful. For several days he spent most of the time lying in the shade; before the week had ended, however, he was hobbling around, even attempting to run on his stump. On the third day after the accident I saw him with the aid of his wings jump upward and snatch cicadas from the branches of shrubs. He was soon back at most of his old tricks. The end of his leg sometimes got a bit sore from overuse, but the callus was soon thick enough to stand much abuse, and before a month went by he was again visiting at the homes of the farm neighbors. Often he stayed away all morning and traveled several miles. Once he frightened a housewife by bringing her a live snake, he apparently enjoyed her excited screams. He usually stayed out in the brush all morning, seldom coming into the house until noontime. If the boys tried to find him he was very clever at eluding them by crouching low under the bushes.

MARSHAL SOUTH

Marshal South was born in South Australia in 1889 and moved to the United States as a young man. In 1930 he built and moved into a primitive adobe shack, which he named Yacquitepec, atop remote Ghost Mountain in California's Anza-Borrego Desert State Park. Along with his wife, Tonya, and children, he lived there with few modern conveniences and in close harmony with the land until 1947. More than one hundred essays and poems based on his life there were published in the widely circulated Desert *magazine from 1939 to 1948. His columns were, according to the magazine's founder and publisher, Randall Henderson, "the most popular feature in the magazine." A biographical study edited by Diana Lindsay,* Marshal South and the Ghost Mountain Chronicles: An Experiment in Primitive Living *(2005), includes the complete collection of South's writing for* Desert. *South died in the southern California mountain town of Julian in 1948, not long after leaving Ghost Mountain for good.*

In the following essay, "Desert Refuge 3 (October 1941)," South enthusiastically describes a rare desert rainfall.

Desert Refuge 3 (October 1941)

There is rejoicing upon Ghost Mountain, for once more all our water cisterns are brimming over. Even the "lake," as we term the big cement lined excavation that will some day be an additional reservoir, is half full. For three days ago, after a long "torture of hope" during which heavy thundershowers marched in complete circle around us—drenching the lowlands and mountains within a mile and leaving us bone dry—the rain gods relented.

There was drama in their storm-sending too. And rebuke. For we had gone to bed dispirited and, let us confess it with shame, rather a little angry and full of complaint. All the day long we had sweltered in the hot heavy atmospheric breathlessness which precedes a desert deluge. And all day the sullen thunderheads had banked around us, glooming the sky everywhere except for one clear spot, seemingly not greater than the area of our mountain, that hung directly over our heads. And it had rained. In sheets and curtains of slashing grey the forked flails of the lightning had ripped waters from the heavens—to north, to east, to west, to south. But not here. "All we get," said Rider bitterly, sniffing the damp fragrance of the breeze that came up in the evening from the distant lowlands, "—all we get is the *smell.*"

So we had gone to bed disgruntled, trying hard to be philosophic in the knowledge that sooner or later our turn would come. But, after the manner of frail humans, not succeeding very well.

But the grey, still whiteness of the next dawn crept across the desert to the hollow rumble of heavy thunder. Far off, but approaching. Hardened by so many previous disappointments, I noted it with a drowsy mental shrug. And went to sleep again.

"Daddy! Daddy! Wake up! It's *raining!*"

I woke with a jerk—in more senses than one—for Rider's hand was upon my arm and he was shaking me vigorously. "Real rain," he said breathlessly. And, as I sat up, fighting the sleep from my eyes, I heard Rudyard's shrill refrain piping from the next bed: "Yes, weal *wain! You* gotta get up an' fix the 'pouts! Hurree Daddy! *Hurree!*"

So I got up hastily—Rudyard is very definite in his commands and has all the authority of extreme youth—and rushed out to "fix the spouts." The spouts and water gutters always have to be "fixed"—that is, swept out and cleaned—immediately before a rain. It is a last minute chore that no previous planning can avoid. The reason is that the Ghost Mountain packrats, secure in the truce of brotherhood which reigns at Yaquitepec, long ago decided to use our gently sloping house roof each night as a dance floor. Which would be all right, for they are lovable little animals, if they did not also use the water gutter as a check booth in which to park bits of cholla, the dry rinds of cactus fruits, mescal pods, dead juniper sticks and all the thousand and one other classes of trashy "valuables" which packrats—exactly on the order of humans—lug along with them and regard as "very important." And which they conveniently forget—in the water spout. It isn't any good to clean the spouts the night before, except as to a reduction in labor. For the next morning will find a new collection.

"Hurree, Daddy! Hurree!" shrilled my imperious taskmaster, racing, a bare-skinned sprite, after me into the dawn, "Hurree! *Hurree!*"

And there was need of hurry. For ominous big drops were already plunking in wide-spaced intervals upon the iron roof. Far aloft we could hear that weird, sinister rushing sound—like the churning of a great wind—which is the advance message of released rain masses already plunging downward towards a thirsty earth. Rider was dashing here and there, closing shutters, dragging dry firewood into the kitchen and setting the innumerable pots, pans, pails and jars that are his own personal water-catching outfit, beneath the run-off and drip point of every inclined flat surface not connected with the main gutter system. We all worked fast. But I had barely tossed out the last bit of cactus joint and given the clean metal gutter a final wipe with the damp cloth when the deluge struck in a blinding white fury. Junipers, rocks, ocotillos and tall podded mescals blotted suddenly in a sheet of falling water. Rider and I reached shelter in a spume of stinging drops that seemed to tingle with the electricity of the forked fire that of a sudden split the sky overhead with a deafening crash. Rudyard bolted in at our heels like a little drenched duck, water pouring from his tangled brown curls. Then it *rained!*

The storm lasted half a day, with the first fury—when the water fell in solid curtains, succeeded by scattered and dwindling showers. The day previous Rider and I had collected a bundle of yucca leaves, with which to put a new seat in one of our chairs. A lucky circumstance, for now there was an indoor job all ready to hand. So, while Tanya sat in the window seat explaining the rain in complicated baby-talk to wide-eyed little Victoria, I and my two eager assistants hunted up awls and began to shred the long green bayonet-like leaves into quarter inch strips. The big leaves shred readily, following the lines of the fibers that run from butt to tip. When the leaves are fresh cut these strips are very pliant and braid or twist easily; but if they are dry they quickly can be gotten to the right condition by soaking in water. When we had accumulated what looked like a sufficiency of strips—in spite of Rudyard's "assistance," for he has rather elastic ideas at present on what constitutes a quarter-inch width—Rider brought the forlorn chair that needed fixing. It was in a bad way, never having recovered from the time when Rudyard invented a dramatic game called "Beeg fire on Bwoadway. Peoples jumping into net." He had used the canebottomed chair as the "net," jumping lustily into it from the height of a box set upon the table. The drama had been suppressed quite suddenly by unimaginative grown-up "police." But not in time to help the chair much.

Well, what is one chair seat, anyway, in the scheme of things. Kingdoms, we have been told, have crumbled for the want of a horseshoe nail. And heaven alone knows how many automobile classics have been lost to unprogressive speed drivers who neglected to use "whoozawizz" spark plugs. So why worry about a chair seat? I cut out the old torn bottom and began to braid in the new one. We have found that braiding, after the manner of the South Sea Islanders with their coconut fiber, is the best treatment for yucca.

It is a fairly long job to braid the seat into a regular sized chair, braiding as one goes and lacing the completed cord back and forth, basket fashion. So by the time the rain was over and the children were racing up and down through the puddles in the hot sunlight that had broken through the scattering clouds the chair was about finished. Not a brilliant example of weaving, perhaps, but something that would serve well enough. I clipped the last strand end, set the chair by the table and went out to sniff the fragrance of the damp, rejoicing, sun-sparkled desert. Tanya and Victoria were already out, sitting on the damp rock step beside the "lake" watching Rider and Rudyard sailing their long-stored boats. Over the crests of the clean-washed juniper trees winged, like migrating fairies, a wide scattered drift of gauzy-winged flying ants. "Plush bugs," as Rider calls the bright many-legged, fluffily scarlet little round insects which appear mysteriously after warm weather rains, were already ambling about underfoot.

It is true enough, as Kipling said, that "Smells are stronger than sounds or sights to make your heartstrings crack...." But of all the scents that can stir up haunting memories and sheer delight for the human nostrils I know of none half so potent as the fragrance of the desert after rain. It is something too deep and subtle for description. If you know it you know what I mean. And if you have never lifted your head and drunk in the winey, aromatic fragrance that wells from the grateful earth and stretching leagues of wasteland after a heavy shower, you have missed something—missed one of the greatest and most mysterious thrills that the wilderness holds. To stand in the midst of a sunlit, rain-washed silence and drink deep of this prayer of thanks, welling up like incense from plant and shrub and rock and spiney thorn to the Great Giver of all Mercy, is a moment—a sacred moment. One stands awed, listening to one's own humble heartbeats. Thus stood our dusky brothers, the "savages" of the dim, fled yesterdays. With them the Great Spirit was something *real*—not an empty thing, blurred in a tinsel mockery of Sunday clothes and stereotyped ritual and hollow words.

But rain at Yaquitepec means not only water in the storage cisterns. It means mud. Mud is a valuable thing. So long we have been without it. Or, having it, have had it in little dabs—as much, maybe, as one can obtain from a pint of wash water, or from the frugally saved unused portions of a brew of tea. Such dabs take a tedious time in making a showing upon an adobe wall—though we can point to considerable areas of the mud walls of Yaquitepec that were built in just such piecemeal fashion.

But now is one of our widely spaced periods of abundance when we revel in mud. There is mud upon my hands as I write, and mud upon my feet. Dried mud that has been imperfectly scraped off. Rider, still working at outdoor jobs, is pleasingly decorated all over with wet, clayey signs of toil. Rudyard has mud in his hair. Purposely put there, we discovered later, as a result of his having remembered a story told some time ago by a visitor to the effect that the Apaches plastered their heads with mud as a hair tonic. Even Victoria has had her innings. For seizing an opportunity, she crawled off her rug and into a gooey batch of adobe which I had just trampled to the right consistency. In the ensuing cleansing operations, to the accompaniment of lusty yells, Tanya got well mudded too. So that makes it unanimous.

But a lot of new wall has gone up, built with a shovel and a trowel and the plain bare hands, and in breathless haste—racing against the swift soaking away of the surface water in the open pools, from which we take it. In some of the pools that have been previously well trampled by our bare feet—after the manner of the old buffalo wallows on the plains—there is still water standing. So for a day or so yet there will be mud—and wall building. Then again operations will stop. Thus, in such fits and starts, goes our building. Woefully primitive, of course. But we are shamelessly unashamed of the method. And it is likewise quite true that the house isn't finished as yet. Nor are we anxious to have it so. "Finished" is an ominous word, reminiscent, somehow, of the practice of sending elegant young ladies and young gentlemen to an elegant "finishing school." Too many things, now, in this era's progressive set-up, are regarded as "finished." And a lot of them frequently are.

PANAMINT RUSS

Asa Merton Russell, better known as "Panamint Russ," was a mid-twentieth-century miner in the Death Valley region. In 1930 Panamint Russ built a small stone cabin, now known as the Geologist's Cabin, at the base of Manly Peak in Death Valley National Park. From there he mined for gold for many years. He was reputed to have discovered and then forgotten the location of a high-grade claim of gold. His story is one among many of the ongoing search for lost and legendary gold mines in the California deserts.

The following article, "Life on the Desert," was written by Panamint Russ and published in Desert *magazine in 1955. It gives a colorful and exciting description of his life as a gold miner.*

Life on the Desert

My cabins are the base of 7200-foot Manly Peak in the Panamint Mountains, Death Valley, California. I started building them in 1930, the same year I found the gold high up the side of old Manly and began my mining operation, now a long horizontal shaft into the side of the mountain.

From my favorite chair in front of the cabins I have a 90-mile view across the valley. My locust trees are the only ones within 125 miles, as far as I know. Just looking at the way they defy wind, storm, rain and snow gives me inspiration. Their bark is unscarred, well-balanced, leaves emerald green, but how they survive I'll never know.

The concord grapes are doing well, too. Twenty-five years ago coming through Riverside, California, I stopped at a nursery and bought a half dozen bare-root size, wrapped them in a newspaper, laid them on the running board with a wet gunny sack and today they are 20 feet of beauty.

I had hardly finished cleaning out my spring, a yearly chore, and settled back in my spring steel patio chair (some tenderfoot brought it out five years ago and forgot to take it back) when the black pickup truck with the white lettering drove up to pay respects on a routine inspection.

It was tall kindly-faced Matt Ryan, the National Monument ranger, and his wife. I'd always missed him before. It was the first time we had met. His headquarters is at the other end of the monument and I am only here two weeks every year for my assessment work.

We had a chat about the baby burro stealers, the bad sports who kill game and leave it by the springs to stink, the roving high graders, the city fools who lose their heads and wander from their car as soon as they are lost, about many things. He thought I had a very nice camp and mill-site and asked how the mine was coming along? Was I working alone?

He knew the answers, it's the same as last year and the years before that. The miners were enticed into the airplane factories. With a pair of pliers and a screw driver they had a cost-plus paradise, and still don't realize it's over. Just sit around waiting for the call that doesn't come, instead of going back to the hills.

The sun was already behind Manly, so after a cool drink of that natural granite filtered water, Matt had to shove off. In parting he remarked: "Russ, as long as you show Good Faith, that's the main requirement."

After the black pickup crossed Butte Valley it looked like a speck and disappeared from view. I relaxed and wondered how much Good Faith one man is supposed to have.

That ranger didn't know that I had plenty of Good Faith long before they decided to make this a national monument—had to. And I have never lost that Faith even though there have been a few setbacks. I am now 60, so if I finish my mine tunnel in the next five years, we'll retire together, my Faith and me.

But Matt's decree got me to thinking how many times I have come face-to-face with Good Faith out here on the desert:

It's been ten days now, since I have seen anyone, a fellow was supposed to drop the mail and some supplies off four days ago. He may have had trouble, because at that time we had a twister. The wind blew down the canyon so hard it raised the roof off the shack. Pulled the nails out three inches. I got some heavy wire and wired it to the floor. Just finished when the wind changed, came back up the canyon, started raising the other side of the roof, cracking rafters. I cut the clothesline and wired that side down just in time. Was undecided whether to move to a nearby cave in the rocks or stick it out. I stuck it out. In the morning the wind died down. I spent the morning picking up and bringing back to camp everything that was loose and moveable.

I wondered if the elements were trying to run me off, or just annoy me. Next day the wind blew again so hard it was impossible hardly to go outside. Three more lost days with nothing accomplished. I needed Good Faith.

Last year the burro, "Jubilee" we called him, was shot or stolen. I had raised him from a colt and just got him to where he could pack and be of some help. Now my back has to take his place.

The year before that I was caught in a cloudburst about 9 p.m. on the side of a mountain. Had to drive by instinct: water all over the road, rocks bouncing over the top of the car, both front tires flattened, roads washed out. I was two days trying to get to town. It took Good Faith.

The year before that—all alone as usual—I was trying to pry out a boulder in the path the wheelbarrow must use and double-ruptured myself. That meant the hospital and recovery.

No matter how good a desert driver you are, things will happen on these dirt roads. A sharp rock sticking up out of the sand took the plug off the crankcase and then knocked the plug out of the gas tank. The new broom I had along made some plugs until I could get some new ones, but I lost my gas and oil, and before reaching camp the car buried itself up to the hub caps in the soft sand and silt. I had to cut sagebrush and lay it for a road to get clear.

Another year, if I recollect right, the tunnel was in about 150 feet deep—too far for a wheelbarrow. I needed an ore car, ties and rails. Had a fellow helping me that year. He forgot to shake his trousers in the morning (a desert must!) and a scorpion bit him before he had his suspenders adjusted. So he quit and I was along again with my Good Faith.

The fellow who was doing the packing by contract to get my stuff up to the tunnel had five burros. He always slept on the ground by the spring. Chewed strong Black-Jack plug tobacco. He was about asleep one night when a snake crawled over his face between his upper lip and nose. He's not sure what kind it was but he swallowed his tobacco and got so upset he couldn't sleep or eat, so he pulled out for a couple of weeks. These delays are costly but we must have Good Faith.

Well, I tried to hire a couple of fellows to help me. Had no luck. A mine owner at Tecopa said, "Russ, there just isn't any extra help around. But to help you out, you take two of my muckers for a week or so." I had to guarantee them two weeks' work, board and room.

My tunnel is one mile up the side of Manly Peak, a winding burro trail the only way up. These two fellows couldn't even climb up that trail with the water, powder, drills and lunch. So I had to keep my agreement and give them two weeks' work around camp. Another two weeks lost!

Pack rats run all over my cabins. Some of them have been stealing cotton from the bottom of the mattress I am sleeping on and I couldn't figure out where they were taking it. All I knew was they sure made a lot of noise getting it through the springs. I just bought six new traps and went out to the car to get them. As I raised the rear compartment door, there was my cotton. The packrats were building a new nest among my extra powder, fuses and caps. All they have to do now is finish their nest with matches, set one off, burn up my transportation and I will have a 65-mile hike back to town—and my Good Faith will be shattered for this year.

But out here, there's always a blue sky, good pure water filtered by Nature through lime and granite rocks, smogless air, no 50-cent parking lots, fresh sage and pinyon pine. I shouldn't ask anyone for help, because it's my problem. But I have Faith, and it will pay off!

I started the tunnel five years ago. I was married and had a son. My wife said one year, "I am going to break you of that desert habit of yours. Every year you go to Butte Valley and stay too long. I warn you, if you stay over two weeks this year I will divorce you!"

I had good intentions of returning in two weeks but circumstances were against me. Some fellow and his family had been working some tailings on an old dump and were just ready to pull out. Their grub was all gone. He warmed up the motor of his truck, but forgot he had drained the radiator the night before. When that cold spring water hit the hot motor it cracked the head. They were stuck and I had the only car within 65 miles.

I agreed to see them through. Gave them my grub and started for Trona, via Ballarat. At noon a cloudburst hit the Slate Range and all roads were washed out. It was several days before I could get over the roads. Then there was no motor head at Trona. Had to send to Los Angeles. So by the time I returned to camp, helped install the head and did some necessary work, it was three weeks before I arrived home in Los Angeles. My wife was gone and had taken my son. I was divorced. My boy, now 24, has never seen the tunnel his Dad has been working on for so long.

The day I found that high grade ore among the boulders and decided to drop down the hill a thousand feet and drill that tunnel, I had to have a lot of Good Faith.

The tunnel is now 330 feet in solid rock, all driven by hand. I still have 150 feet to go, but I figure I can finish it in another five years, if I have Good Faith!

SALLY ZANJANI

Sally Zanjani, born in San Francisco, is a historian whose academic work focuses on the roles of nineteenth- and twentieth-century women pioneers and miners in the West. A resident of Nevada since childhood, Zanjani teaches at the University of Nevada, Reno, and is the author of several books, including A Mine of Her Own: Women Prospectors in the American West, 1850–1950, *published in 1997.*

The following excerpt from a chapter in that book, "Listen and the Mountains Will Talk to You," chronicles the life and adventures of Mary Elizabeth White. White, also known as Panamint Annie, was born in 1910 in Washington, DC. Half Iroquois and a doctor's daughter, she moved to Death Valley in 1931 and lived and mined in the region until her death in 1979.

from *A Mine of Her Own*

After her arrival in 1931 when she was twenty-one Mary Elizabeth White realized that she could not earn her living by cooking at dude ranches where none existed, so she took up prospecting. She spent as much time as possible soaking in the hot springs at Shoshone and drinking in the healing warmth and stillness of the desert. Soon the bleeding lesions in her lungs had healed, and she had regained her beauty: strong clean features, straight black hair cut short to cause no trouble, and skin with the bronze flush of her Iroquois mother. In the years to come she would climb mountains all day in her search for mineral, arm wrestle any man who thought he could play with her, and muck out the ore in her mines as though she had never been ill. When her cure was complete, she briefly returned to Colorado in 1936 and married a cowboy named Bryant, who died when she was seven months pregnant with their child. Panamint Annie returned to the desert, where her daughter Doris was born in 1938. Over her lifetime Annie gave birth to eight children, only four of whom survived.

Soon the name Mary Elizabeth White became a mere technicality; other prospectors had started calling her Panamint Annie, because she reminded them of the original Panamint Annie. The story of this prototypical Death Valley prospector has unfortunately been lost with the memories of these old men, but they used to say that Mary Elizabeth reminded them of the original in several ways: she came from "back East," she prospected, she was a "rough gal," and she sent her daughter away to school.

Although Doris stayed with her paternal aunt in San Bernardino during several school terms, her recollections provide a fascinating glimpse of the life Panamint Annie lived in the 1940s and probably during the depression years before Doris's birth. Home was a 1929 Model A truck, later an old army ambulance. Annie was apparently the first woman prospector to outfit the flatbed of a truck with beds and a tent, turning it into living quarters much like the wagons that had served many prospectors of an earlier age. High in the Panamints, where Annie had claims, winter nights sometimes turned cold enough to freeze the hot-water bottle they used to keep warm. When Doris fretted about where their next meal was coming from, her mother would tell her, "Don't worry about it. Between me and God, we'll take care of it." And a meal always appeared. Sometimes it was a stew from a snake or rabbit Annie had shot. Sometimes the cuisine was not very haut, as when the rabbit was diseased (a food most prospectors would scorn). On those nights Annie would tell Doris if a grub could survive cooking, it had to be "a better being than we are." Once in a while Annie herself went to bed without eating but never hungry enough to sell her keepsake, the first vial of gold she had ever mined.

A good deal of the time Panamint Annie and Doris teamed up with other prospectors. Death Valley ranger and historian Kari Coughlin estimates that at least two hundred people prospected and mined in the mountains of the Death Valley country during these years. Special legislation permitted continued mining after the creation of Death Valley National Monument in 1933, and the west side of the Panamint Mountains lay outside monument boundaries. Although Doris recalls twelve or fourteen women prospectors, mostly working with husbands, Annie was the only woman to join a prospector "family," an informal group of eight or ten, on a comradely basis with no romantic involvements. The family camped in the mountains, sending someone into town for supplies every two to four weeks if they could scrape together the money. Despite limited physical contact with the outside world, radio had greatly reduced the prospector's isolation. On their odyssey across the desert in the summer of 1914 Josephine and George Scott did not learn of the outbreak of World War I until a passing traveler told them; during World War II Annie's Death Valley family gathered each evening around a radio to listen to the news and move the thumbtacks representing the positions of the opposing armies around on a map.

In the division of domestic tasks among family members, even this tough and unconventional woman accepted a fairly traditional role as laundress, baker, and healer. Everyone shared the cooking over a communal campfire. Annie made special treats, baking cinnamon rolls and, most memorably, frying potato chips on the occasion when an intrepid vendor sold the family a fifty-pound sack of potatoes. She also did the laundry, probably out of consideration for her companions, having observed that "men hate to wash." In return they gave her extra water and food. When first aid was necessary, she administered it with firm yet gentle hands fully capable of healing fingers slashed to the bone in an accident, setting broken bones, or delivering a baby, as she had done more than once, a hint of the doctor she might have become.

Doris was treated like a little treasure by eight strict but kindly uncles called Old Man Black, One-Eyed Jack, and so forth. No one ever swore in her presence, including her mother, who was known for "language that would blister the ears of a drill sergeant." The uncles insisted that Doris have her sponge bath each evening, even when they ran so low on the water they brought up in large galvanized tanks from the spring at the base of the hill that they had to go without the prime necessity of a miner's existence—his coffee.

The main business of the family, of course, was prospecting. Each pursued this on his own and so intensively that sometimes it continued into the night by lantern light. When someone made a strike, he would "come dancing in." After the family heard the news, absolute discretion was the rule. No loose talk at the assayer's shop. No spreading the word through the bars on a drunken spree of celebration. Mining claims belonging to anyone in the family were never discussed with outsiders, no matter how drunk one became. Living as they did, far beyond the reach of the law, where a deputy sheriff might be glimpsed once a year or so, the family guarded against claim jumpers on their many unrecorded claims by patrolling for one another and "watching our own backs."

Although the older family members saw the thirties and forties as safer than the days when desperadoes had been known to rob a prospector of his gold, his claim, and his life, the desert itself remained as deadly as in the time of Lillian Malcolm and Mrs. Riggs. Nonetheless, Panamint Annie had no tales to tell of perils narrowly survived, because she avoided dangerous situations and scrupulously followed her own rules for desert survival. Whenever Doris returned from another term at school, her mother would rehearse her in the catechism: "Anything that looks like a rope coiled, stay away. Never put your hand on a rock without looking. So many people get disoriented. Fix your stationary block wherever you're going. Whether it's this mountain or this rock, always remember that if you pass it more than twice you're going the wrong way. Never leave a vehicle, because the distance is so vast that you can't comprehend it." Annie's invariable summer uniform consisted of jeans, T-shirt, light shirt, and high miner's boots to guard against snakebite. (Only Klondike Helen Seifert coexisted comfortably with snakes; she said it pleased her to encounter anything alive in the desert and always allowed snakes to slither on their way in peace.) Annie, by contrast, carried a gun solely for the purpose of shooting rabbits for dinner and snakes. Aside from rattlers, only one thing alarmed her: when she lost her teeth in old age, she refused to wear dentures for fear she might swallow them.

Annie's prospecting methods were a combination of instinct and looking for the colors according to the mystical rule the old-timers had taught her: "Listen and the mountains will talk to you. They'll tell you where the gold is if you listen." Annie listened and tirelessly hunted a wide variety of minerals from gold to uranium. Always excited by the unlimited possibilities awaiting her, she would be out before daylight, with her hammer on her belt, a holed-up shovel, and her pick in hand, not returning until after dark. Doris tagged along at her heels. When her mother said, "I think that mountain has something in it," and commenced to walk back and forth over the ground before zeroing in on the spot where she would take samples, Doris settled down to play with lizards and horned toads. When Annie developed a mine, she did the timbering, blasting, and mucking herself and made her own candles for underground work in order to save money on kerosene.

Occasionally, she made a good strike worth as much as ten thousand dollars, but, as Doris recalls, "she blew it as fast as she made it." Supporting her children had top priority, next came debts to be paid, a stake for her next prospecting trip, necessities long postponed such as new tires for the truck, and quiet charities such as shoes and food for Indian friends. Then came the binge drinking and, above all, the gambling. Annie never touched liquor until she turned forty-five, after which her three-day benders became part of Death Valley folklore. One time she broke her back in a fall and started driving to Las Vegas for medical attention; the trip took her three days because she painfully dragged herself into every bar along the way. Her back healed with no ill effects beyond an inability to raise her arm above her head, but on another occasion, her drinking lost her the best gold strike she ever made. When she came into town for supplies, she went on a prolonged binge with a group of men. After the alcoholic fog cleared from her mind, the place where she found gold had vanished from memory. "I know it's up there," she would say. "I'll find it. You wait." But she never did. Perhaps even a bonanza would scarcely have altered her way of life, because she would have gambled away the proceeds at blackjack or keno. Doris, who grew up with a longing for certainties, saw her mother's gambling as an intrinsic part of the prospector's mindset. Prospecting itself was a gamble. Waking each morning, as Annie did, with an unreasoning belief that this would be the day she would finally "hit it" in the mountains seemed to Doris the very essence of gambling.

Panamint Annie may have lied about her mines, like many a prospector before and since, though she would insist that she merely "exaggerated," but she always repaid her grubstakers with scrupulous fairness. Indeed, she was one of the few women prospectors known to work on grubstakes and to receive some of them from another woman. Annie split her ten-thousand-dollar strike with this woman, the owner of a grocery store near Death Valley Junction. When her prospecting yielded nothing, Annie endeavored to repay her grubstaker all the same by repairing the transmission on the woman's car. Annie also had a woman partner for some years, Mrs. Frederica Hessler, a Georgia schoolteacher who had inherited the entire town of Rhyolite from her brother, after the booming mining camp of the early twentieth century had turned into a ghost town where wind moaned through the ruins of massive gray stone walls. On a nearby mountain Panamint Annie and Frederica worked together at their gold mine, a site that today is being extensively mined for great profit.

As these working partnerships suggest, Annie's relationships with other women were based on friendship and cooperation rather than competition. She knew and liked Louise Grantham. Although she bristled whenever she saw a woman standing helplessly by the road unable to change a tire, Annie harbored no scorn for women who had led conventional lives and lacked her ability to survive in the wilderness. In Doris's words, "Mom always had a philosophy that each individual has their own capabilities, whether you use your mind or your body, your strength, or God-given talent. She never looked down at anybody." Nor did conventional women hold their skirts aside from her as they did from Lillian Malcolm. Her encounters with visitors to Death Valley demonstrate this. When Annie appeared on one occasion in her last years, unwashed, bizarrely clothed, giving full vent to her opinions in sulphurous language, and preceded by her reputation as one who would "bum around with anyone who had the price of a bottle and didn't much care what was in it," a middle-class woman sightseer warmed to her at once. Somehow the sheer force of Annie's charismatic personality swept all before it. One fervent admirer saw Annie and John Wayne as the last immortals in the West.

Panamint Annie usually dealt with problems by denying they existed, and child care was no exception. She strapped her baby girl Mary Ann onto her back and took her down into the mines much as her Iroquois forebears would have done and as she herself had probably done with Doris when the child was too small to remember. At other times Annie neutralized the perennial incompatibility of prospecting and child care by having Doris watch the younger children. All the same, when her youngest son, Bill, born the year she turned thirty-seven in 1947, reached school age and she could not arrange to send him away, she felt obliged to move down from the mountains into a makeshift cabin in Beatty, a small settlement across the hills from Rhyolite, where he could attend school. During this period she mined nearby with Frederica Hessler and prospected intensively throughout the area.

The passing parade of Annie's husbands and live-ins showed one common characteristic: all shared her passion for mining. "Some of them would live, sleep, and die mining," commented Doris. On the whole, these liaisons appeared neither long-lasting nor deeply felt. "Some of them treated her nice. Some of them didn't. There was nothing unusual about them," was Doris's dry assessment. And all had to fit in with Annie's freewheeling way of life, which was "I do whatever I want to do when the mood is on me." Perhaps mining engineer Bill Skogli, with whom Annie worked a mine near Shoshone and went on long prospecting sojourns in the hills, stood out a little from the rest. At least, when he left to mine in Arizona and neither wrote nor returned, Annie took sufficient notice to make a false report that he had stolen her car in order to find out what became of him. When she learned that Skogli had died from a heart attack, she said in her usual stoic way, "I really cared for him but he just goes off and dies on me."

Annie would not have classified that false report as a lie, nor would anyone else in Death Valley, where she was known for her raw honesty and a kind of emotional absolutism. She never mentioned people she disliked: they simply did not exist. She buried pain deep within herself. When little Mary Ann fell and died from a concussion at age two, Annie was so grief-stricken that she forebade anyone to mention Mary Ann's name again. Only once did Doris ever see her mother cry, and that was at Mary Ann's funeral. After that day Annie's sorrow stayed inside, breaking out only when her little son Bill struck his head in a fall and she went temporarily berserk with fear that it could happen all over again (Bill suffered no ill effects). Usually, she brushed difficulties aside with the maxim "Tomorrow will take care of itself." More than twenty years after her death it was her unashamed and undiluted candor, her way of stripping things down to the stark essentials the way she had stripped her life, that people remembered most: "She wasn't afraid to tell you what it was, no matter how bad it hurt."

Although she never acknowledged pain, hunger, grief, or obstacles, Annie voiced strong opinions on a wide variety of subjects, sometimes telephoning the Nevada governor with unsolicited advice. Equality for women was a strong theme of hers, and none could doubt that she had lived what she believed, ever since she took her fateful motorcycle ride. She thought, and often said, a woman could do any job a man could do, the only difference being that a woman could not lift quite so heavy a load. She strongly favored equal pay for equal work and as strongly opposed female dependency ("Don't wait for a man to do something for you, or you'll be waiting for an eternity"). She raised Doris to be what she believed all women should become—independent. Like many women prospectors before her, she saw the wilderness as a proving ground for independence. When Doris once burst into tears and said she hated the desert, Annie laughed long and hard and told her, "Doris, if you only knew—and you will someday. When you're in a city, everyone does things for you. You got people taking care of running water, electricity, picking up trash for you. When you're out on your own by yourself, there's no one to do it for you. You have to learn to do it yourself." Annie's belief in woman's equality also encompassed sexual behavior and justified her many lovers: "It's no big deal. Men do it all the time." She was, in many ways, a thoroughly modern woman in an archaic way of life.

Deep pleasure in the wilderness remained a powerful element in prospecting from the beginning, but Annie, more than any woman except perhaps Fannie Quigley, exemplified the prospector as naturalist. Although she acknowledged that the desert was "the worst thing in the world on humans," she knew the land and all the living things on it as no one else did and treasured them, from the young buck glimpsed on a mountain ledge ("Now you show me a more beautiful sight than that!") to the swift-moving roadrunner and the nesting eagle. When Doris once picked a bunch of wildflowers, her mother cautioned her to take only two or three, because the flowers are "God's decoration for the desert." It was said that she knew every bird's nest from Death Valley to Tonopah, and Death Valley rangers and naturalists had a high regard for her knowledge of the flora and fauna. As the *Los Angeles Times* put it, "she probably knows more about the National Monument than any person alive." Annie herself was modest: "I haven't seen it all, but I'm gaining on it." In her last years, when the strength that had formerly moved boulders failed her, she still went up Titus Canyon, where she had often mined, to see the view and hear the mountains talk. "The desert is the sea," she used to say. "It moves just like waves in the ocean."

Toward the end, crippled by arthritis, ravaged by cancer until she lost nearly half her normal weight, and unable to mine, she scraped a living by selling homemade jewelry and other souvenirs from her station wagon. But she still spoke of prospecting: "As soon as I get a stake, I'm going out in the hills and make another strike." There was another dream she talked about from time to time: "There's places in this world no man has ever been, and I'm going to find it." Born earlier, she would no doubt have become the mountain woman and explorer on new frontiers that she was so well suited to be. She died in September 1979 and was buried in the little cemetery at Rhyolite. When the shaggy Joshua palm over her grave bloomed the following spring for the first time in living memory, Doris saw the shining white blossom as a kind of benediction.

ANONYMOUS

The Small Tract Act of 1938 authorized the lease of up to five acres of public land in the California desert for use by private citizens as homes or weekend cabins, or for recreation or other personal purposes. The majority of the land made available was located in Twentynine Palms, Victorville, Apple Valley, the Morongo Basin, and the Coachella Valley. The numerous cabins that quickly arose in those regions became known as "jackrabbit homesteads," and many of the original cabins are still in use today. Excerpts from two jackrabbit homestead memoirs, written by Elizabeth Crozer Campbell and June Le Mert Paxton, follow this selection.

The overwhelming public response to the 1938 leases, at giveaway prices, was reported in the following news article, published in 1945 in Desert, *a popular magazine produced by Randall Henderson from 1937 to 1958 that featured articles about travel, history, mining, geology, exploration, and life in the southwestern United States and northwestern Mexico.*

Stampede for Jackrabbit Homesteads Deluges Los Angeles Land Office

A 1945 version of the land rushes of former years was staged in Southern California during February when applicants for "Jackrabbit Homesteads" deluged the United States land office in Los Angeles with 1500 applications for 5-acre leases on public lands in Twentynine Palms, Victorville, Morongo and Coachella valleys.

Registrar Paul B. Witmer and members of the Land Office staff in the Postoffice building in Los Angeles had long waiting lines of people who came either to get information or make applications.

In a majority of cases the applicants filed sight-unseen, making their selections from maps in the government files.

The land office personnel tries to make it clear to each applicant that the lands are not designed as a source of livelihood, and in most instances are several miles from the nearest water and power-lines.

The procedure is for the applicant to fill out a blank, the limit being five acres for any American citizen except that both husband and wife are limited to one filing. The application is accompanied by a $5.00 filing fee. Final approval of the application is given in Washington and requires from two to three months. When approved, the applicant must pay the first year's lease at $1.00 an acre. Thereafter for the five-year term of the lease the cost is the same—$5.00 a year for five acres.

The Izac law under which public lands are opened in this manner permits the sale of the 5-acre tracts. However, the regulations set up by the department of interior have so far restricted all applicants to leases. The department is now considering plans for selling the land outright to lessees after cabins are built or other improvements are made on the property. There is no requirement as to residence on the land, however, such as was necessary under the old homestead laws.

Witmer stated that additional lands will be made available as required to meet the demands of the public. "There are 8,000,000 acres of public land in Southern California," the registrar stated, "and most of it is on the desert. Those desiring to make applications should understand that this is not farm land. It is mostly sand and rocks—and the government makes no promises as to water or other improvements."

ELIZABETH CROZER CAMPBELL

In 1924 Elizabeth Crozer Campbell (1899–1971) moved to Twentynine Palms, California, with her husband, Bill, a World War I veteran who was ill from the effects of mustard gas poisoning. They were among the many people who have relocated from other parts of the country to southern California's deserts in search of better health. At first they camped in a tent at the Oasis of Mara, a longtime residence of the desert Indians and now the site of Joshua Tree National Park headquarters, and the couple later homesteaded 160 acres. Campbell wrote a memoir of her life there, The Desert Was Home: Adventures and Tribulations of a Desert Homesteader, *which was published in 1961. She and her husband were also accomplished amateur archaeologists who worked for Los Angeles's Southwest Museum of the American Indian.*

The following excerpt is from "Thorns," a chapter in The Desert Was Home.

from *The Desert Was Home*

Thorns

Most unsettled land goes through the same cycle. First, Indians who do not want trappers or cattlemen, then, cattlemen who do not want homesteaders, and finally homesteaders who love quiet, open spaces and do not want subdividers, billboards or hot dog stands.

When we first camped on the desert it was apparent our presence didn't create enthusiasm. This we felt might be due to the fact that gossip had it that we were "officers." It might have been because our car looked a little better than the average old wrecks navigating these parts, that we didn't drink, or that my husband wore his old army shirts, trousers and puttees. We tried to be pleasant and mind our own business, but it was evident that many people felt animosity toward us.

It was not entirely personal. If one homesteader became firmly rooted, more would settle, and, as barbed wire fences spread, the cattle would have to be removed to other districts. The less inhabited the region and the worse the roads, the more ideal for cattle range or bootlegging, and homesteaders were no help to either.

No one encouraged homesteaders, and it was hard to obtain information about the land, for those who knew cannily shut their lips like clams. Fortunately for us, we had a friend in the Land Office who was glad to give us all the information he could about the district.

In spite of dark looks and innuendoes we camped on at the springs, showing no intention of being ousted by lack of cordiality. It was only a few nights before we moved to our homestead that the first sign of open hostility appeared. We were taking our bedtime baths in the tent when two shots were fired just close enough to graze the canvas. These were followed by a derisive shout in the distance. Presumably it was done to frighten us away. With Bill's makeup, had we not already made up our minds to move, those shots would have been sufficient cause to keep us camping at the springs for years. But we wanted to homestead more than just to live roughness down and, God knows, we had roughness enough on our own place. By the time we were dressed and out of the tent no one was around, though the darkness would have covered anyone's retreat. We never discovered who shot at us, though we had our suspicions.

I crawled into bed that night with all the tenderfoot uppermost in my soul. This was to be my way of life, I thought, never knowing when someone was going to shoot at me. I voiced objections to my husband.

"You aren't going to let roughnecks scare you out, are you?" he said witheringly, and it became our motto for several years to come. No, your self respect demanded that you mustn't let roughnecks intimidate you. I comforted myself that in a couple of days we would be on our own land and away from this public water hole where all kinds of people stopped.

After we were established on our homestead, I was alone and resting on my bed one morning when two men appeared at the tent flap. I knew who they were, and admired neither, so I sprang to my feet and picked up my gun. One seemed fairly sober, the other had been in a fight and obviously was recovering from a terrible spree. His eyes were bleary and face bruised, nose swollen and blood was all over his shirt and trousers. His story was that he had become drunk on so-and-so's liquor, beaten up his wife and finally been licked to a pulp by another man who was trying to defend the woman. Supposing Bill to be an "officer," the bloody one wanted him to arrest the man who had trimmed him.

"Get out!" I yapped. "Get off the place and if you want to see my husband don't come back till you know he is here!"

Unsteadily they retreated and wavered across the desert in their car. So here it was again, I thought; well, you couldn't let roughnecks frighten you out of camp, as Bill had said.

Shortly afterwards the two men returned, and pulled their car up beside Bill, who was armed.

"Keep right on driving," he ordered in a voice I had never heard him use, a voice of cold steel. "We'll have no drunken brawlers on our homestead, and besides you got what was coming to you. Don't ever set foot on our place when you're drunk, and don't stop your car. Keep right on moving." One glance was enough to show Bill meant business, and they drove away, never to return. But word went out now that the new homesteader was "hard-boiled," and hostility toward us grew.

Following this, someone with a high-powered rifle fired bullets over my head when I walked from one place to another on our land. It was probably the same person who shot at our tent, and after he drifted away the shots stopped.

All new homesteaders in our district met more or less of the same thing. If you were rough and liked evil ways, the old gang let you alone, but if you had a spark of anything in you that stood for law and order, attempts were made to drive you from your land. Homesteaders just weren't wanted.

As I look back on those days, I realize how little it took to start a row, and I can understand it, for only our self control prevented us from acting unreasonably about some things as others did.

No one thing was greater cause for minor feuds than roads. All over the desert for miles about were "roads." These wound in every direction and, though merely two wheel ruts, connected mines with water holes or windmills with the outside world. Tracks that had been followed for years were considered public property, and were regarded in the light of county roads. Homestead law permitted patented land owners to fence their property, regardless of what tracks zigzagged across their land, the only requirement being that a cleared and passable track must be put around the fence no matter how many other roads went to the same point. But would any old-timer travel those new roads, often better than the others? He would not! All that was needed to make a man see red was a few strands of barbed wire and a "No Trespass" or "Private Property" sign. Out would come the cutting pliers and down would go the wire, while the occupant of the rickety desert car drove in state over the tracks he had always traveled. It was *his* desert, anyway, and he wasn't going to be thrown off it by some blankety-blank homesteader.

Then the feud would be on. The homesteader would call on the trespasser, tell him what he thought, and finally put up a sign on his fence. "John Doe, you are hereby warned that if you set foot on these premises you will be shot on sight," signed Jim Blank. Shootings did occur some years after our advent, but no actual deaths over homestead rows took place in the early days, though two men of our acquaintance had notches on their guns from some years back. Yet here they were, wandering about the desert, and out of jail.

If "hate signs" didn't relieve the feelings of a feudist, he might resort to a "hate board." This was a board through which nails had been driven, designed for burial in a desert road rut, nails pointed upward, with intent to puncture an enemy's tires.

I think the signs were what infuriated people. We too fenced our land, across which ran a road that had been in use for years, but we advanced slowly to the fray. First we put up posts and let them stand without wire till people were used to that, then a brush barrier and finally the wire. Not till we had been on the desert for years did we ever put up a private property sign, and we had almost no road or fence troubles.

Someone would make a remark, which would be enlarged in repeating it, and the war would be on; usually over nothing very much.

One of our neighbors had some baled hay stolen, and he did not know who took it. "I guess," he began saying, "someone's black goat was hungry."

Just how the story started that we were the ones who said it I suppose we will never know, but we were in ignorance of the whole affair.

Now, only one homesteader owned a black goat, and he didn't live so far away. Not long before, he and his wife had come to call, and had proposed stealing pipe for us from a mine up in the hills. They could procure all our plumbing for us, naming a price. Bill told them flatly that he would have nothing to do with theft, and rose, terminating the call. Realizing they had tackled the wrong people, they were no doubt annoyed with us. We thought we would see no more of them, but soon they were back with a dead calf in the truck. "Mayn't we give you some veal," they ingratiatingly urged. We all knew the calf had been rustled, in fact they openly said so.

"No," said my husband again, "we're funny that way, we never have anything on the place that doesn't belong to us."

"Well, how about some for the dog then?"

Patiently Bill explained that we didn't feed the dog stolen meat either, and crestfallen, they turned away. How badly they wanted to incriminate us!

At the time of the goat episode they drove to our place in their truck and pulled to a stop where we were standing. Immediately they burst out with what they thought of us for saying that "somebody's black goat was hungry." In the course of the conversation the man, nursing a shotgun, poked the barrel out of the car and into my face. That was too much for Bill. Seizing the gun barrel, he reversed it toward their faces.

"Be more careful where you point your guns," he growled, in that smouldering voice I had learned meant business, "when you come here."

They were finally convinced that we had said nothing about a hungry goat, but the hay owner had. There was an anticlimax of half apologetic remarks, and they drove away, leaving a smell of brimstone and sulphur on the air.

At once we called on the hay owner, he of the gun with notches. "It would be nice," Bill said, "if you would tell Brown who *did* say 'somebody's goat was hungry.'"

Angrily the man seized a three-foot mining spoon and swung it in the air at my husband. Bill snatched the spoon and roared, "Don't you threaten me, Mat!" Sitting in the car, I thought I would see murder enacted before my eyes. My heart beat like a trip hammer. Then a strange thing happened. Mat, who had notches in his gun, dropped his jaw, stared at my husband and *wilted*. I think some secret admiration stole into his heart for the man who had dared to defy him. A change came over his expression, and soon both men were chatting reasonably about the goat.

The end of the story I do not know, for Mat left the valley the next day, selling out his homestead to a neighbor. And the owner of the goat also departed, serving time in jail for one thing or another, his cabin shut and locked. It is my opinion that most feuds start over nothing more reasonable than this.

Some years later we met Mat in a blacksmith shop in town. His face beamed. He threw an arm about Bill's shoulder and drew my arm through his, giving it a friendly squeeze, so plainly no rancor remained.

"Let's sit in your car and chew the fat awhile," he said. His head was as shiny as the proverbial billiard ball, all of his upper front teeth were out, and tobacco juice stained the corners of his mouth, but his face was alight with friendship. "Do you know what I'm agoin' to do when I git rich?" he asked. "I'm goin' to git some new false teeth, git me a blue serge suit, and git me a wife."

Bill was never aggressive, but he had the courage of his convictions, and was absolutely fearless. Though he knew that for awhile he was the only man in our district standing openly for the right, he took comfort from his philosophy of life, which was simply that one right thinking man should be able to accomplish more good than a dozen wrongdoers.

Two factions gradually came into being, one, law-abiding that stood with Bill, the other which found it convenient to keep the country unsettled and discourage law and order. And the unlawful element was in the majority. Anyone who showed a spark of integrity or was a friend of ours was our "satellite" and was the recipient of jeers or affronts. Our sheriff had been in power for more than twenty years, and that is too long for some sheriffs. Besides that, he was related to the cattleman who had the grazing privileges in our district. That was the root of the whole trouble.

As I ride around the desert today I think there is much to be said for the cattlemen, however. Cowboys didn't exterminate mountain sheep, or destroy the desert growth. They didn't carry away desert holly or cactus to die in somebody's garden; they never set fire to fine old Joshua trees, hundreds of years old, just to see how good a torch they would make. I go to the beautiful palm groves around the water holes and see the boles of the stately old palms mutilated with initials and names for all to read—the names of *homesteaders*, not cattlemen or cowboys. There are, of course, some homesteaders who cherish the desert, and try to conserve its growth and wild life, but the many who do not have left their ravages like locusts. When we were open range the desert was wild, beautiful and clean. No ancient trailers or cheap buildings peppered the landscape; no ugly poles stood out to mar a treeless land; the old desert wasn't littered with bottles and tin cans, and papers weren't blowing in every direction.

Do I sound like a matriarch mourning the good old days? Well, maybe I am a trifle. Small desert towns can be attractive, but most of them are not. I appreciate the fact that I now live in a Judicial Township, and that the days of universal gun toting are past. I too have electricity and telephone service brought to me by ugly poles, but I know the desert was lovelier without those things.

The cattlemen were always decent to us. We were never forbidden to picnic beside their windmills or water holes. Perhaps their worst fault was the subtle shielding of any element that might have kept the country wild, but it is hard to tell how much they did of that. They suffered, too, at the hands of homesteaders. Their calves were rustled, their cattle frightened away from water holes, and their valuable bulls shot with no more excuse than someone claiming to be afraid of them.

The coming of homesteaders and barbed wire fences means the ousting of cattle, and I know how the cattlemen felt. When I look at the dunes from afar, the dunes I will always love and know that I can no longer roam over because they are homestead land, private property in short, I think the same sadness of spirit, the same senseless rage possesses me that must have made the cattlemen boil when we early settlers were determined to homestead.

Periodically Bill, with a few of the better thinking people, made jaunts to the county seat, a two hundred mile round trip, asking for law and order, in the hope that the constant dropping of water would wear away the hardest stone. But the plea that the hand of the law might reach out and touch us now and then roused little response. San Bernardino County, one of the largest in the west, was still considered cow country. If one must go and live in a cow county, then expect to bear the consequences, was the inference. Once in the presense of five witnesses the sheriff lost patience entirely. "If you want protection so badly," he snapped, "then go and get guns and defend yourselves." So that is exactly what people did, and the gun rack in our cabin loaded with shot guns, rifles and riot guns dated from those days. For some years after our arrival everyone carried guns. Some of them openly as we did, others inside their shirts or trousers. No one showed any inclination to be the first to disarm.

JUNE LE MERT PAXTON

June Le Mert Paxton was a popular desert poet and writer who lived for many years, beginning in the 1930s, on a desert homestead in the Morongo Basin, located in the high desert of California near Joshua Tree National Park. She is the author of Desert Peace, and Other Poems *(1946) and the memoir* My Life on the Mojave *(1957).*

The following excerpt, "A Letter to My Daughters in Hollywood, California," is from My Life on the Mojave.

from *My Life on the Mojave*

A Letter to my Daughters in Hollywood, California

Yucca Valley, California
July 6, 1932

Dearest Evelyn and Adelaide,

Now that I have been here a few weeks, I will try to give you a more detailed account of the place. As you, too, well know, your father is anything but favorably impressed, and I could just hear him saying to you girls when he returned home, "You have never been in such a God-forsaken country in your life! How anyone could file on a homestead, much less try to live there, is beyond me! The twenty-three miles of rocky road after one leaves Highway 99 are a nightmare! The only happy person in all that country must be the chiropractor in Devil's Garden, who could make a good living by resetting the disjointed vertebrae of hapless travelers!"

Your dad's reactions are not hard for us to understand. There is very little outside his home which is of any interest to him, so breaking up our family circle, even for a short time, makes a difficult situation. But, girls, the place has its nice features in spite of its remoteness from civilization. The valley, from this point, somewhat resembles an elongated pie tin, with the San Bernardino and Little San Bernardino Mountains fluting the outer rim. To the southwest, high above the lower ridges, is Mount San Jacinto. This towering peak we see from our yard, as well as Mount San Gorgonio to the west. Both crests, I hear, are often covered with snow far into the summer.

There is also more shrubbery around here than one would expect to find on a desert, and the sand is a gravelly sort, with real sand dunes only here and there. The closest dune hides our place from Warren's Well, which is a little to the south and west; the ranch house is within walking distance if one does not mind crawling under the barbed-wire fences or encountering the strange-faced cattle that stare, unfriendly, at him.

And that reminds me! Please send up a pair of jeans and walking shoes. Oh, yes, and several boy-size shirts for blouses. The thorny bushes tear at my garments, and the wind at times has no regard for a lady's skirt. Bring your own hiking clothes when you come, because the mountains that rim us on the north are just back of the shack, and you will enjoy the climb to the topmost peak. From the top one can see for miles and miles over the valley.

The Joshua trees, named by the early Mormons, are numerous; in fact, all about us are groves of these weird-looking trees. I have also learned the names of many odd local shrubs, such as Spanish Dagger, creosote, catsclaw, juniper, and varieties of the sage and cactus.

Some days a stray burro loiters around to keep Old Kate company, and we will try to have them both here when you come, so that you can take a ride. You will enjoy the picnic-style suppers that John prepares out of doors these warm evenings. He has made a little tin stove on which, over a scrapwood fire, he warms left-over food and heats the water for dish washing. Our Spot is tiny, yet he barks at, and chases away, the cattle when they roam too near the house. He does not seem to mind Old Kate. I can imagine that up here one learns to make friends with every creature, furred or feathered.

Sometimes I go with John to Matty's little store about three miles to the west, and then on to Tom Heard's homestead where we can buy buttermilk. The Heards have lived here since about 1913. Late one afternoon we went through an old sandy riverbed to see Bill McClung. I think Bill is a bachelor; he lives alone. There must be springs up there, because the little house is surrounded by trees, and he has a pond where huge frogs jump in and out of the water and croak horribly! John thinks that if he takes me around the valley I will be less homesick.

In the mornings, after I have had a wonderful sleep and the sun comes in bright and early, I am on top of the world. When the afternoon stretches out into a long day, however, I often feel not so sure of my venture in this almost unknown country. At such times all of this experience seems to be a part of a weird play, and I, who have come upon the scene, cannot remember my cue, much less the lines!

I am very eager to have you girls come to see this place and am hoping that you both will be good sports; that will be a great help to your lovin'

Mother

PEARL BAILEY

Pearl Bailey, born in Virginia in 1918, was a famous American singer and entertainer. During World War II she toured the country performing for American troops. She debuted on Broadway in 1946 in St. Louis Woman, and for the rest of her life she toured, recorded albums, and appeared on stage and film. She died in Philadelphia in 1990.

In 1954, seeking a place to enjoy time away from her busy career, Bailey, along with her husband, famed jazz drummer Louie Bellson, purchased Murray's Ranch in Apple Valley, California. It became the only dude ranch in the area that catered to African Americans. Bailey writes about her beloved retreat in the following excerpt from The Raw Pearl, *an autobiography she published in 1968.*

from *The Raw Pearl*

Possibly the best thing to do when a project has ended is to take a trip. So, when *House of Flowers* closed after about five months, we decided to take a rest. Don, Peetney, Mr. Reilly (my dog), Louis and I piled in the car and headed west. Our destination was Vegas or California. We traveled at a snail's pace. We cooked outdoors but didn't sleep out. I'm not quite that much of an outdoor girl.

Off we started at 10 A.M., four musketeers on vacation. Our first goal was to get to Chicago in a hurry, because we figured the wide open spaces didn't start until we were past the Windy City. Stopping at every Howard Johnson's along the way didn't help our speed, so by midnight we had only reached Harrisburg, around 160 miles from New York. Next day we vowed we'd make it for sure. Where do you think we reached? Bucyrus, Ohio—so we did a little better than the day before. When we finally got to Chicago, all of our budget money was eaten up or spent for souvenirs. We had to cash checks and begin again.

It was a lovely trip, with Peet gathering wood and me as "chef." Lou and Don had the pots and pans to do. Oh, how they grumbled! When we arrived at Kingman, Arizona, at the edge of town there was a fork in the road. One side went to Vegas, the other to California. We chose California.

I looked like a blown-up balloon by then, I was so fat. In fact, we all should have hid in some local forest—we'd lived it up, and we looked like real lovable bums.

When we got to Victorville, I asked at the Texaco station whether there was a place called Murray's Dude Ranch around Apple Valley. I remembered it from USO days. Well, everyone knew and loved Mr. Murray, so the man gave us directions and off we went. When we finished following those directions we were sitting on top of a hill. There was one family up there. We began all over. It was hot and we were hungry. We were about to give up and return to what we called civilization when the place loomed into sight. There under a tree sat this wonderful man, Mr. Murray, that we were to grow to love as a friend and to cry for at his ending.

Our intentions were to stay one day—and that became three, during which time Louis and I decided that California was for us. We drove around looking for just a small spot to buy. Mr. Murray, ready to retire, spoke of perhaps giving up thirty-five of his forty acres. He would keep five, as he said, "to die on." Louis and I had no intention of going into such a deal. Whatever would we do with that much space? It sounded ridiculous to us. There were two musicians staying there, the only boarders. They kept telling us the old man wanted us to have the place. Finally, we spoke to "Boss" (everyone called Mr. Murray that), and told him that not only was it too much space, but it was also a case of spending that kind of money.

He said, "I'm not interested in the money now. I don't have to worry about that because I've owned the ranch outright for so long. It's just keeping the place clean and right that bothers me. Just give me something to let me know you'll be back, something to bind the agreement."

Lou and I had no other argument. That spot represented peace and contentment, so we okayed the deal. The next day Mr. Murray walked up the slope to the huge swimming pool and stood there gazing into nothingness, and he wept. He didn't know Lou and I saw him. Then he came to the cottage and said, "I'm ready," and off we went to the bank. I think even the people in there wept a bit when the paper was signed. Mr. Murray was an institution.

We left and went back to New York to get ready to head for the desert. We named the ranch the Lazy B, and there were questions as to why. Aha! Get this reasoning: I told Louie people would pass by and look at this huge ranch and say to themselves, "There are actors in there, leading a glamorous life. I wonder what those 'lazy B's' [dirty word] are doing there." Also, B is for beaver, and let me enlighten you: we worked like beavers from sunup to sundown. That's where guests and relatives and I sometimes clashed; they would come to visit Lou and me, and right away would want to sit Louis and me down in our own home and treat us like stars or something. When will they realize that Mama taught me to keep a clean house years ago, and I've not forgotten it, show business or not. I enjoy doing my own work. Secondly, how could we get relief from our theatrical work if we can't release that tension by wandering around our place getting joy? Thank heavens I've learned to be a good guest in other people's homes, learned it from the ones who came into ours and tried to take over. Folks who go to visit someone and must have something to do should go home and do it.

On the ranch we planted trees (Arizona cypress, the big boys), raked, cooked for a large group, chased a few snakes (and I'm scared of them), pulled weeds, got tumbleweed from around all those acres. All I can tell you is that New York was never like that. There were no lights outdoors on the place when we started. That desert is dark. As far as a telephone was concerned, we had to go to the highway and use a public one. It would have cost a fortune to run a line back there. I was happy because I hate the telephone. They are misused instruments. Don't you love the people who wake you and ask, "What do you know?" I always answer, "Plenty, but I'm not going to tell you this early."

Talk about a hick, that's me. Once I planted flowers all around the thirty-five acres. You can imagine how many times I bent my aching back and, would you believe it, they were all poisonous? A gorgeous desert plant, but if put in the mouth, the harm was done. So, with children around, it wasn't the thing. I had to pull them all up—in 110 degrees yet.

When we moved to the ranch, the main buildings and all the cottages were in terrible condition. Often I have to laugh at the thought of someone dropping in on these theatrical folks looking for "big-shot living" and instead seeing us on this place. We even used iron beds and cots. We took the bitter with the sweet. At least we knew what our dream was for the land.

Once when we had to go East and stopped over at Mama's in Philadelphia, we went up to Germantown to see my sister Eura. She wasn't home, but my nephew Billy, his wife Roz (from England), and their three children, Debra, Felix, and Billy, were there. He explained that things weren't going well with him. His work and a couple of other issues were getting him down, and he wanted to leave Philadelphia. We said, "We are not able to pay a tremendous salary, but you're welcome to come out to the ranch and act as foreman." They were elated, soon got on the Super Chief and came on out.

Now it was a group of city folk on the ranch for sure. There was no kitchen in the big house, but there was a huge kitchen-dining room that had been used when the ranch was working, so we all had room to eat together. I cooked three big meals a day, and we really went to town building our dream. Louis and I had a small house built and made it a lovely home. Billy and Roz redecorated their house. We ordered a telephone and outside lights, and the ranch became a city.

It was a pleasure to see folks drive by and look at what we had achieved. We lived there for nine years, and one day will probably live there again. So much is there—the earth where food can be planted, the lovely sunsets, the air, and all that space to wander away and get with God.

TOD GOLDBERG

Tod Goldberg was born in Berkeley, California, in 1971, and moved to Palm Springs in 1985. He is the author of the novels Fake Liar Cheat *(2000) and* Living Dead Girl *(2002) (a finalist for the* Los Angeles Times *Book Prize); the short story collection* Simplify *(2005); and the* Burn Notice *novels* The Fix *and* The End Game *(2008 and 2009). His short fiction has appeared in numerous journals, magazines, and anthologies, and he also writes for the* Los Angeles Times *and* Palm Springs Life Magazine. *He is the director of the University of California, Riverside, Palm Desert MFA Program in Creative Writing and Writing for the Performing Arts. He lives in La Quinta, California.*

The following short story, "The Salt," takes place in and around Palm Springs.

The Salt

Beneath the water, beneath the time, beneath yesterday, is the salt.

The paper says that another body has washed up on the north shore of the Salton Sea, its age the provenance of anthropologists. "Washed up" is a misnomer, of course, because nothing is flowing out of the Salton Sea, this winter of interminable heat: it's January 10th and the temperature hovers near one hundred degrees. The Salton Sea is receding back into memory, revealing with each inch another year, another foundation, another hand that pulls from the sand and grasps at the dead air. Maybe the bodies are from the old Indian cemetery first buried beneath the sea in '71, or perhaps they are from Tom Sanderson's family plot, or maybe it is my sweet Margaret, delivered back to me in rusted bone.

I fold the newspaper and set it down on my lap. Through the living room window I see Kim, my wife of seven months, pruning her roses. They are supposed to be dormant by now, she told me yesterday, and that they are alive and flowering is nothing short of a miracle. Much is miraculous to Kim: we met at the cancer treatment center in Palm Springs a little over a year ago, both of us bald and withered, our lives clinging to a chemical cocktail.

"How long did they give you?" she asked.

"Nothing specific," I said. The truth was that my doctor told me that I had a year, possibly less, but that at my age—I was 72 then—the script was likely to be without too many twists: I'd either live or I wouldn't. And after spending every afternoon for three months hooked to an IV, I wasn't sure if that was completely accurate. What kind of life was this, I wondered, that predicated itself on waiting?

"I'm already supposed to be dead," she said. "How do you like that?"

"You should buy a lottery ticket," I said.

She rummaged into her purse then and pulled out a handful of stubs and handed them to me. "Pick out one you like and if you win, we'll split it."

We live together now behind a gate in Indian Wells and our backyard abuts a golf course that my knees won't allow me to play on and that my checkbook can't afford. My yearly pension from the Sheriff's Department is more suited for the guard gate than the country club. But Kim comes from money, or at least her ex-husband did, and so here we are, living out the bonus years together. At least Kim's hair has grown back.

I pick the newspaper back up and try not to read the stories on the front page, the colored bar-graph that details the Sea's water levels from 1906 until present day, the old photos of speedboat races, the black bag that holds a human form, the telephone poles jutting out of the placid water. I try to read page A-3, where the other big local news stories of the day are housed safely out of sight from passing tourists: *A broken water pipe has closed the Ralphs in Palm Springs. A dead body found in Joshua Tree identified as a missing hiker from Kansas. Free flu shots for seniors at Eisenhower Medical Center.*

"Morris," Kim says. "Are you feeling all right?"

I look up from the paper and see that Kim is standing only a few inches from me, worry etched on her face like sediment. "I'm fine," I say. "When did you come in?"

"I've been standing here talking to you and you haven't even looked up from the paper," she says.

"I didn't hear you," I say.

"I know that," she says. "You were talking to yourself. It would be impossible for you to hear me over the din of your own conversation." Kim smiles, but I can see that she's worried.

"I'm an old man," I say.

She leans down then, takes my face in her hands, and runs her thumbs along my eyes. "You're just a boy," Kim says, and I realize she's wiping tears from my face. "Why don't you ever talk to me about your wife? It wouldn't bother me, Morris. It would make me feel closer to you."

"That was another life," I say.

"Apparently not," she says.

"She's been gone a long time," I say, "but sometimes it just creeps back on me and it's like she's still alive and in the other room, but I can't seem to figure out where that room is. And then I look up and my new wife is wiping tears from my face."

"I'm not your new wife," Kim says, standing back up. "I'm your last wife."

"You know what I mean," I say.

"Of course I do," Kim says.

The fact of the matter, I think after Kim has walked back outside, is that with each passing day I find my mind has begun to recede like the Sea, and each morning I wake up feeling like I'm younger, like time is flowing backwards, that eventually I'll open my eyes and it will be 1962 again and life will feel filled with possibility. What is obvious to me, and what my neurologist confirmed a few weeks ago, but which I haven't bothered to share with Kim, is that my brain is shedding space, that soon all that will be left is the past, my consciousness doing its best imitation of liquefaction.

I go into the bedroom and change into a pair of khaki pants, a buttoned-down shirt and a ball cap emblazoned with the logo of our country club. In the closet, I take down the shoe box where I keep my gun and ankle holster and for a long time I just look at both of them, wondering what the hell I'm thinking about, what the hell I hope to prove after almost 45 years, what exactly I think I'll find out there by the shore of that rotting sea but ghosts and sand.

Dead is still dead.

I find Kim in the front yard. She's chatting with our next-door neighbors, Sue and Leon. Last week, Leon wandered out of his house in the middle of the night and stood on the 15th fairway shouting obscenities. By the time I was finally able to coax him into my golf cart—this was sometime around three in the morning—he'd stripped off all of his clothes and was masturbating furiously, sadly to no avail. That's the tragedy of getting old and losing your mind—that switch flips and everything that's been sitting limply beside you starts perking up again, but you can't figure out exactly how to work it. Today, he's smiling and happy and seems to have a general idea about his whereabouts, but seems blissfully unaware of who he is, or who any of us are.

"I can't thank you enough for the other night," Sue says when I walk up.

"It's nothing," I say.

"He was happy to do it," Kim says. "Any time you need help, really, we're just right here."

"His medication…Well, you know how it is. You have to get it regulated. I wish you'd known him before all of this," Sue says, waving her hands dismissively, and then, just like that, she's sobbing. "Oh, it's silly. We get old, don't we, Kim? We just get old and next thing you know, you're gone."

Leon used to run some Fortune 500 company that made light fixtures for casinos. They called him the King of Lights, or at least that's what he told me once in one of his more lucid moments. But today he's just a dim bulb and I can't help but think of how soon I'll be sitting right there next to him at the loony bin, drooling on myself and letting some orderly wipe my ass.

"I have to run out," I say to Kim, once Leon and Sue have made their way back to their condo.

"I could clean up and come with you," she says. "It would just take a moment."

"Don't bother," I say. "I'm just gonna drive on out to the Salton Sea. See what's going on down there. Talk a little cop shop."

"Morris," she says, "if I go inside and look in the closet, will I find your gun there?"

"I'm afraid not," I say.

"You're a fool to be running around with that thing. Do you hear me?"

"I hear you," I say.

We stand there staring at each other for a solid minute until Kim shakes her head once, turns heel, and walks inside. She doesn't bother to slam the front door, opting instead to close it quietly, which makes it all the worse.

In the spring of 1962, I took a job working for Claxon Oil and moved, along with my young wife, Margaret, to the Salton Sea. Claxon had hired me to be the de facto police for the 500 people they'd shipped into the area in their attempt to find oil beneath the Sea, a venture that would prove fruitless and tragic. At the time, though, Claxon was simply concerned about keeping order: they'd already built an Army-style barracks and were busy constructing seafront hacienda homes for the executives who'd oversee the dig and, presumably, the boomtown that would come once the oil came spouting out of the ground. My job was to provide a little bit of law, both with the working men (and families) and the Mexicans and Indians who populated the area. There'd already been three stabbing deaths in the past year—two roughnecks and one Mexican—and it didn't seem to be getting better.

I was only 28 years old then and had spent the previous three years trying to figure out how to get Korea out of my head. I served two years in Korea during the war and another five trying to conjure a better future for myself by reenlisting until it seemed pointless, before finally returning to Granite, Washington, where I'd grown up. My father was the sheriff there—as I would later be—so he hired me on to be his deputy. It was reasonable work until a young woman named Gretchen Claxon went missing from the small fishing resort on Granite Lake. I found her body, and the man who'd done unspeakable things to it, a few sleepless weeks later. I'd like to say that I was honest and fair with the killer, a man named Milton Stairs that I'd gone to elementary school with, but the truth is that I nearly killed him: I broke both of his arms and beat him so badly that he ended up losing the ability to speak. I was rewarded with a job offer from Gretchen's grieving oil baron father and a salary well beyond my comprehension.

A year and a half later, Margaret would be dead from ovarian cancer and I'd be back in Granite.

But today I'm standing on the other side of a stretch of yellow caution tape, though this isn't a crime scene, watching as a rental security officer stands guard over a patch of dirt while two young women and a man wearing one of those safari vests brush rocks and debris away from a depression in the earth. The Salton Sea laps at the edge of the sand, the stench rising from it as thick as mustard gas. The two women and the man are all wearing masks, but the security guard just keeps a handkerchief to his face while in the other hand he clutches a clipboard. It's not the body that smells—it's the sea, rotting with dead fish, sewage runoff and the aroma of red tide algae.

Forty years ago, this was roughly where Katie Livingston had her little bar and café. At night during the week, the working men would sit at the bar drinking the stink out of their skin, but on the weekends the LA people would drive in with their boats and water skis and, eventually, speedboats, and would come into Katie's looking for authenticity. More often than not, they'd leave without a few teeth and, on occasion, without their girlfriends, wives or daughters. They thought the Salt would be like an inland Riviera. They thought we'd find oil and prosperity and that a city would rise from the fetid desert floor.

Thirty-eight years ago, Katie's bar slipped into the sea. Thirty-five years ago, Katie's home followed suit. Shortly thereafter Katie followed her bar and house, simply walking into the water with a bottle of wine in her hand, drinking big gulps all along the way. They never did find her body, but that was okay: her entire family watched her walk into the sea, bricks tied around her ankles. It wasn't a suicide, her son wrote to tell me, because she'd been dead for at least three years, but more a celebration of the Salt. All things return to it.

At some point, however, memory becomes insufficient in the face of commerce and space. These bodies that keep pulling themselves from the sea are a hindrance to something larger and more important than an old man's past: real estate. The Chuyalla Indians intend to put a 26-floor hotel and casino here and then, in five years, one hundred condos. They intend to fund a project that will eradicate the dead—both the people and the lingering fumes of a sea that was never meant to be—and once again tempt the folly of beachside living in the middle of the desert.

I'll be dead by then myself. Or at least without the ability to know the difference.

The security guard finally notices me and ambles over, his gait slow and deliberate, as if traversing the twenty feet from the body to the tape was the most difficult task of his life. "Can I help you?" he asks, not bothering to remove the handkerchief from his face. I'd guess that he's just a shade under sixty himself, too old for real police work, which probably makes this the perfect job for him.

"Just came down to see the excitement," I say. I open my wallet and show him my retired sheriff's card. He looks at it once, nods, and then from out of his back pocket he fishes out his own wallet and shows me his retirement card from the Yuma, Arizona PD. His name is Ted Farmer, he tells me and then explains, as former cops are apt to do, the exact path he took from being a real cop to a rental cop. When he runs out of story, he turns his attention back to the body in the sand.

"Yep," he says, motioning his head in the direction of the grave. "Lotta fireworks. My opinion? They should just leave the bodies where they are. No sake digging them up just to move them somewhere else." We both watch as one of the female anthropologists carries a hand and wrist over to a white plastic sheet and sets it down across from another hand and wrist. "That first hand? Still had rings on it. That sorta thing messes with your head. It's dumb, I know. Lady's probably been dead 50, 100 years, more fertilizer than person. But still."

About 200 yards from the shore a small aluminum boat with a screaming outboard motor trolls back and forth. I can just make out the outline of a shirtless man sitting at one end, a little boy at the other, a long fishing pole bent between them. When she was at her sickest, when it was apparent that the only salve for her illness was the belief that tomorrow could only be better, when we'd begun to live in increments, separating the positives into single grains of sand, Margaret was certain that when she was well (never if) that we'd have a houseboat on this inland sea, that our lives would be lived rarely touching land, that each morning we'd pick up anchor and find another destination, another view of the sun-charred Chocolate Mountains.

When she passed, I gave her that.

I look now at the bones being sluiced from the ground and know, of course, that it's not Margaret. Oh, but she is here, holding my hand as we walk from Katie's and dip our toes into the water, the air alive with laughter behind us, music wafting through the thick summer air, Chuck Berry singing "Johnny B. Good" into eternity. Her hair is pulled from her face and she's wearing a v-neck white t-shirt, her tanned skin darkening the fabric just slightly, a scent of vanilla lifting from her skin. It's 1962. It's 1963. It's today or it's yesterday or it's tomorrow.

"You okay, pal?" Farmer says. I look down and see that he's got a hand on my chest, steadying me. "Drifting a bit to stern there."

"Not used to the heat anymore," I say, though the truth is that I feel fine. Though my perception is dipping sideways, it does not bother me. Seeing the past like a ghost is a welcome part of my new condition and if it brings with it a few disorienting side effects, I suppose I'm willing to make the trade. Farmer fetches me an unused bucket from aside the dig, turns it over, and directs me to sit, which I do. After the horizon has straightened out, I say, "I used to be the law out here, if you can believe that."

"When was that?"

"About a million years ago," I say. "Or it could have been fifteen minutes ago."

Farmer winces noticeably and then just shakes his head, like he knows exactly what I mean. We watch the anthropologists going about their work in silence. It becomes clear after a while that the two young women are actually students—graduate students, most likely—and that the man in the funny vest is the professor. Every few minutes he gathers their attention and explains something pertaining to what they've found. At one point, he goes back to the white plastic sheet and lifts up a leg they've pried from the earth and makes sure his students have made note of an abnormality in the femur; a dent of some kind.

"You know what I think?" Farmer says. "Guys like us, we've seen too much crazy shit, our brains just don't have enough room to keep it all. Pretty soon it just starts leaking out."

"You're probably right," I say.

"I guess I've seen over a hundred dead bodies," he says. "Not like this here, but like people who were alive ten minutes before I got to them. Traffic accidents and such, but sometimes I'd get called out on a murder, but I was mostly a low-hanging-fruit cop, if you know what I mean. I tell you, there's something about the energy surrounding a dead body, you know? Like a dog, it can just walk by, take a sniff and keep going. Us, we got all that empathy. What I wouldn't give to lack empathy."

The two women lift the trunk of the body up out of the dirt. There's still bits of fabric stuck to the ribcage and my first thought is of those old pirate books I used to read as a kid, where the hero would find himself on a deserted island with just the clothed skeletons of previous plunderers lining the beach. How old was I when I read those books? Eight? Nine? I can still see my father sitting on the edge of my bed while I read aloud to him, how the dim light on my bedside table would cast a slicing shadow across his face, so that all I could make out was his profile. He was already a sheriff then himself, already knew about empathy, already had spent a few sleepless nights on the beginnings and endings of people he'd never know intimately, though he was only 28 or 29 himself. Thirty-five years now he's been gone, and still I grieve for him. You never stop being somebody's child, even when you can see the end of the long thread yourself. Maybe that's really what Kim finds absent: It's not simply Margaret that calls to me in the night, even when the night is as bright as day, it's all that I've lost: my father, my mother, my brother Jack, who passed before I was even born, but whose presence I was always aware of, as if I lived a life for him, too. How many friends of mine are gone? All of them by now, even if they are still alive. And here, in the winter soil of the Salton Sea, the air buttressed by an ungodly heat, I remember the ghosts of another life, still: these bodies that keep appearing could be mine; if not my responsibility, my knowledge, my own real estate.

I tell myself it's just land; a place, like any other place, made of air and dirt and water. My mind has ascribed emotion to a mere parcel of a planet: It's the very duplicity of existence that plays with an old man's mind, particularly when you can see regret in a tangible form alongside the spectral one which visits periodically.

"They bother looking for kin?" I ask.

"Oh, sure," Farmer says. He waves his clipboard and for the first time I notice that it's lined with names and dates and addresses. "We got some old records from back when Claxon was out here detailing where a few family plots are and such. Claxon kept pretty good record of who came and went, but this place has flooded and receded so many times, you can't be sure where these bodies are from. Back then, people died, they just dug a hole and slid them in, seems like."

"That's about right," I say.

"Anyway, we get a couple visitors a week, like yourself."

"I'm just out for a drive," I say.

"What did you say your name was?"

"Morris Drew," I say.

Farmer flips a few pages, running his finger down the lines of names, and then pauses. "I'm sorry," he says quietly.

"So am I," I say.

I drive south along the beaten access road that used to run behind the marina but now is covered in ruts and divots, the pavement long since cracked and weathered away, plant life and shrubs growing between vague bits of blacktop. Back when Claxon Oil still believed life could take place here, they built the infrastructure to sustain a population of 100,000, so beneath the desert floor there's plumbing and power lines simply waiting to be used, a city of coils and pipes to carry subsistence to a casino, one hundred condominiums, tourists from Japan. They'll bring in alien vegetation to gussy up the desert, just as they have outside my own home on the 12th hole, they'll install sprinklers to wash away the detritus of fifty years of emptiness. There are maybe 600 people living permanently around the Salton Sea, Ted Farmer told me, more if you count the meth addicts and Vietnam and Gulf War vets out in Slab City, the former Air Force base that became a squatter's paradise.

I stop my car when I see the shell of the old Claxon barracks rising up a few hundred yards in the distance. To the east, a flock of egrets have landed on the sea, their slim bodies undulating in the water just beyond the shoreline. They've flown south for the winter but probably didn't realize they'd land in the summer. I can make out the noise from the lone boat on the water, but can't see its shape, can't see the man and the child and the hope for fish.

The barracks themselves are a swiss-cheese of mortar and drywall, to the point that even from this distance I can see the sparse traffic on Highway 80 through the walls, as if a newsreel from the future has been projected onto the past. Farmer warned me not to go into the old building, that transients, drug addicts and illegals frequently use it for scavenging and business purposes.

It's not the barracks that I'm interested in; they merely provide a map for my memory, a placeholder for a vision that blinds me with its radiance. It's 1962 and I'm parked in my sparkling new Corvair, Margaret by my side, and we're scanning the scrub for the view she wants. It's not the topography that she cares about; it's the angle of the sun. She tells me that anyone can have a view of the mountains, anyone can have a view of the sea, but after living in the Pacific Northwest her entire life, she desired a view of the sun. She wanted a morning room that would be flooded in natural light at dawn, that would be dappled in long shadows in the afternoon, that at night would glow white from the moon. All my life I've lived in clouds, she said, I think I deserve a view of the sun. She got out of the Corvair right about *here*, I think, and she walked out into the desert, striding purposefully through the tangles of brush and sand while I watched her from the front seat. She was so terribly young, just 23, but I know she felt like she'd already lived a good portion of her life out, that she needed to be a *woman* and not the girl she was. She always told me that she envied the experiences I'd had already in life, that she wished she could see Asia as I had, wished that she knew what it felt like to hold a gun with malice, because to her that was the thing about me that was most unknowable.

Late at night, she'd wake me and ask me what it felt like to kill someone, to know that there was another family, somewhere in North Korea, who didn't have a father. She didn't ask this in anger, she said, only that it was the sort of thing that kept her awake at night, knowing that the man sleeping beside her, the man she loved, had killed before. I'd see Margaret get old before time had found its bearing with her, long before she was done being a girl. I would see her bald and shapeless, her bones the density of straw. I would hear her beg for me to use my gun without malice, to save her from the suffering of her very body, to relieve the pressure that boiled through her skin, the slow withering of her veins, the viscous loss of self that would turn her into something foreign and angry, her voice the consistency of withered crabgrass, begging me, begging me, pleading for a mechanical end to an unnatural death sentence. I would see her in the moonlit glow of the Salton Sea, her body slipping between my hands into the deep, murky waters.

I step out of the car and see Margaret, her face turned to the sun, motioning for me to join her, to see what she sees. She calls out for me to hurry, to join her there in the spot that will be our home. Did this happen? I'm not sure anymore, but at this moment it is true, it is now, it is then and it is always. Surrounding me are nearly a dozen old foundations, this tract of desert bifurcated by the phantom remains of paved roads and cul-de-sacs. From above, I imagine the land surrounding the old barracks must look like a petroglyph left by an ancient civilization.

I triangulate myself with the sea, the mountain and the barracks and then close my eyes and walk forward, allowing my sense memory to guide me, to find the cement that was my home. But it's useless: I trip over a tumbleweed and nearly fall face first into the ground. Sweet Christ, I think, it's lucky I didn't break my hip! How long would it be before anyone found me? Hours? Days? Weeks? With the heat as it is, I'd die from exposure before Kim even noticed that we'd missed the early bird at Sherman's Deli. So instead I walk from foundation to foundation, hoping that the layout of the Claxon Oil Executive Housing Unit makes itself clear. I picture the payroll manager, Gifford Lewis, and his wife, Lois, sitting on their patio drinking lemonade, their baby frolicking between them in a playpen. I picture Jeff Morton, sitting in his backyard, strumming a guitar he didn't know how to play. I picture Sassy, the Jefferies' cocker spaniel, running across the street to our scratch of grass, her tail wagging in a furious motion. I picture myself leaning down to pet Sassy and the way the dog would lick up the length of my arm, her tongue rough and dry from the heat, and how I would step inside and get a water bowl for the dog, and that the dog would sit and wait patiently for my return and then would lap up the water in a fuss, drops of water flying from the bowl and catching bits of sun so that each drop glowed a brilliant white.

I try to see the world as it was and as it is now, try to find what used to be my home, what used to be my life, try to locate the Fourth Estate of my memory:

a dry reporting of fact. You lived *here*. You slept *there*. You made love and you witnessed death and you mourned and you buried your wife in the simple plot Claxon provided and allowed behind your home and you carried your wife's corpse—because that's what it was, it wasn't a body anymore, not with the dirt and the sand and absence at all of any kind of reality, any kind of relevance beyond what you'd emotionally ascribed to it—to her final resting spot because that was what she asked of you, not to allow her to rot in the desert, to give her a perpetual view of the sun and the water, to let her float free of the pain, because that's what she wanted you to give her, a release from the terrible pain, the numbing end of her life that felt, she said, like a kind of slow-acting poison that killed her one inch at a time. That is *across the way*. And you see the end of your own life, don't you? You feel the creeping dread that you've beaten that same slow poison yourself but have found another, more insidious invader. And what will you do about it, Morris Drew? Why did you bring that gun with you? What will you do?

When I get back home, I find Kim sitting on our back patio, her eyes buried in a magazine, golf carts moving in a steady stream past her as dusk has begun to fall. She doesn't see me, so for a time I just stare at her. I imagine what she might look like with the fine lines around her eyes smooth, her gray hair blond, her skin thick and healthy instead of thin and stretched like parchment. The trauma of memory is that it never forgives you for aging. What would Margaret look like to me today? Would she be an old woman or would she be young in my eyes, perpetually 23 years old? The other trauma of memory is that it can absolve you of reality if you let it, and the reality is that I've come to love another woman, finally, a fact I'm not ashamed of.

"I'm home," I say.

"I know," Kim says, not looking up.

"I didn't know if you heard me come in," I say.

"Morris," she says, absently turning pages, "your footfalls have the delicacy of a jackhammer. I heard you walk from the garage, through the kitchen and all the way to your spot there by the sliding door. There are no secrets between tile floors and you."

I laugh for the first time all day and it feels good. I sit down beside Kim and put my arm around her and pull her close. "There's something I need to tell you," I say.

Kim looks up at me and I see the young woman she must have been. I've seen photos, of course, but you never truly see someone in a photo. You see what they looked liked, but not who they were. Fear shows you all the colors in a person's skin. "Are you sick?"

"I am, but that's not what I want to tell you," I say. I reach down and lift up my pant leg and show her my empty holster. "I almost killed myself today. I'm not proud of that, but I wanted you to know that it won't happen again."

"Jesus, Morris," she says quietly.

"I threw that gun into the Salton Sea," I say. "Gave it a proper burial, said some prayers over it and even cried for a time, but I'm not gonna let it take me from you."

Kim sighs and I know that if I look down I'll find her crying, so I stare instead at the long shadows crawling into the bunkers on either side of the 12th hole, at the last glimmers of sunlight that eke over the rim of the San Jacinto Mountains, at the green shards of grass that grow just beyond our patio. I watch as lights flicker on inside the condos across the fairway from us and I think that where I am now, at this very moment, with my wife beside me, with a hint of cool in the breeze that has swept by me, the smell of jasmine light on its trail, *this* is the memory I want to live out the rest of my years with. A moment of silent perfection when I knew, finally knew, that I'd found a kind of contentment with who I was, who I'd been, and what I'd tried so desperately to forget. I am not surprised, then, when a strong gust of wind picks up from the east and I make out the faint scent of the Salton Sea, pungent and lost and so far, far away.

PAULINE ESTEVES

Pauline Esteves was born in 1924 near Furnace Creek, California. She is a tribal elder and former tribal chairwoman of the Timbisha Shoshone Indians, who have lived in their ancestral homeland in and around Death Valley for centuries. Esteves was nine years old when government officials began establishing a national park at Death Valley, and her tribe subsequently endured decades of pressure from park officials to move away from the area, which they refused to do. After years of struggle, the Timbisha Shoshone Tribe received federal recognition in 1983, and on November 1, 2000, the Timbisha Homeland Act was signed into law, transferring more than 7,700 acres of land in California and Nevada to the tribe.

The following excerpt is from a speech Esteves delivered to Congress in 2000 that was vital to the passage of the Timbisha Homeland Act.

Shoshone Indian Homeland Statement to Congress

The word "timbisha" refers to a red material found in the Black Mountains not far from our tribal village at Furnace Creek. Our ancestors, the Old Ones, used this material, called ochre in English. They would use it like paint on their faces, to protect them and heal them. The Old Ones believed that this material, "timbisha," strengthened their spirituality.

Our people, the Timbisha, are named after this material and so is our valley. The term "Death Valley" is unfortunate. We refrain from talking about death. Instead, we refer to "one who it has happened to." Even more importantly, this is a place about life. It is a powerful and spiritual valley that has healing powers, and the spirituality of the valley is passed on to our people.

Our people have always lived here. The Creator, Appü, placed us here at the beginning of time. This valley, and the surrounding places that the Old Ones frequented, is "tüpippüh," our Homeland. The Timbisha Homeland includes the valley and the nearby mountains, valleys, flats, meadows, and springs.

Then others came and occupied our land. They gave us diseases and some of our people died. They took away many of our most important places. The springs...the places we used for food. The places we used for our spiritual practices. They didn't want us to carry on our religion or our ceremonies or our songs or our language. The names of our places became unknown to some of our people.

We never gave up. The Timbisha people have lived in our Homeland forever and we will live here forever. We were taught that we don't end. We are part of our Homeland and it is part of us. We are people of the land. We don't break away from what is part of us.

THE CHANGING DESERT—
A NEW TYPE OF BEAST

JOSEPH E. STEVENS

Joseph E. Stevens was born in 1956 and is a graduate of Princeton University. His book Hoover Dam: An American Adventure *(1988) provides a history of Hoover Dam, beginning with the mid-nineteenth-century visions of Dr. Oliver Meredith Wozencraft and geologist William Phipps Blake (whose journal entries of desert exploration are excerpted elsewhere in this anthology). The completion of Hoover Dam, in 1935, ultimately allowed for settlement of the deserts and inland regions of the Southwest.*

In the following excerpt, Stevens traces the beginnings of the dream of harnessing the Colorado River.

from *Hoover Dam*

In the nineteenth century, explorers began to probe this 2,000-square-mile wasteland, mapping its contours, sampling its soil, and speculating about its bizarre topography. They found that the desert was bounded on the east by a sixty-mile belt of wind-blown sand dunes, which they called the Walking Hills, and on the west by the rocky palisades of the Vallecito and Santa Rosa mountains. Within this outer ring the valley swept downward to the Salton Sink, three hundred feet below sea level, where midday temperatures surged past 120 degrees Fahrenheit and hot winds scorched away any hint of moisture.

This was some of the most desolate territory on earth, a place where even reptiles would not go, yet in 1849 and 1850 caravans of fortune hunters, on their way to California to dig for gold, dared to traverse it. One of the argonauts was Oliver Meredith Wozencraft, a 35-year-old New Orleans physician who had abandoned his family and his medical practice and set out for the gold fields. He arrived at Yuma in May, 1849, weak from cholera and the rigors of his ride across Texas and New Mexico, but grimly determined to push on to the Golconda he was sure awaited him in California.

The four-day journey across the Colorado Desert was always hazardous, but in May, when the heat was near its zenith, the trip was virtually suicidal. The people of Yuma desperately wanted a doctor, and they urged Wozencraft to postpone his crossing until fall, when the temperatures moderated. They hoped that once he recovered from his illness and resumed practicing medicine he would forget the gold rush and stay in Arizona. But Wozencraft would not be deterred; joining a group of similarly gold-crazed men, he crossed the Colorado River on a ferry and rode off on his mule into the sand dunes. The wind and sun soon took their toll, and two members of the group collapsed. At last perceiving the folly of what they were attempting, the others turned back, all but the heat-befuddled doctor, who rode on until he reached the east bank of the Alamo Barranca, one of the dry channels cut by the Colorado in its flood stage. There he tumbled from his saddle, crawled to the brink, and stared in a delirium at the dusty, down-sloping riverbed that led north toward the Salton Sink.

"I felt no distress whatever," Wozencraft wrote of his ordeal. "I was perspiring freely and was as limber and helpless as a wet rag. It was an exhilirating experience....It was then and there that I first conceived the idea of the reclamation of the desert." The idea was a hallucination, a vivid dream of water rushing down the barranca and flowing across the desert to green fields and oases of palm trees. The dream might have died aborning if one of the doctor's companions had not arrived at that moment with a full water bag; the man revived Wozencraft, hoisted him onto his mule, and led him back to Yuma.

Wozencraft soon recovered from the effects of sunstroke, but he could not forget the mirage he had seen on the edge of the barranca. He questioned the people of Yuma and was told that when the Colorado was at flood stage, water did indeed run down the Alamo and New River channels and other smaller arroyos into the Salton Sink. He looked again at the heavily silted banks of the river and at the desert plain below and was struck by a simple yet wholly original idea: if he could dig a ditch through the riverbank to the Alamo Barranca, gravity would draw the waters of the Colorado down into the desert, causing the fertile soil to sprout the green paradise he had seen in his delirium.

Wozencraft eventually traveled to California, but by then he had forgotten the gold rush. In the cool climate of the Pacific coast he dreamed of the burning desert and began to work to make his idea of reclamation a reality. Large-scale irrigation was virtually unknown in the United States in the middle of the nineteenth century, but with common sense and the help of a surveyor named Ebenezer Hadley, the doctor devised a plan for dredging the Alamo Barranca, linking it to the Colorado, and diverting the water through side canals so that it would nourish fields throughout the desert valley.

The feasibility of his plan was soon confirmed by a government geologist, William P. Blake, who surveyed the Colorado Desert for a railroad route. "The Colorado River is like the Nile," Blake wrote. "If a supply of water could be obtained for irrigation, it is probable that the greater part of the desert could be made to yield crops of almost any kind. By deepening the channel of the New River...a constant supply could be furnished to the interior portions of the desert."

Blake's report was not entirely optimistic, however; it cautioned that diversion of the Colorado might lead to uncontrolled flooding of the Salton Sink. Wozencraft, now completely enthralled by the commercial potential of his plan, chose to embrace Blake's comments about an American Nile and ignore the warning of potential disaster. He busied himself establishing a land company and getting himself elected to the California legislature, where he planned to drum up support for his irrigation scheme.

During the gold rush years, the doctor generated little interest among fellow lawmakers for his reclamation proposals, but he never stopped lobbying or accumulating political favors. Finally, in 1859, his persistence paid off. The legislature passed a bill giving him title to ten million acres of the Colorado Desert and its blessing for his irrigation scheme. The bill also provided that California's congressional delegation would work to secure federal rights to the land for Wozencraft.

With an empire almost in his grasp, the doctor hurried to Washington, where for three years he lobbied the House Public Lands Committee, marshaling every bit of evidence and bringing forward every cooperative witness he could find to support his project. On May 27, 1862, his land bill was reported favorably to the House by the committee chairman, but the majority was unconvinced; as Wozencraft watched in dismay from the gallery, the bill was killed.

Penniless and heartbroken, he returned to California to resume the practice of medicine, but the alluring vision of a garden in the desert continued to haunt him. As the years passed, he convinced himself that the legislation awarding him ten million acres had failed not because of its merits but because Congress was preoccupied with the Civil War. In 1887, at the age of seventy-three, he returned to Washington and again presented his land bill, only to see it derided in committee as the "fantastic folly of an old man." Several days later, Wozencraft, now broken in body as well as spirit, died of a heart attack in a Washington boardinghouse. The visionary was dead, but his dream of irrigating the Colorado Desert survived and was seized upon by a new generation of would-be empire builders. The triumphant realization of this dream, the fabulous results it produced, and the incredible natural disaster it unleashed ultimately would lead to the construction of Hoover Dam.

MARC REISNER

Marc Reisner, born in Minnesota in 1948, was a writer and environmentalist who worked for the Natural Resources Defense Council. In 1986 he published Cadillac Desert: The American West and Its Disappearing Water, *a landmark book on the complex and often controversial policies, history, and practice of water management in the West. Reisner died in 2000.*

This excerpt, taken from the chapter "The Red Queen" from Cadillac Desert, *depicts the building of the Los Angeles aqueduct, an epic and controversial engineering feat that took six years to complete. When finished, in 1913, the aqueduct imported into southern California a continual flow of water from California's Owens Valley, on the abundant eastern watershed of the snowy Sierra Nevada, 223 miles across the Mojave Desert. The aqueduct remains in use today.*

from *Cadillac Desert*

The aqueduct took six years to build. The Great Wall of China and the Panama Canal were bigger jobs, and New York's Catskill aqueduct, which was soon to be completed, would carry more water, but no one had ever built anything so large across such merciless terrain, and no one had ever done it on such a minuscule budget. It was as if the city of Pendleton, Oregon, had gone out, by itself, and built Grand Coulee Dam.

The aqueduct would traverse some of the most scissile, fractionated, fault-splintered topography in North America. It would cover 223 miles, 53 of them in tunnels; where tunneling was too risky, there would be siphons whose acclivities and declivities exceeded fifty-grade. The city would have to build 120 miles of railroad track, 500 miles of roads and trails, 240 miles of telephone line, and 170 miles of power transmission line. The entire concrete-making capacity of Los Angeles was not adequate for this one project, so a huge concrete plant would have to be built near the limestone deposits in the grimly arid Tehachapi Mountains. Since there was virtually no water along the entire route, steampower was out of the question and the whole job would be done with electricity; therefore, two hydroelectric plants would be needed on the Owens River to run electric machinery that a few months earlier had not even been invented. The city would have to maintain, house, and feed a work force fluctuating between two thousand and six thousand men for six full years. And it would have to do all this for a sum equivalent, more or less, to the cost of one modern jet fighter.

The workers would have to supply their own hard-shelled derby hats, since hard hats did not yet exist, and even if they had the city couldn't afford them. They would live in tents in the desert without liquor or women—although both were available nearby and ended up consuming most of the aqueduct payroll. They would eat meat that spoiled during the daytime and froze at night, since the daily temperature range in the Mojave Desert can span eighty degrees. Nonetheless, the men would labor on the aqueduct as the pious raised the cathedral at Chartres, and they would finish under budget and ahead of schedule. If you asked any of them why they did it, they would probably say they did it for the chief.

The loyalty and heroics that [William] Mulholland [head of the Los Angeles Department of Water and Power] inspired in his workers were a perpetual source of wonder. For six years he all but lived in the desert, patrolling the aqueduct route like a nervous father-to-be pacing a hospital waiting room—giving advice, offering encouragement, sketching improvised solutions in the sand. In sandstorms, windstorms, snowstorms, and terrifying heat, his spirits remained contagiously high. Pilfering, which can add millions to the cost of a modern project, was almost unknown. Although the pay was terrible—Mulholland simply couldn't afford anything more—he initiated a bonus system that shattered records for hard-rock tunneling. (The men were in a race with the world's most illustrious tunnelers, the Swiss, who were digging the Loetchberg Tunnel at the same time.)

Throughout the entire time, Mulholland showed the better side of a complex and sometimes heartless character. If he wandered through a tent city and discovered that a worker's wife had just had a baby, he would stop long enough to show her the proper way to change a diaper. He would sit down and eat with the men and complain louder than anyone about the food. In lieu of newspapers, his wit was breakfast conversation. Once, when a landslide sealed off a tunnel with a man still inside, Mulholland arrived to check on the rescue effort.

"He's been in there three days, so I don't suppose he's doing so well," said the supervisor, a mirthless Scandinavian named Hansen.

"Then he must be starving to death," said Mulholland.

"Oh, no, sir," said the supervisor. "He's getting something to eat. We've been rolling him hard-boiled eggs through a pipe."

"Have you?" said Mulholland archly. "Well, then, I hope you've been charging him board."

"No, sir," said the flustered Hansen. "But I suppose I should, eh?"

And Los Angeles loved Mulholland even more than the men, because its reward would be infinitely greater than theirs—to the thirsty city, he was Moses. And he was that greater rarity, a Moses without political ambition. When a move was afoot a few years later to run him for mayor, Mulholland dismissed it with a typical bon mot: "I would rather give birth to a porcupine backwards than become the mayor of Los Angeles." But nothing that William Mulholland ever said or did quite matched the speech he gave when, on November 5, 1913, the first water cascaded down the aqueduct's final sluiceway into the San Fernando Valley. It had been a day of long speeches and waiting, and the crowd of forty thousand people was restless. Mulholland himself was exhausted; his wife was very ill, and he had slept only a few hours in several nights. When the white crest of water finally appeared at the top of the sluiceway and cascaded toward the valley, an apparition in a Syrian landscape, Mulholland simply unfurled an American flag, turned toward the mayor, H. H. Rose, and said, "There it is. Take it."

It was the high point of Mulholland's life and career.

NINA PAUL SHUMWAY

Nina Paul Shumway was born in Nebraska in 1888. She was the daughter of Coachella Valley pioneer W. L. Paul, who played a key role in launching the valley's successful date farm industry, which thrived throughout the twentieth century but has been overshadowed by urban expansion and skyrocketing land values. In 1909 Shumway's father purchased a quarter-section of desert land near present-day Indio and moved there with his family, including the newlywed Shumway and her husband, Harry. She published her autobiography, Your Desert and Mine, *in 1964. She died in Palm Springs in 1984.*

In the following excerpt, a chapter from Your Desert and Mine, *Shumway writes in a humorous and upbeat manner about coping with the difficulties of early desert farming.*

from *Your Desert and Mine*

Let the Desert Rejoice

Evenings and mornings grew cool. We no longer had to wrap ourselves in wet sheets to get a night's sleep. Though the sun was still hot and dazzling it was comfortable in the shade. The mountains no longer radiated heat like great piles of simmering slag. The air currents that flowed from their high crests refreshed the parched land like streams of water. The pale gray face of the desert brightened with the golden bloom of rabbit-brush. And all living things rejoiced in a general revivescence.

The angry, itching rash of prickly heat that had plagued me all summer, sluffed its scabs and faded out like a bad dream. Infused with new ambition, Mama and I unpacked wraps and blankets, cleaned house, started cooking all our meals, and settled down to normal family life.

Two pretty girls delivered milk to us every morning, driving in from their father's ranch on the edge of town. The wife of another rancher, more distantly located, brought butter, eggs, and dressed poultry once a week when she came in to do her trading.

Several of the town ladies called and invited us to church and Ladies' Aid. Among these was the minister's wife who enlisted my talents in a projected performance of "Thompson's Hired Man," rehearsals to begin when she had signed up enough volunteers to round out the cast. There was an ice-cream social to raise funds for something or other and announcements were posted inviting everybody to a dance to be held at the schoolhouse. Altogether the social prospect was promising, in spite of that watermelon party.

Papa, always a dynamo of energy, kept Harry on the go in their quest for more information on local agricultural conditions, prospects, and practices. Sometimes Mama and I went along to stimulate our hopes with the accomplishments of established pioneers and gather ideas for the house we were planning to build at the Esperanza. These occasions were infrequent because they necessitated using the hack instead of the single buggy and as we had not yet succeeded in procuring a driving-mate for Roaney we must endure the slow, lumbering pace of Belle and Queen.

Nevertheless we succeeded in making several encouraging contacts. One of these was with "the turkey lady"—a rugged character who was successfully raising birds for the Inside markets and gave us explicit (if to me somewhat embarrassing) information on the details of breeding for best results.

In marked cultural contrast was our visit to the J. P. Read Ranch. Mr. and Mrs. Read, a delightful elderly couple, had already been in the Valley five or six years. Upon their arrival they had begun planting date seeds supplied by the government. In their garden of several thousand young seedling palms a number were beginning to bear early-ripening fruit which Mr. Read, an enthusiast at seventy, proudly pressed us to sample.

From this, Papa's original interest gained fresh impetus. And I learned a few rudimentary facts about date culture. Dates could not be budded or grafted like most fruits. This left two methods of propagation. The ideal one was to plant the offshoots or suckers of choice varieties which were invariably of the same sex and fruit-type as the parent palm. But since these must be imported and cost from six to twenty dollars apiece, a prohibitive price for most ranchers, and in any case were unobtainable in quantity, there was no alternative but to raise the palms from seed.

It was true that dates did not come true to seed and that after giving them three to five years of care, about half the palms turned out to be males, which of course bore no fruit, and of the remaining half only a small percentage could be counted on to produce fruit of a marketable quality, still, in Mr. Read's opinion, date growing could be proved profitable. Why not, since this Valley with Imperial and a limited area of Arizona were the only places in the United States where dates could be grown commercially and our annual importation averaged about five-hundred-thousand dollars?

It took no persuasion to win Papa's championship of any method, however arduous, which would produce a date garden. And he left Mr. Read's quite convinced that the ranchers could do nothing better than secure seed from the government and plant it as fast as possible.

Equally conclusive was the evidence we gathered on house-building. So far as we could discover, all substantial ranch dwellings in the Valley conformed to a single plan. Based on the idea of shutting out the glare and providing airy sleeping quarters, the rooms were entirely surrounded by wide screened porches supplied with canvas "flaps" which, by an arrangement of ropes and pulleys, could be raised for shade in lieu of awnings, or battened down for protection in a sandstorm.

Convinced by its universality that this design was a signal success, we adopted it. How were we to know that the porches were not only veritable dirt-catchers but rendered the rooms dark as cells, and that the "flaps" would so vociferously live up to their name?

With that almost pathetic anxiety somehow to make up to Mama for having brought her to this harsh country, which I now perceive in so many of Papa's attitudes, he set out to "make this the finest house in the Valley." It was to be plastered inside and out for better insulation. Dressing rooms with built-in wardrobes were to be part of the sleeping porches. And there would be a bathroom with a porcelain washbowl and tub. As the house, like the well, was to be a joint project, the bathroom and the livingroom with its big stone fireplace would be used in common, though each family would have its own cooking, eating, and sleeping quarters on opposite sides of the house which would face north toward the road.

Aided by a carpenter from Banning named Eastman, who was to do the building, the lumber footage was figured and a list made of all needed materials. The order was sent to the Consolidated Lumber Company of Los Angeles, a wholesale firm with whom Papa could deal through his connection with the Sheridan Lumber Company. And on November 24, 1909 they made shipment. The bill amounted to what now seems a ludicrous sum—$344.52 exclusive of freight charges to Coachella. This of course was substantially less than the cost would have been at retail prices.

With the construction of the house under way we were more than ever impatient to have the well finished and get started clearing and grading the land. To tear out the brush and disturb the ground before we had water to flood it, was only to ask for trouble. With the first wind our top soil with its precious humus would be piled up against the nearest eastward windbreak. But we could and did get a crew of Mexicans started cutting the mesquite clumps and grubbing out the stumps on the twenty acres which was to be the first improved unit of the Esperanza.

We were now definitely embarked on our project. Papa's old joke about going down to make the desert bloom had become too factual to be funny. We were engaged in the actual process of fulfilling literally the ancient prophecy: "The wilderness and the solitary place shall be glad for them; and the desert shall rejoice and blossom as the rose...And the parched ground shall become a pool, and the thirsty land springs of water; in the habitation of dragons, where each lay, shall be grass with reeds and rushes."

I should like to be able to say that I perceived the tremendous import of our undertaking and was properly impressed by it. But honesty compels me to admit that any such breadth of vision was still beyond me.

SUSAN LANG

Susan Lang was born in 1941 in Hollywood, California. From the age of three, she was raised on a 160-acre homestead claimed by her mother in 1930. Located in the Pipes Canyon area of the high desert, near Pioneertown, California, it was a popular filming location for many western movies. Lang is the author of Small Rocks Rising *(2002),* Juniper Blue *(2006), and* Moon Lily *(2008), all semi-autobiographical novels inspired by her mother's life, particularly her challenges as a single mother in the rugged desert canyon. For the past twenty years, Lang has taught English at Yavapai College in Prescott, Arizona, where she lives.*

The following excerpt is from Juniper Blue.

from *Juniper Blue*

By the time Ruth reached the lower desert, the turbulence latent in the high desert had fully manifested. All through Devil's Garden she had to hold tight to the steering wheel to keep the Model A from being pushed off the road by gusts embedded within a steady but strong sidewind. She had rolled up the windows and latched tight the windshield, but that didn't keep out the grit they had to breathe. Even the dirt road in front of her lifted its surface to meet her as she drove over it. Especially bad were sections where the road was torn up for the new Colorado River pipeline. Ahead of her, the air in the lower desert was so choked with dust that she could barely make out the mountains lining the pass she would soon turn into. Then she would have to point her vehicle straight against the wind's force. Palm Springs, usually visible ahead to her left as the road turned right toward San Bernardino, had been completely obscured by a thick curtain of blowing sand.

Sand pinged the tin hood and fenders of the Model A, pocking the windshield glass. Ruth gunned the car up the hill where the road turned right into the pass, pushing the gas pedal to the floor to give the vehicle enough momentum to reach the top, she and the two children coughing as they went. Ruth assumed that once they started down the other side and reached the paved highway, the dust in the air would lessen because she would be driving on blacktop. Yet when she went over the top of the hill and across it, the high viewpoint offered no glimpse at all of highway at the bottom. The only thing visible below was a boiling soup of green-brown dust.

For the first time, she wondered if she should turn back, questioned seriously if she would be able to get through the murky tempest that lay between her and the road up Black Canyon. She had been through dust storms before, but nothing like this one that pitted her windshield with pebbles and made the air more impossible to breathe with each mile she drove.

The problem didn't lessen when she reached the paved highway. The tarred surface in front of her was barely discernible beneath the steady layer of low-blowing sand streaking straight toward her over the road. The headwind slowed the Model A to a crawl, but the reduced speed did not enable her to see any more of the road ahead. Now she understood that she would have to turn back, at least for the time being. The distance out of the dust storm was much shorter back the way she had come.

Ruth began looking for a turnout, where she could safely swing the car around, but couldn't see well enough where the pavement ended and the soft shoulder sand began. Not wanting to get trapped in the treacherous sand trap beside the road, she continued struggling slowly forward, all the while searching for a widened edge of paved road. Without disaster, she managed to get the children's shirts up over their noses, telling them to breathe through the cloth, though her own shirt would not stay put there without her holding it, and she needed both hands to hold the wheel steady against the vicious gusts. Even taking a hand off the wheel occasionally to wipe grit from her eyes was risk enough to stir up butterflies in her belly and shoot streaks of heat into her armpits.

When at last, in a small pocket of near visibility between gusts, Ruth caught a glimpse of a wide turnout across the road, she swung the car around toward it. She'd been lucky to see the road widening, she realized, as a wave of sand closed her view about halfway through her turn, and the air became even murkier than before. She trusted the vision she'd had and made her turn. Changing direction brought the wind behind her, which kept pushing the car to greater speed than she dared on a highway she could barely make out. Only a change of texture when blowing surface sand met pavement then left it again gave any clue to where the road was, and sometimes even that was hard to make out. It puzzled her, though, that instead of the low-blowing sand streaking from straight down the road behind them, the sand now crossed at an angle. Had there been a slight but sudden shift of wind?

The children's coughs had become less frequent, though she found herself choking and hacking and clearing her throat constantly as she drove on, straining to keep the vehicle on the highway, while she waited for the hill that would lead her away from the lower desert. She could see nothing farther than a few feet around her now in the darkening brown-green thickness. It seemed, after a while, that she should have reached that hill some time ago, but without any visible external references, she found it hard to trust her own perceptions. Time seemed to have been carried off on the wind along with the rest of the world. Ruth was reminded of a day years ago when she struggled through a snowstorm, and time and direction became as unreal as they were now. At the moment up, down, sideways were all the same, and she started to wonder if maybe she had climbed the hill without realizing it, since she had no clear reference point. Had she simply not noticed the car rising above level ground? That seemed entirely possible under these conditions, even probable.

A moment later Ruth noticed a diffuse light somewhere just ahead to the right of her. She assumed what she saw was the headlights of a vehicle coming toward her—another traveler about to enter the tempestuous landscape that had forgotten its stationary place—and she meant to stop and warn whoever it was not to go on any farther toward the pass. Her decision to turn back had been a wise one; already the air was lightening slightly, and she was relieved that the sand traveling over the pavement had thinned from a steady stream to scattered trickles snaking their way across. Through blurred air, she was now able to make out creosote and other brush several yards away from the highway. Visibility improved enough for her to realize, as she came closer to the light, that it was both singular and stationary. She had expected the highway to curve toward the light to her right, but it did not, and she found herself driving past the dim glow that came from somewhere just past her range of visibility. Had someone driven a vehicle off the highway and become stuck in the sand? Entirely puzzled, she considered stopping to investigate, but looked over at J.B. and Maddie, who now slept leaning one on the other. It seemed she had no choice but to keep her vehicle moving on the pavement. Pavement?

Realization came to her at the same moment that she saw another dim light, this time to her left, with another close following it. Then ahead of her more lights yet, and the double lights of an actual vehicle coming toward her on this road. A road that should not be paved at all—if she really had turned and gone up the hill. Somehow she had lost her way. But how had she become so disoriented as to not realize that fact the moment the sand had cleared enough to see the pavement under it? And where had she gone then? Just where was she?

The air cleared a bit more as she drove farther, though the thick haze of dust retained a strong hold. And the sky seemed especially dark. She could make out the sides of a huge mountain whose top was lost in the darkness overhead, and now Ruth understood that she had driven into the shelter of its cove. Blurred buildings began to appear on each side of her, most of white stucco, restaurants and shops and even some small hotels. She saw another Model A in the lane ahead of her and a truck coming her way. Beside the highway, a line of palm trees waved frantic fronds. Other fronds lay along the side of the road. She began to understand and vaguely recognize this place, though she had come here only once, and that was several years ago. The clearing she'd seen when she turned must have been the road to Palm Springs, Ruth realized with great relief, and she had wandered into Ben's mythological Springs.

It wasn't such a bad thing, she told herself, as accidentally as it might have happened. It got her out of the main dust storm—and it occurred to her that if she found someplace to spend the night here, it would be easy enough to go on to Black Canyon in the morning. If the storm ended.

Other than a few passing cars, Ruth saw no one out on the street as she drove through town, though most shops and restaurants appeared to be open. It was no wonder the streets were so bare, with all the wind and dust in the air, an occasional palm branch flying across the sidewalk like some giant bird. She drove through town until buildings became more scattered along the road, then turned the Model A around and headed back the way she had come, searching for a hotel for the night. She had liked the name of one she had passed—the Cricket Club. Such a whimsical name for a hotel. She liked the clean lines of the place too, without fancy signs or elaborate buildings.

The hotel front did seem more classy when she actually pulled up into its driveway and killed her engine. But quietly classy, she thought, as she looked at the two stately palm trees guarding the door, the green, green lawn, and beds of colorful flowers bordering the building. Wind had speckled many of their petals over the grass and layered the concrete drive with soft ribbons of color. A gust of wind rocked the car as she looked down at the children waking beside her. J.B. sat up, rubbed his eyes and looked at her. Both eyes were ringed with dark grit. Maddie's were too, and she had a smudge of mustard on her cheek. Ruth tipped the rearview mirror to look at her own face, and burst out laughing.

"We look like a family of raccoons," she said. She turned the mirror toward them, had them stand and see their faces.

J.B. laughed at the sight of himself. Maddie made claws of her hands and began hissing and imitating the low chatter of the animal, the sound eerily accurate.

"This isn't Black Canyon," J.B. said after he'd peered outside the car windows. He looked over at Ruth, puzzlement spreading over his sleepy face.

As she wiped their faces, Ruth tried to explain how she'd thought she was turning back toward home, away from the sandstorm, and ended up in Palm Springs—at least she thought it was Palm Springs.

"I want to go see Grandmother Siki," Maddie said, with that adult clarity she affected when she deemed it necessary. "I want to go to Black Canyon."

"We can't get there now, I told you. The sandstorm is too bad. Maybe we can stay here and go in the morning." But already Ruth was reconsidering that plan. The three of them were dressed for Black Canyon. Even their one change of clothes was not appropriate for this place. She had no tolerance for the kind of snobbery their worn denim and faded cotton might encounter in a place like this. Maybe she should drive farther, even closer to the mountain, and see if there was someplace less classy.

Before she could decide, a man came out the door of the hotel and hurried toward the Model A. Ruth swiped at the edges of her eyes to get off more of the grit as he approached her window, which she unrolled as he stood there, his short hair blown forward to frame his face, like some kind of carnival mask. She smiled, despite her predicament, at the sight.

"Are you with the movie crew, Madam?" the man asked, his eyes registering but not judging her appearance. A gust rushed in, shoved hard against the passenger window to get back out.

Ruth shook her head. "We had to turn back in the storm," she half shouted, indicating the children with her eyes. "Now I'm not sure what to do."

"I suggest you come inside, then. The wind's still quite strong. The radio reported a hurricane along the coast of Baja." He looked over at Maddie and J.B. "Would you like some help?"

"Thanks, but no," she told him, pushing at the door to open it. The door would barely budge against the wind. The attendant reached over and yanked it open, then held it back so it didn't slam into her as she got out, reached in and slung her knapsack over her shoulder. She went around and pulled open the passenger door for the children. It took all her weight to shut it again.

"Where are we going?" J.B. yelled as she tugged the two of them through the wind toward the hotel's front door, where the attendant stood ready to do battle with gusts and open it for them.

"Out of the wind," she shouted. They reached the building just as a huge blast spattered them with debris. She was surprised to see a group of people gathered behind the big glass window watching their approach.

Once she had the children inside and the door closed behind them, the three of them were greeted by several of the crowd, who seemed to be relieved they'd made it inside. Ruth was surprised to see such small-town, neighborly concern in a highbrow place, and she warmed to them enough to describe in detail her harrowing attempt to make it through the pass in the storm. "But we're not with the movie crew," she finished, the thought suddenly occurring that such a misperception might be behind all the attention.

"Of course not. Nevertheless, we hope you'll remain with us until the storm passes," a woman with a massive gold necklace said, while others were murmured in agreement. The woman gave Ruth and the children's appearance a once-over, then said politely, "You needn't worry about paying under these circumstances."

Ruth laughed, allowing pleasure at the friendliness to override the insult—what else would the woman expect of such ragamuffins. Especially in these hard times. "Don't worry," she said. "I can pay for a room if we stay the night. But thank you. I appreciate your concern, all of you." She stopped then, puzzled at the way people appeared to be smiling at her. "Is something funny here?"

The tall woman who spoke before stepped forward and opened a palm toward Ruth, clattering the numerous gold bracelets on her arm. "The flower petals," she said, her smile broadening. "They're all over you. It's quite striking—and lovely."

Ruth looked down at herself to find tender speckles of deep and pale blues and others of violet and white, pink and yellow scattered over her arms and chest. Maddie and J.B. were also sprinkled with color. Ruth reached down and plucked out a purple petal caught in her son's dark hair.

"You look more like fairy people than victims of a fall windstorm," the woman said. "A princess and her two attendants." Ruth liked her strong face with its high cheekbones, tanned skin and expressive hazel eyes. She was older, but ageless at the same time. "Either way, welcome to the Cricket Club refuge, though there will be no cricket today."

Ruth was just about to say something about the playful name of the place, when Maddie gave out the most perfect imitation of a cricket chirping Ruth had ever heard. She felt her face heat as the girl kept it up, one hand placed over her mouth. The crowd around them first looked at Maddie in amazement, then burst out laughing.

"What a wonderful sound joke," the woman said. She bent her knees and lowered herself to Maddie's level, long linen skirt and all, so she could look directly into the girl's face. "That's so clever of you." Holding out her hand to Maddie, she said, "I'm Esther Golden."

"I'm Maddie," Maddie said, just as formally.

"Maddie who?"

The girl stood a moment considering. "I don't know," she said, finally. "Maybe Cricket."

That got another laugh from the crowd and from the woman, who now rose, heavy jewelry and all, as agilely as she had descended. "What a superb child," she said. She offered Ruth her hand. "Esther Golden. The Cricket Club is my hotel."

Ruth accepted the hand. "Ruth Farley." She looked around for J.B. and found him by the wall, running a hand along the stucco surface as he examined the room's high ceiling. "And this is my son, J.B.," she said. Everywhere, it was always Maddie with her weird little ways who got the most attention. J.B. was even more interesting—yet his quiet introspection brought him less notice. Sometimes Ruth couldn't help but wonder if it was his darker skin that kept eyes away from him.

"Yes," the woman said. "Such a serious boy."

Ruth opened her mouth to take issue with the woman's statement, when light flashed from the walls. At the same instant thunder boomed through the room as if it would fling the stucco structure to kingdom come. No one said a word, but they all turned to look out the big bay window just as another thudding clap came, this time stabbing a wavy knife of light into the ground across the highway. A huge gust swept leaves and flower debris against the windowpane and held them squirming along the glass before sweeping it clean again. Ruth noticed a few splotches of water remained on the pane.

"My word!" the woman exclaimed. "We're likely to lose our electric if this keeps up." She called the attendant over and told him to bring the kerosene lamps out of storage. Turning to one of the middle-aged men in the crowd, whom Ruth had noticed because of the cream-colored suit he wore, Esther said, "You'd best give up on the rest of your entourage, George. Most likely they won't be arriving today. If the pass is as bad as this woman says, they'll not get through any better than she did." She turned back to Ruth.

"You, sweetie, I'd like to have stay in the Willow Suite. Joey here will show you the way. Bring her one of those lamps too, will you, Joey?" she said, as another bolt of lightning lit up the window and thunder crashed down around them, rattling glass panes and the vases beside them.

"Dinner will be served at seven, come hell or high water," she called as the attendant led Ruth and her children from the lobby into a hallway, "and high water is likely." By now rain was washing over the glass sheets that formed one side of the long corridor, while bursts of thunder and flashes of light followed one after another.

Their rooms were large, with a living space and divan in addition to the bedroom, very much like the Biltmore's, except for the glass door leading to a large patio, which at the moment had a fast current washing leaves and debris across it. Excited exclamations turned Ruth's attention from the rain to the bathroom, where Maddie and J.B. had had their first-ever sight of a real toilet and a porcelain bathtub.

Ruth demonstrated the use of the toilet, and the children delighted in copying her waste-letting. Then she stripped them, filled the large tub with hot water that miraculously waited behind the faucet with the red center dot, and all three of them wallowed and played there, splashing and laughing as if they had never seen such a thing before—which of course two of them hadn't—until thunder became only a constant and tinny grumble against the huge mountain that helped shelter the town from the virulent effects of the sandstorm. By then the water in the tub was nearly as cold as that coming down outside.

Dinner seemed relaxed enough for the few guests who had made it to the hotel before the storm strengthened, though for Ruth the white tablecloths, candles, and china dishes served to underscore her own now clean but rustic attire. The only thing that enabled her to relax was the glass of ruby-colored liquid in front of her, and she found it difficult not to gulp it down at once and ask for more. It was truly the most delicious-tasting wine she'd ever had. Of course, for years during prohibition, she'd been able to get only the crude wine and moonshine whiskey that Matt Baxter's brother furnished. She remained clear-headed enough, however, to understand from conversations around the large round table that Esther's brother was the producer of the film that the missing crew had been coming to shoot. He and a few others had driven out from Hollywood the evening before.

Ruth half listened to the moviemaking talk—dominated for the most part by the producer brother, who had more words to say than anyone she'd ever seen—and immersed herself in food without speaking for most of the meal. It had been some time since she had tasted such a thing as slices of roast beef floating in gravy, accompanied by potatoes au gratin and asparagus dribbled with yellow sauce. She ate the food out of politeness and curiosity—and considerable hunger. She was attuned to the tastes of goat and game and of the wild plants that grew around the zucchini and tomatoes she planted in her garden. She'd come to prefer them as she had the mesquite-pod flour she'd learned to prepare in Black Canyon. It greatly enriched the flavor of the breads she made. For special occasions, she even added coarse flour made from dried sumach berries. Though she had stocked up on canned food, she wouldn't use any until the foods from the canyon ran out, as she knew they would when winter came on in earnest.

J.B. and Maddie certainly had never seen such foods. They sat at a table near the window, cautiously picking small bites of the exotic meal. The two of them had spent the hour before dinner bouncing up and down on the beds in the hotel room and now seemed content to quietly observe the adults.

Ruth was just swiping the last of the gravy from her plate with a slice of bread when Esther turned toward her during a brief pause in the director's monologue and firmly announced—in a voice obviously meant to put an end to the man's verbal domination, "Now tell us about yourself, Ruth. My instincts tell me you have an interesting story."

Caught completely off guard, Ruth almost choked on the bread she'd jammed into her mouth. But she swallowed and smiled, took a drink of water from the crystal glass, then patted her lips with the cloth napkin—as if this were the way she behaved at every meal, as if she really had china plates and cloth napkins in her canyon when clay bowls and the back of a hand served just as well.

"Oh, I've nothing exciting to tell. No interesting story, really."

Esther laughed. "A fairy princess with two wood sprite children appears in the middle of a storm, telling us she comes from the faraway wilds to the north and is on her way to the wilder wilds of Black Canyon. And you say nothing's unusual or interesting about that? It sounds like a plot for one of George's movies. One of the better ones, before he turned to westerns."

"Now, Esther," the director said. He turned to look at Ruth for the first time during the dinner. "Yes, do tell us your story, before my sister writes your script herself."

"Only because you're so wrapped up in your own importance, George." Esther wagged a finger at him.

"Don't be unkind, Esther. Your brother has a lot on his mind with the new project," the woman sitting beside the director said. She reached over and squeezed George's arm. "And the stories he tells are fascinating."

"Well, he doesn't have to lord it over everyone else. He misses a lot that way, Barbara. In fact, I'll bet a new Ford that this young woman has a more interesting tale to tell us than any of his." Esther turned toward Ruth expectantly. The others looked at her also, though not expectantly, but as if they were waiting for her to make a fool of herself—which seemed quite possible. Ruth found it hard to appear collected with a torrent of blood rushing to her head. She was sure everyone could feel the heat her cheeks radiated.

"So, tell us, young woman, just where it is you live up there in that wilderness country. And *why* you live such a place?" Esther's brother demanded.

"Because I want to live there, that's why." Ruth set her chin and continued on in the only way she knew how to respond to such idiotic patronization— a word she'd gleaned recently from her novels, but an attitude she'd long encountered. "And as for *where* I live, I doubt that you'd have any idea if I told you."

Esther's delight was registered in her melodic laughter, though others at the table were brought up short by Ruth's curt reply.

"I know more about the area than you seem to think, young lady," the producer countered. Ruth thought she detected high color in his cheeks as well. "It so happens I'm involved with a project we're building smack out in the middle of that wilderness. A rocky area past that little town, Oak Valley or whatever it's called."

Could he mean the movie set? Could this flabby city man be responsible for those huge machines that were fast scraping away the rocks and wild plants, she wondered? She pictured the clouds of dust released by the dozers as she drove by the place. Machines scraping away all remnants of earlier residents so someone could make a pretend version of a place that had never existed? Looking at the man across from her in his cream-colored suit and silk shirt, Ruth could almost understand how and why such stupidity could be happening. The thing that startled her, though, was the fact that she had just left that place, only to have a ferocious sandstorm drive her here to encounter the agent responsible for its making. The thought sent a chill up her back.

"Near Juniper Valley, you mean? Perhaps a movie set?" Ruth kept her tone casual. She sipped at her water, glanced at her empty wineglass. She could use something stronger than water at the moment.

"Not your ordinary movie set. A genuine frontier town—and it will be a permanent town. All Frontiertown buildings will be functional for the people who live out there." Golden's face lit up with enthusiasm as he went on describing the place. "The Dirty Dog Saloon will be a real bar, now that prohibition has ended. The General Store will have actual groceries. We designed some of the set buildings as living quarters, like the Land Survey Office and Dr. Brown's Medical Office."

"What people who live out there? Nobody lives out in the Rimrock. There are a few homesteads left around the periphery, maybe, and in Juniper Valley, but..."

"You build the town and the people will come," Golden announced proudly.

"Meanwhile your dozers are scraping away an archaeological site."

"So I've been told. Merely a few broken pieces of primitive pottery. Nothing compared to what I'll build in its place."

"Enough, George. You've taken over again. But you, Ruth, you haven't told us just where you live out there," Esther insisted once again, waving him quiet with clatter of bracelets.

This time Ruth welcomed the distraction from Golden's abomination. The idea of it made her sick at heart. Yet she wasn't so innocent herself. At the Coconut Grove, she'd sampled again the hollow world she'd homesteaded to evade. Now she felt as if she had somehow been contaminated, brought back Hollywood on her clothing to violate her desert and canyons. David Stone had been the first warning of intrusion and led her to the rest. She was only now beginning to understand the extent of it, not so much an intrusion as a full-scale invasion.

For the moment, Ruth put aside her concern over what the area might become and went on instead to explain how she'd found her homestead, describing her struggles to build and remain there. She told them about midwifing her nanny's two new kids, about the bobcat she chased off with a broom when he tried to escape with one of them, how the cat came back later that night and took them both. Deliberately vague about just where her place was located, she hoped that would keep Glory Springs inviolate. She left off also any mention of how the children came to be and how all that fit into her story. Ruth sensed that Esther was on the edge of probing, but the woman didn't push, seeming satisfied that Ruth's story was as unusual as she had promised. Other guests, including Golden himself, became quiet and listened, though his expression was one of puzzled amusement.

When she could, Ruth escaped to her room with Maddie and J.B., both of whom fell asleep almost as soon as they lay on their soft mattress. Ruth spent hours awake, listening to the hard rain that was still coming down steadily. She couldn't stop worrying about the effect such a movie town could have on the small desert community and on the wild landscape she loved, cringing again over her own intrusion there—the buildings and windmill she had put on the barely touched land of Glory Springs. How different was that from what she hated to see others doing? As much as she had despised the tin cans that cowboys had hung from the pines to catch pitch before she homesteaded the place, she herself had put much more than a few tin cans on that land.

She remembered the strong sense of destiny she'd felt at her first sight of Glory Springs, the passion unleashed when she found her place. In what way was that different from the passion shown by Golden for his foolish Frontiertown? She remembered what Thomas had told her about the changes the people of Black Canyon had seen, remembered what Jim had said about civilized encroachment in the area. And suddenly she wasn't sure of anything, of her own role and purpose, of what was right for herself or for the rest of the world—which kept turning out to be far more complicated than she had imagined. The only thing she was sure of was that she wanted to be with her friends in Black Canyon tomorrow, wanted to pound mesquite pods on the *na'dai* and bake the wild flour into bread, to drink tangy liquid made from that mesquite as they sat around the campfire while night came on. She wanted to eat with her friends the sweet flesh of plants and small wild creatures—some she had never dared ask about. All she could do was lie awake and wait for morning so she could get on her way. Only when she heard the rain dwindling away enough to allow her to make the trip did she let herself fall asleep for a short while.

STATE SENATOR DAVID G. KELLEY

In 2002 the California State Senate enacted a resolution authored by Senator David G. Kelley that designates a five-mile stretch of Interstate 10 near Indio as the Doctor June McCarroll Memorial Freeway (Res. Ch. 105, Stats. 2000).

June Hill Robertson McCarroll (1867–1954) moved to the California desert in 1904 and practiced medicine throughout the Coachella Valley area, including the area's Indian reservations, from 1907 to 1916. In 1917, after she was run off the road near Indio by a careless truck driver, McCarroll came up with the idea to hand-paint a white line to divide the road into two lanes to help prevent accidents. In 1924, thanks to McCarroll's extensive efforts, the state of California adopted her idea, which eventually came to be used as a safety measure on roads throughout the world. That first roadway eventually became U.S. Highway 99 and is now Indio Boulevard, running adjacent to Interstate 10. The senate's resolution honoring her invention is printed below.

California Senate Concurrent Resolution No. 58— Relative to the Doctor June McCarroll Memorial Freeway

This measure would dedicate the portion of Interstate Highway Route 10 near Indio in Riverside County between the Jefferson Street and Indio Boulevard interchange and the junction with State Highway Route 86 to the memory of Doctor June McCarroll, and would specify that this portion of Interstate Highway Route 10 shall be known as the "Doctor June McCarroll Memorial Freeway." This measure also would request the Department of Transportation to determine the cost of appropriate plaques and markers showing that special designation and, upon receiving donations from nonstate sources covering that cost, to erect those plaques and markers.

WHEREAS, Doctor June McCarroll first arrived in California in 1904, when she moved to Indio in order to place her ailing husband in a health camp for persons infected with tuberculosis; and

WHEREAS, It was in Indio where Doctor June McCarroll acquired the name "Doctor June" and traveled at first by horse and buggy and later by horseback, in order to practice medicine; and

WHEREAS, In 1907, Doctor June practiced medicine on five Indian reservations and later became the doctor retained by the Southern Pacific Railroad to treat its employees in the Coachella Valley; and

WHEREAS, In later life, she expressed regrets that younger doctors were seemingly unable to function without modern hospitals and other conveniences when she had sometimes operated on kitchen tables, explaining "I would clear off the table, tie the patient down, and administer the anesthetic"; and

WHEREAS, Doctor June is also credited with starting the first library in the Coachella Valley; and

WHEREAS, Although Doctor June McCarroll has a reputation in the Coachella Valley based on her practice of medicine and dedication to her immediate community, she is also known for her role in initiating the painting of centerlines upon streets and highways; and

WHEREAS, In 1924, after she and the Indio Women's Club and the California Federation of Women's Clubs proposed it, the idea of painting a centerline on state highways was adopted by the California Highway Commission, and the first white stripe was painted by hand with a paintbrush, on Indio Boulevard, coincidentally, on the street where Doctor June McCarroll was then living; and

WHEREAS, This idea has saved thousands of lives since its early beginning over 75 years ago; and

WHEREAS, It is appropriate that the portion of Interstate Highway Route 10 near Indio in Riverside County between the Jefferson Street and Indio Boulevard interchange and the junction with State Highway Route 86 be dedicated in the memory of this remarkable person; now, therefore, be it

Resolved by the Senate of the State of California, the Assembly thereof concurring, That the Legislature hereby dedicates the portion of Interstate Highway Route 10 near Indio in Riverside County between the Jefferson Street and Indio Boulevard interchange and the junction with State Highway Route 86 to the memory of Doctor June McCarroll; and be it further

Resolved, That this portion of Interstate Highway Route 10 be officially designated the "Doctor June McCarroll Memorial Freeway"; and be it further

Resolved, That the Department of Transportation is requested to determine the cost of appropriate plaques and markers, consistent with the signing requirements for the state highway system, showing this special designation and, upon receiving donations from nonstate sources covering that cost, to erect those plaques and markers; and be it further

Resolved, That the Secretary of the Senate transmit copies of this resolution to the Director of Transportation and to the author for distribution.

KATHERINE AINSWORTH

A longtime resident of the Palm Springs area, Katherine Ainsworth was born in 1908. She was married to writer Ed Ainsworth, a well-known journalist who wrote for the Los Angeles Times. They were friends with and patrons of the artist and writer John W. Hilton, who lived nearby. In 1978 Katherine Ainsworth wrote The Man Who Captured Sunshine: Episodes in the Life of John W. Hilton.

In the following excerpt from that book, Ainsworth writes of Hilton's surprising association with General George Patton during World War II. When Patton came to the desert near Indio and Desert Center, California, to prepare his troops for the North African invasion in 1944, he sought "desert rat" Hilton's advice on how to cope with desert conditions. A reflective essay on the desert, written by Hilton himself, appears earlier in this anthology.

from *The Man Who Captured Sunshine*

News of war came to John Hilton in Hidden Springs, a remote camping site where Zane Grey spent many hours writing. Hilton was camped with Randall Henderson, publisher of the Desert Magazine, and other members of the Sierra Club. Hilton had entertained the group at the campfire the previous night. Suddenly young Randy dashed into the camp dodging rocks and palm trees as he shouted, "Japan has bombed Pearl Harbor."

"Randy volunteered for service the next day and I never saw him again," recalled Hilton. "News that America was again involved in war shattered the tranquility of our desert and shook us up as it did everyone in the nation. Pretty soon rumors were being spread about as to the role the desert was to play in the war effort. Some folks said there was to be a training camp installed somewhere, but I didn't pay any attention."

One night shortly after the bombing of Pearl Harbor, John Hilton and his neighbor Russ Nicoll were having dinner at the Indio Hotel. He noticed that at a nearby table a group of army officers were studying some maps and talking intently to the owner of the hotel. In a few minutes, the proprietor came over to Hilton's table and said, "These men are on General Patton's staff. They are trying to locate a good place for a training camp and want to meet someone who knows the desert. I told them nobody knew more about it than John Hilton. Can you come over and suggest a suitable place?"

John went and after being introduced to the men, sat down at their table and looked at the map before them. It was an ordinary map put out by the Automobile Club of Southern California. John listened as they outlined their requirements and then studied the map for a few moments. He picked up a pencil and drew a large circle near Mecca, east of Indio.

He told them that he thought they would find what they were seeking in that general area. The land was reasonably level and well-drained, there was an adequate water supply available from the L.A. aqueduct, and the highways were nearby. The men thanked him and left.

Hilton thought no more of the matter until government surveyors came down to the desert and were quickly followed by large tractors and other earth-moving equipment. Within months Camp Young was established and General Patton and his tanks and army recruits arrived to train for the North African campaign of the war. When the camp was well established, General Patton called upon John's knowledge and familiarity with the desert terrain.

Hilton was invited to dinner out at Camp Young and General Patton presented his special guest with a map. It was the same map on which John had drawn the circle that night at the Indio Hotel. On it the engineers had drawn a small rectangle which was about the shape of the camp, which fitted inside the circle. On the map, Patton had written, "Thanks, John, you saved us a lot of trouble."

This map Hilton still cherishes as much as he would a Presidential Citation.

It was a matter of great satisfaction to John that this great general valued his knowledge of the desert enough to call upon him for advice. Patton wanted him to accept a job with the army, but Hilton's long years of childhood indoctrination with the loving, pacifistic teachings of his church would not permit him to yield on this persuasion. Then, too, he wanted to remain free of any obligations and entanglements with the governmental red tape. He decided that whatever information he possessed which could be helpful he would gladly give, but when the demands increased, his painting time was pushed aside.

These phone calls for advice became more and more frequent. Usually when he arrived, he would find Patton studying a map on a table and the general would ask, "What is this terrain like? Is there any cholla or barrel cactus in this area?"

If Hilton said "No," then Patton would say, "Good, then we can use this area for night maneuvers."

Then he would go on to explain, "I want my men to take just as rough a beating as I can give them in as near the situation they will have in North Africa, but there isn't any cactus in North Africa so I don't want them wasting time pulling cactus spines from each other's ass."

From time to time John went on exploration trips with the General, or "Old Man" as he was affectionately beginning to refer to him in his mind as did most of the soldiers.

"I think I got to know the Old Man pretty well, if anyone ever did, on those trips," Hilton said. "One time we went out across the Colorado River into Arizona, inspecting the area. We were traveling in two Jeeps—two officers and myself and the General and two drivers. We stopped on a mesa and the Old Man said, 'Let's stop right here on this level place for the night.'

"I said, 'General, in my opinion this is not a good place to camp. This is the time of year for sidewinders and this spot is between two deep washes full of ironwood trees and a lot of brush, and it's just close enough that sidewinders would be possibly crawling from one wash to another, and coming through our camp.'

"The General just snorted, 'Oh, bosh, I don't see any sidewinders.'

"Well, who was I to argue with a general, so we set up camp," continued Hilton in recalling his experiences with Patton.

"We soon had a roaring ironwood fire going and all of a sudden one of the drivers got a strange look on his face and I followed his eyes. There, coming out of the darkness right towards us, was a sidewinder. There's a thing about campfires and sidewinders—they are attracted by the light. It fascinates them like it does a moth.

"Well, that little old sidewinder came twisting from side to side right in and I saw that by now the General was watching, too. Just as that sidewinder got fairly close to the campfire, the Old Man pulled out one of his famous pearl-handled revolvers he always wore and let him have it.

"That night the two drivers slept in one of the Jeeps. I looked kind of longingly at them, but the Major, the General, and I slept on the ground in sleeping bags. After Patton stood in that famous stance of his, legs spread apart and hands on hips glaring off into the darkness, and roared out, 'Those damned snakes won't crawl into a sleeping bag!' we meekly had followed his action and timidly slipped into our bags.

"Sure enough, the snakes didn't bother us. Guess even sidewinders knew better than to defy Old Blood and Guts Patton. But the next night he said sort of casually, 'Hell, John, you pick the camping place.'"

Hilton continued in a contemplative mood, "I think I got to see a side of Patton few people ever saw—the tender side. He could get very emotional about a beautiful sunrise or sunset. He liked to sit around a campfire at night and when I played my guitar and sang, he joined in with the other fellows and sang in his high pitched tenor voice.

"Funny thing about Patton's voice. For such a big, imposing man, his voice oddly enough was almost effeminate. I often thought he probably wouldn't have sworn so much if he'd had a deeper, more masculine voice. Sometimes he recited poetry. He was especially fond of the poems written by Don Blanding, and once he recited some of his own poetry."

Hilton was privileged to go on maneuvers with Patton and watch him ride, standing up in his Jeep and screaming orders in his high pitched voice and robust profanity. He was a figure of a man Hilton would never forget. Patton's reputation for toughness was greatly misunderstood by many people, but his theories proved out to be correct and saved many a boy in combat.

"He put winter uniforms on his men in the middle of the summer desert heat," recalled Hilton, "on the theory that if they could stand it here, they would find North Africa a breeze. He also insisted they have their shirt collars buttoned and ties neatly tied. Once he told his men, 'I don't give a damn what you do when you get over there. Take off your shirts, or anything you damn please, but right now you're going to dress and look like American soldiers. The ones of you who can't stand it, better drop out right now, and get sent to Nova Scotia or some other damned cool place. We'll be rid of you and you won't be a drag on us later on.'"

Patton's rigid insistence upon proper military dress got him into an embarrassing situation which he loved to tell on himself. He was returning from a lecture given at a club in Indio when he spotted a young man high upon a pole. This young man appeared to be in a slovenly uniform and without a tie.

The general commanded, "Stop this car. I wish to speak to that man."

The driver backed the car and Patton roared out, "Soldier, what's your name? Which division is yours? Why the hell aren't you in proper military dress with a tie?"

Without stopping his climb up the pole the man answered, "My name is Andy Jones. I'm not a soldier. I work for the telephone company and you can go straight to hell, General."

Each time Patton related this story, he would guffaw and slap his thigh with enjoyment.

"On one bivouac," recalls Hilton, "I saw proof of his great concern for his men—a side Patton kept successfully hidden from the public. His men had orders to dig slit trenches and sleep in them during that night's maneuvers. I saw Patton walking up and down the rows of slit trenches checking up on his sleeping soldiers. Suddenly he began swearing and screaming and kicking some of the sleeping men. The men were beastly hot and they preferred to sleep outside the trenches in the hope of catching any stray breeze.

With each kick of his highly polished boot he screamed, "Get your asses in those trenches. If you can't obey orders here, you damned fools, how the hell do you plan to obey them in combat?"

"Oh, he was tough, all right," laughed Hilton and then soberly added, "but his every thought was the safety of his men. He was careful not to make them take anything they wouldn't have in North Africa. Once he said to me after some parents had written whining notes of protests to Congressmen, "I want these men to go through hell in the training process. I want them to get more tired. I want them to get just as beat as a man will in combat. That's the only way they will be tough enough to survive this goddamned war."

The "Old Man" appreciated John's efforts and knew what a personal financial sacrifice he was making. There would be packages of meat or butter or sometimes cans of gasoline dropped off at the door of Hilton's modest little desert home. John steadfastly refused to become an employee of the army, despite Patton's urging.

"He was in our home just one time," said Hilton. "We were having a little party and he came and stayed for just a few minutes, but you know, that man had great charm. He shook hands with everyone in the room and had a personal comment to make to each person. Then he had to leave, his duties were so pressing. I don't know when he managed to sleep."

General Patton had a lighter side to his nature when he dropped the martinet facade he favored so much of the time. With such a man as this, stories about him were rampant. One of the best, according to Hilton, was when the Women's Club of Indio wrote him a letter and said he should stop the soldier boys from whistling at the women on the streets of their town. He responded by asking permission to attend their next meeting. The ladies were all aflutter when in came this tall, handsome, and awe-inspiring general to talk to them. In his brisk military manner he explained to them that the boys who were whistling in the streets were just exactly like the sons, nephews, grandsons, brothers, and even husbands of the women right there in the Women's Club and were probably whistling at the gals somewhere else on some other streets.

"Now, I'm a great disciplinarian," he growled at the admiring ladies, "and can stop this if you want me to, but it seems to me that we have a lot of necessary rules already without trying to make other kinds. If I were you charming ladies, I would never worry as long as some man whistled at me in the street, but if I walked down the street and no one whistled, then I would really start worrying."

"Needless to say," laughs Hilton, "he had no more trouble with this situation."

As Patton got new officer groups to train at Camp Young, Hilton was asked to take some of his Kodachrome slides and go out and give an indoctrinational lecture on the desert. Years before when the first of the new color film packs arrived down on the desert, Hilton bought most of the supply and started taking pictures of various desert scenes, showing plants, animals and snakes, pointing out which of these were poisonous and which were not.

"The tarantulas, of course, are horrible looking things and sometimes I'd even take a live one along and handle it to show them it was harmless," Hilton recalls, "Then I'd show them the pictures of really poisonous things to look out for, and near the end of my lecture I showed the picture of a man who had actually died of thirst on the desert. It was not pretty; it was horrible. I sneaked in old Queho lying there, partly mummified from the heat and dryness of the desert with his empty canteen lying beside him. I showed this gruesome picture to impress upon these youngsters the necessity of always checking the supply of water in the cans before taking Jeep rides out into the far desert or even if they were just going into Indio.

"These really were survival lessons and to soften their impact I slipped in a group of pictures I had taken years before. These were pictures of nude models posing on some of the desert and dunes. I would say, 'Well, of course, you can't tell what you'll find on this desert,' and then as a gag I would flash one of these pictures on the screen.

"Well, one night I was about three-fourths through my lecture when there was a stir at the back and every man stood up, and in comes the Old Man. I got to thinking about those nudes and wondering how he would take it, but there wasn't any way of slipping them out of the box, so I went on with the lecture and when I got to them I shoved them through the projector as fast as I could.

"About the third nude I heard General Patton's high pitched voice yelling, 'Mr. Hilton, if you don't mind, would you start over with those last few pictures? I think some of my younger officers would like to study them from a tactical standpoint.'"

Trying to keep up with these lectures, writing his monthly article for the *Desert Magazine*, running his little shop, keeping the supply of paintings on hand, and with practically no money coming in, it was getting harder and harder for John. The Hilton family was suffering and could not possibly have existed without the generous informal honorarium of foodstuff coming from the military base. When the Old Man took Hilton aside and spoke to him confidentially and, without mentioning any specific time or date, indicated that the men were trained and ready for combat, and that they would undoubtedly soon be moving out, John was tempted to accept the offer of becoming a warrant officer with the 3rd Tank Corps. He began making tentative preparations for this move.

Then one morning he received a phone call which was to change the course of his life.

CÉSAR E. CHÁVEZ

César E. Chávez (1927–1993), a Mexican American farm worker, labor leader, and civil rights activist, was born in Yuma, Arizona. He cofounded, along with Dolores Huerta, the National Farm Workers Association, which later became the United Farm Workers (UFW). In the early 1970s, the UFW organized a number of successful strikes and boycotts, many of which took place in the agricultural Imperial and Coachella Valleys, to protest for higher wages and better working conditions for farm workers employed by grape and lettuce growers.

Chávez delivered the following speech in Coachella, California, at a rally for support of workers' rights.

Speech at Coachella, 1973

Brothers and sisters in the field—for you are our brothers and sisters of the fields, just as your mothers and fathers were the brothers and sisters of our mothers and fathers. We are the people of the fields, the people who have made the trees and the vines the work of our lives. We know together what the truth is and you know, as we know, that there was never anyone to care about us until the Farm Workers Union came. *Only* the Farm Workers Union gave us new hope and new dreams for our families and our children. You know that and we know that and nothing can ever change it. Before the Farm Workers Union we had only each other to care for. It was our common problems and our common suffering that gave birth to this union, our own union, the Farm Workers Union. All around us were those who said that it could never be done. Everywhere people said that the growers were too strong for us, that the police would be against us, that the courts would beat us down, and that sooner or later we would fall back into the poverty and despair of our forefathers. But we fooled them. We fooled them because our common suffering and our love for each other and our families kept us together and kept us sacrificing and fighting for the better tomorrow that all of us dream about as we work among the vines.

Stop and think: What do you dream of as you work day after day in the hot sun? You dream of a nicer home for your wife, a good school for your kids, some dignity and rest for the older ones. Those have always been the dreams of the farm workers, but they never started to come true until we built this union. No one can ever erase from your minds or our minds that there was no one who cared, no one who fought for us, no one who helped hold out to us hope for a brighter tomorrow. It was the Farm Workers Union, *your* Farm Workers Union which did it. Never until the Union came did our wages move ahead.

Think of it: We are the youngest union in America and yet we have *doubled* our basic wage, and even more!

Never was our health or the health of our families given any consideration until your Farm Workers Union. Remember the days when the vines were our toilets. Remember the indignities that our wives and mothers and daughters experienced. Think back upon it, brothers and sisters. It is a brighter day today in the fields and all of us who look deeply into our hearts know that the difference is entirely because of our union—*the first union that ever cared!*

But more than that we need to remember that there was something which made that *union*, for unions are not pieces of merchandise to be bought at a corner store. Our union was born out of our common suffering, our common hopes for our children, and our common love for each other.

Brothers and sisters, that love is still strong in our hearts. We think it is still strong in your hearts. We all must know that to let outsiders come in and destroy that love we have for each other is to destroy what we can make tomorrow mean for our children and our loved ones.

We came as far as we are today through sticking together. We will go even further tomorrow if we remember that under everything else our strength is our love and respect for each other.

WILLIAM LUVAAS

*William Luvaas was born in 1945 in Eugene, Oregon. He has published several novels,
including* The Seductions of Natalie Bach *(1986) and* Going Under *(1994), and one
short story collection,* A Working Man's Apocrypha *(2007). His writing has won many
awards, including a National Endowment for the Arts Fellowship in 2006. He teaches
creative writing at San Diego State University and lives in Hemet, California.*

Luvaas's forthcoming novel, Welcome to Saint Angel, *from which the following
excerpt is taken, deals with development run amuck in one of the fringe communities
that borders the California desert and the urban Inland Empire.*

from *Welcome to Saint Angel*

Those of us foolish or spendthrift enough to grow a lawn out here meet once a
month at Ches Noonan's place in Two Horse Flats (used to). Truth be told, I
stopped growing mine years ago. Ches's house low and labyrinthine cavernous,
half an air conditioned acre, hallways spoking off the entry hub like a hotel
complex, cathedral ceilings in rooms fronting the pool, walls of glass. Cactus
garden out front the only desert friendly touch—and Penny Noonan's all season
tan. Ornamental gravels in pie-shaped wedges around the cactus garden, color
coordinated, seems like, by the folks who did St. Angel Family Mortuary. One
and a half acres of grass with desert looming all sides, like a plantation set
plunk in the middle of Arabia. Or a movie set. Like a studio came out and laid
turf over the sand. Folks passing on the county highway plow tire tracks in
hot asphalt slowing to look: sprawling ranch house, Olympic-size pool, palm
trees...all that grass. Ireland gone west. What my ex-wife Sondra would call
a conundrum. Out here conundrum is our way of life. Life and death forever
rubbing against each other. In March, life pulls a little ahead, vernal pools fill
up and we get the wildflower bloom. Death catches up by mid-May when we
get our first triple digit days.

"What we do is we pump water up from deep earth and spray it on the
ground," Ches explained at our last Christmas party. "What don't evaporate
trickles into the sand. You can hear grass roots sipping for all they're worth."

"What's it sound like?" Esther Johnson asked. Esther's always been a bit
goofy.

"Why, like a kid slurping with a straw at bottom of a glass."

Sam Jenson scowled. "Bull crap. I been here fifty-six years, I never heard no
slurping."

"You don't have any grass roots to slurp, Sam."

Sam worked a disdainful toe into Ches's thick turf. "Hell no I don't. I'm water wise and desert smart. You won't catch me spraying no water or evaporating it off my swimming pool neither."

"Some of us have dreams," Esther said primly.

"No damn snowbird is got nothing to teach me."

"How can I be a snowbird when I live here year round, Sam?"

"Up here." Sam tapped his head and turned to Ches. "I'll tell you something else, Mr. King of God's Creation, they dug a well over to Hungry Man Canyon deep enough they hit peterfied wood. That's a lesson ri'chere: redwoods drank too much water. Look where it got us. Waste not want not. You and your upnorther ideas is a blot on the desert."

"Doesn't seem to stop you from enjoying my pool."

"Our pool! You fill 'er up with water belongs as much to me as it does to you."

Ches got a hoot out of that. He called to Penny, entertaining a group of Silk-Setters across pool. "Sam says we're running a public swimming pool here."

"I do believe it." Penny frowned at the motley crew of us surrounding her husband. Ches smoothed down the divot Sam had taken in his lawn with the toe of a huge shoe. It's a grass growing principle out here: entropy makes the slightest inroad, you have gophers, bushy-tailed ground squirrels, red ants, moth larvae, drought—God knows what all—trashing your lawn.

We began as the DESERT LAWN GROWING ASSOCIATION: DLGA. But the Desert Links Golf Association threatened to sue us for appropriating their acronym. So we switched to DESERT GRASS GROWING ASSOCIATION: DGGA—until long hairs started leaving petitions to legalize marijuana at our county fair booth, and Nora Proudhon, some of them, threatened to quit, fearing people would mistake us for a pot grower's association. So we changed again to DESERT GREEN LAWN ASSOCIATION: DGLA.

From the start, folks at DGLA saw me as their mascot. Me and Sage Littlefeather (representing the *La Cienega del Diablo* nation), among the few members whose income is under six figures. A mascot, as everyone knows, is what its team condescends to but doesn't wish to be. I had a lawn at first. Kept it green for a time, until grass woke me up mornings with its screaming, anticipating the day's heat. There's some can't hear it. Dogs and black birds can. I've seen gophers emerge from their burrows at height of the racket and scamper off down the county road. When members protested I was setting a poor example, I argued that grass still remains grass in its dry state. "Half the hills in California are covered in sere grass. I'm holding up tradition." Members voted to keep me in. Littlefeather and them suggested I paint my brown grass green, like they do in modular estates over in Hemet. No way, Jose!

With Sage, Hailey Sahlstrom, Sam Jenson, Rob and Daphne Thompson, I founded the Honesty Faction of the DGLA (later known as the Dirt Faction) and challenged Ches for president. "You notice they're all leftover hippies, Indians and desert rats," I overheard John Sylvester complain one meeting. I walked over and asked which category I fit into. "Goodness!" Clover Abernathy said, a hand pinned to chest of her flowery yellow blouse as if rooted there. "Why, I'd call you our resident hippie hanger-on, Sharpe." Sylvester smugged his chin.

"That's better than what my ex used to call me."

Ches laughed. "What was that?" Large man, Ches, a peculiar pink diamond-shaped wart tip of his nose. Huge feet: 16 triple D. I know shoes, once worked as a salesman at Walksmart over in Escondido. Sondra called it my "foot period"—previous to my "armpit period." Body basic employment.

"You don't want to know, Ches, and I sure as hell don't want to tell you."

"Goodness," Clover said.

I lost the presidency, but won a free lifetime membership for running.

Nobody knows what all Ches does for a living. There's real estate development, dot-com stuff, people say. "He's in shifty money," Sam Jenson believed. "What people make without making nothing at all." Ches sits on the Municipal Water Board and down at Western Enterprise Bank, half the boards in the county. He was county supervisor, but gave it up once developments started going in. Stayed long enough to give developers the green light to hijack our way of life. All in all, Ches should rightly be over among the wealthy in Palm Desert. But he prefers it here in St. Angel among "real people." We've got those in ready supply. Desert doesn't give you much choice but be real. Dirt grit sweaty.

Sam Jenson was the true local item. His yard dry and dead as long as anyone can remember, even his yucca and prickly pear looked thirsty. So Sam should've been the natural leader of our insurrection. But there's no sane person on earth would follow Sam. Come winter, he twined Christmas lights in among dried honeysuckle vines on the cyclone fence out front of his place; nights, it looked like honeysuckle had returned to life. He didn't believe in lawns but came to meetings each month to keep us honest. Regular *rattus aridus* of the true variety. His skin brown as a paper sack and lizard coarse. Sam wore the same filthy khaki shorts and long-sleeved shirt winter and summer. Ches would remind him to shower before going in the pool, believing Sam came to rinse off: "When does he ever see water?" But I knew Sam came for free beer. He gave up on water years back. Hardcore realist, Sam.

His vintage Airstream trailer parked on its lonesome out Yucca Road, surrounded by a tumbledown chainlink fence, half oxidized and turned to powder in the sun. What Tinkerspoon dubbed "the finest Chevy junkyard in Southern California." His place lit up like the Times Square Christmas tree Thanksgiving to

Easter. "Nine-hundred-seventy-six bulbs in six colors, half blinking," Sam would tell you. Not including sprays of icicle bulbs hung from Tamarisk, fence, rusted car chassis—anything that drooped or ledged. Half Sam's Social Security check went for replacing bulbs. Ches aligned his Cessna Skylane to Sam's place flying home nights from meetings in L.A. Penny Noonan once told me I looked to Sam as a role model. Nonsense. Only a lunatic lizard could see Sam as a role model.

If Sam was the self-appointed representative of sand and creosote bush—no doubt Death sent him over as its direct representative to the DGLA—his opposite would be Sage Littlefeather. Closest we have to a pure life force in the valley. Besides Finley. True son of Saint Angel. "Sam's problem is he's standing upside down," Sage once declared. "Your desert world is turned all topsy turvy, like an upside down lake, except the bottom is dry. Dig down and you find slimy green bikini "—she'll open her mouth and swallow it." Not hard to imagine, given sand drifted half way up a retaining wall east side of Ches's place. No element is less impressed by what we do to contain it than sand. Unless it's water.

Point is, all seemed harmless jesting at the time. Who could imagine what it would grow into? Or imagine how far it would go? The weather was still quite reliable then. You could look around and say: All this makes perfect sense, folks making the best of a harsh climate. Could smile at Penny and think: the world is a fine and neighborly place, I have no claim on her nudity.

SHEROD SANTOS

Sherod Santos was born in Greenville, South Carolina, in 1948. He is the author of five books of poetry, including The Pilot Star Elegies, *published in 1999, which won a Theodore Roethke Memorial Poetry Prize and was a finalist for the National Book Award. Santos's poems and essays have appeared in numerous national magazines and journals, and he has been awarded several major fellowships. He is a professor of English at the University of Missouri, Columbia.*

This poem, "Near the Desert Test Sites," is taken from Santos's poetry collection The Southern Reaches *(1989).*

Near the Desert Test Sites

(Palm Desert, California)
for Logan and Renée Jenkins

Unlike almost everything
Else just surviving here
In summer, poison flowers
Flourish in this sweltering
Heat, tangling like blown
Litter in fences around
The trailer parks and motel
Pools, and turning the islands
Pinkish-white between
Divided lanes of freeway,
Where all day long against
The burnished hubbub of U-
Haul trucks and automobiles,
Off-the-road vehicles and
Campers, the oleander shakes
Its brightly polished pocket-
Knives, as at the motorcade
Of some ambassador hurrying
Through a village of the poor.
And every day by late after-
Noon the overwatered lawns
Around the shopping mall

Still burn off brown, their
Pampered opulence upbraided
By the palms' insomniac
Vision of one ineffable apoc-
Alyptic noon. But the smell
Is somehow sweeter than
That makes you think, a dry
Lemon-sweetness, as if some-
Where nearby wild verbena
Has been forced to leaf
By a match held up to each
Bud—and the silo-skyscraper
Holiday Inn at the famous
Resort "Where the Horizon
Ends" could almost be that
Match the way the heat
Sloughs off it like after-
Burn. And yet, because
Of the way the sun in-
Tensifies everything, one
Always has the feeling there
Is much less here than meets
The eye: the halcyon blink
Of a shard of glass, a Lear-
Jet wafted into vapor out
On the tarmac's run, the way
Common quartzstone gives
Off heat which seems to come
From inside itself, and not,
In fact, from that more-
Than-imaginably-nuclear sun
Which every morning starts
Up so illusionless, and every
Evening slow-dissolves
On the blue and otherwise
Planetary hills, like a Valium
Breaking up on the tongue.

REBECCA SOLNIT

Rebecca Solnit was born in 1961 and lives in San Francisco. An environmental and political writer and essayist, she is the author of numerous books, including River of Shadows: Eadweard Muybridge and the Technological Wild West *(2003), which won the National Book Critics Circle Award in Criticism. Her book* Savage Dreams: A Journey into the Landscape Wars of the American West *(1994), examines the history of nuclear testing in the northernmost reaches of the Mojave Desert, where the U.S. government has tested bombs, above and below ground, since the 1940s.*

In the following excerpt from Savage Dreams, *Solnit describes her participation in a 1994 antinuclear demonstration at the Yucca Mountain test site.*

from *Savage Dreams*

My second year at the Test Site I went in with a bunch of anarchist women from San Francisco and Seattle. Two of the northerners became friends of mine later, but I didn't know any of them well at the time. We'd agreed that we would pair off so that no one got abandoned or left at the guards' mercy without a witness, and then we'd hiked northwest up 95 about a mile north of the main gate, so that we'd have time to cover some ground before we were interfered with. I'm not sure what our purpose was—curiosity?—but my own desire was always to walk as long as possible across the land that was off limits. "Reclaim the Test Site," the big American Peace Test action of spring 1988 had been called. Walking claims land not by circumscribing it and fencing it off as property but by moving across it in a line that however long or short connects it to the larger journey of one's life, the surrounding roads and trails, that makes it part of the web of experience, confirmed by every foot that touches the earth.

Actually, that spring afternoon in 1989 the dozen other women and I only got about a quarter mile in, walking in a gully that made it hard to see us from the land, before the helicopters found us, swooping low overhead with men in paramilitary uniforms leaning out ready to jump. If we were conducting our war as a picnic meander, they were conducting their job as a military maneuver. But when the hovering copter got low enough to pelt us with gravel spat from the ground by its gust of air, we ran, and the men leapt out and ran after us. I ran madly in the bad footing of the desert, with its soft patches of sand and crusted-over dust, cobbled stretches, boulders, loose rocks, and low bushes, only slowing down enough to keep pace with the woman with whom I'd paired off. The anarchists were all wearing vivid colors, and I in my dusty khaki regretted that we were so visible. I wondered this time, as I did so many others, whether I could disappear from view if I walked by myself, but solitude was discouraged here—it could be dangerous.

I ran for a ways without looking back, and then I turned my head a little and saw a man in camouflage all but close enough to grab me, far closer than I expected. He must have decided to join another chase, because it seems unlikely that I actually outran him. And running was one of the things that we usually agreed not to do, as it wasn't in keeping with the spirit of nonviolent direct action. Urgent, unpredictable, quick actions threw the security forces into a panic, made it possible for things to go astray.

I gave up easily, letting them handcuff my hands behind my back, but my companion resisted, letting the two guards know why she was here and by what laws she had the right to be here. She cited the fact that the land was stolen from the Western Shoshone in the first place, and that we had permission from them to be here, that she was following the Nuremberg Principles they were violating. Now I can't even remember which of the women she was, only the unwavering conviction with which she refused to cooperate. She refused to walk, too, and so they herded the two of us into another gully and handcuffed us ankle to ankle. One stood guard over us while the other went for reinforcements. The other women were no longer visible. Picture an immensity of flatness populated only by two immobilized women and two men in camouflage, one of whom was rapidly disappearing. There was nothing to say. The Test Site looked exactly like the landscape outside, though we were now unable to stand up in it because of our shackles.

The second guard came back with a third man. While one guard walked behind me to make sure that I didn't attempt to flee, the other two picked her up, each taking one arm and one leg, and carried her. We progressed a couple of hundred yards in this way, when an older, red-faced guard joined our group of five. He snarled at the guards not to indulge her by carrying her. First he got them to drag her by her arms, then he got them to stop going around the obstacles. They began to drag her through thornbushes and over cacti.

He had convinced them to engage in a mild form of torture, and it didn't seem to have occurred to any of them that they could refuse his orders, though it was this kind of mindless obedience that the Nuremberg Principles she cited were made to combat. Finally she gave up and, near tears, asked them to stop. She began to walk, so she wouldn't be dragged. We walked to the dirt road that ran parallel to the Test Site periphery, where a big van was waiting for us, along with several of the other women in our group. The van was there to take us to the huge holding pens the Department of Energy had built a year or so before, next to the main gate. The guards cut off the plastic handcuffs we were bound with and rehandcuffed us with our hands in front, letting the cut pairs lie where they fell. My companion offered me a drink of water from the bota bag they hadn't confiscated, then she took off her hiking boot with awkward double-handed gestures and took out her Swiss army knife. I pulled out as many of the thorns in her foot as I could with the knife's tweezers. Some of them were huge, and one long one broke off deep in her foot.[...]

That year, 1989, the year of the cactus thorns and of Suleimenov's statement, law-enforcement officials arrested 1,090 people for trespassing in one fell swoop, unloaded us into the special cattle pen they'd built for us, left us there with a canister of water and a portable toilet for the afternoon, then loaded us onto buses. They used the same buses for us that they used to transport the workers from their homes in Las Vegas to Mercury, air-conditioned coaches with tinted windows, reclinable seats, even toilets. It was a peculiar experience, sitting on the soft upholstery provided by the Department of Energy, watching the scenery roll by at sixty or seventy miles an hour on the way up 95. My first year there, 1988, they'd taken all 1,200 of us nearly 200 miles north to the remote town of Tonopah, and I had worried that they wouldn't take us that far unless they planned on hanging onto us for a while. It was a long enough ride, on that strange road as the sun set, to imagine many things. But every year they just hauled us north to inconvenience us, unloaded us a few buses at a time, snipped off our cuffs, and told us to go away and not come back. (Usually we were dumped in Beatty—"gateway to Death Valley"—a former mining boomtown that had restaurants with fingerbowls and tuxedoed waiters in 1906, but was more of a corn-dog kind of place by my time, and was the town where the Princesses had been held.) By the end of the ten-day event in 1988, 2,000 people had been arrested from among the 5,000 participants—and no charges were pressed in the vast majority of the cases. It was one of the biggest civil disobedience arrests in U.S. history, and it barely made the local news.

In 1988, the nuclear bombs exploded at the Test Site were named Kernville, Abeline, Schellbourne, Laredo, Comstock, Rhyolite, Nightingale, Alamo, Kearsage, Bullfrog, Dahlhart, and Misty Echo. Most of them ranged from 20 to 150 kilotons (Hiroshima was laid waste with 15 kilotons, Nagasaki with 21), as did 1989's bombs: Texarkana, Kawich, Ingot, Palisade, Tulia, Contact, Amarillo, Disko Elm, Hornitos, Muleshoe, Barnwell, and Whiteface. They didn't make the news either.[...]

There's something profoundly American about getting arrested at the

Nevada Test Site: The very issues are, not cowboys and Indians, but land, war technology, apocalypse, Thoreauvian civil disobedience, bureaucratic obscurity, and Indians, part of the great gory mess of how we will occupy this country, whose questions are as unsettling as its land is unsettled. Then, of course, after being unhandcuffed and thrown out, the obvious thing to do is to celebrate, which in Beatty means going to one of the diner-cum-casinos for drinks and American food. To start the day in the deadly cold of a desert morning, sitting on rocks and drinking coffee, to fill one's water bottle and mill around with friends and acquaintances as the day gradually creeps toward hotness, to sit through a sometimes stirring and often dull rally of speeches and music (folk to punk and back again), to commit the fairly abstract act of climbing under a wire fence that separates the rocky expanse of cactus and creosote bushes from the rocky expanse of creosote bushes and cactus, to be confronted by hired help in the wrong-colored camouflage (as though they, not we, had a use for stealth), to go through numerous pairs of disposable plastic handcuffs as we captives are rearranged, to idle in a sort of cattle pen built just for us, to be escorted after many hours in the sun into a special luxury bus and be given a tour of scenic Highway 95, to be interrogated by hard-faced sheriffettes with piles of teased hair who are irritated by anyone who wants to give a more complicated name than Jane Doe or Shoshone Guest, to be tossed out into a small town, to catch up on one's friends' well-being and head for fast food and ice cream in the middle of the night, to plunk quarters into slot machines while waiting for the food to come, winning the occasional handful of change, to retrace the pointless route as the liberated activists get driven back to the camp, to wander back through the rocks and thorns in the dark to a sleeping bag on hard, uneven ground under a sky more full of stars than almost anyplace else in the world—could anything be more redolent of life, liberty, and the pursuit of happiness?

JULIE SOPHIA PAEGLE

Julie Sophia Paegle was born was born in 1971 in Salt Lake City, Utah, and since 2006 has taught poetry and creative writing at California State University, San Bernardino. A prolific and prize-winning poet, her writing has appeared in many academic journals, and she is currently at work on a long poem series, "Juniper Complex Fire," which focuses on the ecosystems, species, droughts, and fires of the California deserts—in particular, on a major fire event in Joshua Tree National Park in 1999. The following poem, "What the Fire Forgot," is taken from that series.

What the Fire Forgot

(Juniper Complex Fire, Joshua Tree, 1999, Yucca brevifolia)

> Usually requiring pollination by the yucca moth, the Joshua tree is also capable of sprouting from roots and branches, which allows a quicker recovery after damaging flood or fire.
> —National Park Service

First, ions strike, then a blistering gleam
      grows—no glows—in sparks and darts, over stitch, then seam

flares from where lightning ran down itself to earth,
      then the loss of many lambent valleys' worth

of once large lilies having poured to smolder most
      of their bodies. All those trunks—once host

to night lizards, diamondbacks, the desert kit fox.
      Some trunks formed solitary columned stalks;

others fixed a frenzy of branches, each branch a ghost
      wandering between a hard winter frost

and spring flowering. And all those branches—once rests
      for pallid bats, occasional rain, frequent orioles' nests,

the very years, each reclining within each
      branch's half-inch; and above it all, the variegated reach

westward settlers named as if to summon company
     of old tribes for arrayed imploration skyward: Joshua Tree.

Some eddies resolve; the colder smoke now mottled by yucca moths
     returning as if to settle, with several pollen-sodden deaths,

for young last seen seeding ovaries of a colossal monocot.
     But their traffic won't fan recovery as fast as what

the fire forgot: many valleys' worth of roots buried beneath one stretch
     of still glowing rain shadowed loam. Each root is parched,

parted and startled, morphing sand-ward, slow as stone
     cools is each root moving, severed, airless and alone.

KATHERINE SIVA SAUVEL
and ERIC ELLIOTT

Katherine Siva Sauvel (also spelled "Saubel") was born on the Los Coyotes Reservation in 1920 and grew up in Palm Springs. A Cahuilla elder, she is cofounder and president of the Malki Museum, a museum of Cahuilla culture on the Morongo Reservation (near Banning, California), where she lives. During her lifetime, she has made numerous contributions to the understanding and preservation of Cahuilla culture. She has served on the California Native American Heritage Commission, been honored by the Smithsonian Institution and the National Museum of the American Indian, and written several books on Cahuilla culture and language, including Temalpakh (from the Earth): Cahuilla Indian Knowledge and Usage of Plants, *coauthored with Lowell John Bean (1972) and* 'Isill Héqwas Wáxish: A Dried Coyote's Tail *(2004), coauthored with Eric Elliott. (For more information about Eric Elliott, see "Massacre at Desert Hot Springs" in this anthology.)*

In this excerpt from 'Isill Héqwas Wáxish, *Sauvel describes the removal of a barrel cactus garden near Palm Springs to make way for windmills.*

from *'Isill Héqwas Wáxish*

Problems with Windmills

'Et pé' 'íyaxwe' yéwi pé' pé' pén 'áya' chemkí'iw'a' chémem métechem 'énga'
 pé' Séxnga' qáltem, pén 'ípa' 'úmu'.
This is the way it was long ago with the wild plants which we, the
 residents of Palm Springs, used to gather.

Píka' pé' pemyíkawwenelu'.
They used to gather them there.

Memyíkawwe' hísh te té'iy, kúpashmi'.
They used to gather those things, those barrel cactus blossoms.

'Enga' pé' yéwi pé' Mélkish petétewanqa'. Táxswet téwan'a' míyaxwe'.
The white man had a name for that place. The Cahuilla had a name for it.

Né' kíll pensichúminqa. Mélkish petétewanqa Devil's Garden.
I don't remember it. The white man refers to it as 'Devil's Garden'.

Yáqa ku pé': písh petétewanqa Mélkish.
That's what the white man calls it.

303

'Enga' pé' mèetewet wélqa' pé' téwe' kúpachem hémeqi'.
Nothing but barrel cactus used to grow over there.

Péngax pé' pé' pén píka' chemhíchiwenelu'. Michemyíkawwe'
  kúpashmi'.
We used to go over there. We used to gather barrel cactus blossoms.

Méten pé' michemhívinwe'.
We would pick a lot of them.

Pé' pén támiva' pé' 'íyaxwe' neyyíkawmaxish pé' meséxqa'.
And then in the winter my mother, God rest her soul, would boil them.

'Ay mewáxniqa'.
First she would dry them.

Pé'iy 'áchakwe' mewáxniqa'. Pén háni' 'áy mewénqa' ku pé'.
She would dry them thoroughly. And then she would store them away.

'Ay hétex. Pén 'áy máwa' 'áy támiva' 'áy pé'iy pánga' pé' pechúpinqa'.
She'd store them. And then later in the winter she would soak them
    in water.

'Angapa' pé' 'áy 'ángapa' nánvayaxqa'. Pé' michemqwáwe' pé'emi'.
And then they were ready (to eat) again. And we would eat them.

Pé' pé' 'íyaxwe' pé'. Túku' 'áy Mélkish pá' píshqa'.
That's the way it was. And then the white man got there.

Pé'iy mán 'íka' 'úmu' pehóoqqa' wám. Qamíva' pewén'i' wám.
He pulled all of them up. Who knows where he put them.

Yáqa' pé' pé', "Táxstem menmáxik," yáqa'.
He said, "I am going to give them to the Indians," he said.

Né' 'ípika' né'iy níyik pemvukséqaywe'.
They called me here.

Pé' mán hémyaxwe', mán Táxstem písh pem'áyawwenive' pé'iy kúpashmi', píka' písh memyáwichipi' hémkika'.
They asked whether Indians wanted those barrel cactus, (they said) that they were going to bring them to their homes.

"Hèé," níyaqa'. "'Ipika' mík pék haxcheme'e'máxne'," níyaqa' né'.
"Yes," I said. "You all can bring some over here to us," I said.

"Háni' píka' támika' me'máxne'."
"You can also take some to the east (at Torres-Martínez)."

"Pénga' pé' wélqa 'áchakwe'," níyaqa', "támika' Tóoro pén Martíneznga'."
"It grows well," I said, "over there in the east at Toro and Martínez."

CONSERVING AND PROTECTING
THE LAND

REBECCA K. O'CONNOR

Rebecca K. O'Connor was born in 1971 in Riverside, California. She has published eight books, including a romance novel, Falcon's Return *(2002), and* A Parrot for Life, *a guide to raising and training parrots (2007). Her falconry memoir,* Lift, *is forthcoming from Red Hen Press. O'Connor lives near Sacramento, California, where she is director of development for Ducks Unlimited, a wetlands and waterfowl conservation organization.*

The following essay, "Postcard from Above," shows how encroaching development in the southern California desert region threatens wildlife and open space and, in this case, the art of falconry. It was first published in West *magazine in 2006.*

Postcard from Above

Easing my hooded peregrine onto his perch in the back of the truck, I pretend I don't see the man approaching from the east. He is moving with an intent that makes my palms sweat, despite my focus on tying the falcon's leash, despite the cool morning air. I am on farmland without permission, and the man's brisk chin-up-shoulders-back walk tells me the reservoir from which my falcon had just caught a lesser scaup is his. The duck in my vest and the protruding chest of the well-fed peregrine are evidence that we have ignored the sign that says "No Hunting or Fishing." I am about to be kicked off my favorite field.

The man isn't smiling as he approaches. He holds up a hand when I start to apologize. "I'm Mark Draper," he says. "I lease this land." I stutter to interrupt, but he hushes me again. "I'm not going to kick you off. The guys tell me that you've been flying your bird here for three years. I just wanted to introduce myself."

"I would shake your hand, but…" I hold up my hands, smiling. They must look a bit like farmer's hands, imbued with earth and a touch of blood, a little too soiled for handshakes. "Thank you, Mr. Draper. There are so few places that are open enough to fly my falcon, and I love this place."

We stand for a moment to admire it, a simple sod farm in Thermal, southeast of Indio and close enough to the Salton Sea that the breeze leaves the taste of saline on your lips. There's desert to the north, and date palms to the south and west. I sometimes stop to watch the sun rise in a rose glow through the short palms, while the Gambel's quail scuttle from their breakfast of fallen dates to the safety of the scrub. I savor the hour it takes me to get to this reservoir that irrigates the sod, because the best part of the morning is discovering what species of which waterfowl happen to be migrating through, passing to and from the Salton Sea. There is no reason for me to tell Mark Draper that this is paradise.

I live an hour away in Banning, but drive to the fringes of the Salton Sea to fly my peregrine during the falconry season, the cool months of October through February. I've been a falconer for 10 years and have watched the vast vineyards of Ontario shrink to nothing or give birth to concrete structures from their sandy soil. The flats in Hemet and Temecula have given way to suburbia as well. And the field near my house where I trained my falcon last year is now a Wal-Mart Supercenter Store. Coachella Valley is my wilderness, the only place left with enough open space to satiate the peregrine's tremendous appetite for unadulterated horizon. I just don't know how much longer the landscape will last, or where the ducks we hunt will stop for food when it's gone.

Three years ago a friend of mine took me up in his Beechcraft from the airport in Thermal. My falcon was 6 months old and stubbornly flying off on a daily basis to hunt the farmlands without me. I would track his transmitter and find him 10 miles away on a pole, hungry but still scanning the sky. I wondered what he saw that was inspiring enough to power his wings until he was too exhausted to look for a red-headed girl waving from the ground below. I imagined that the possibilities of the landscape and its bounty were buzzing in his head, too much to take in to decide where to hunt. I don't care for the bump and roar of small planes, but I had to see the falcon's view for myself.

It was astounding, this tremendous expanse of green checkerboard agriculture garishly stretching through the subdued hues of the desert. Every color change was marked by a blue pond tucked in a corner or hidden in the center of a plot of artichokes, stretch of cilantro, carrot patch or table-grape vineyard. I'd seen only a tiny portion from the ground on the days that I followed the waxing and waning signal, the bread-crumb beeps that indicated the falcon's path. From the sky I finally understood his winged excursions. I could see from above why the larks liven this land with their flashes of yellow breast, why the mallards stop to dip their heads in the sparkling water, and why the jackrabbits venture through the budding vineyards. I didn't want to come down, but like the falcon I had to land eventually.

These days my falcon rarely indulges in capricious flights across the farmland, and I'm grateful but concerned. I don't want to imagine the changing aerial view—or the implications. I don't want to wonder where the waterfowl will go when ponds become swimming pools. I would rather pretend that Mark Draper's sod farm will remain our sanctuary.

Next to the pond, Draper peers at the falcon and the Brittany puppy in the back of my truck. I explain that this is the peregrine's third season and the puppy's first. The dog swims to encourage the ducks' flight off the water, and sometimes the peregrine catches one. I tell Mark that I have to follow the falconer permit regulations, carry a duck stamp and the same license as any hunter with a shotgun. I roll my eyes in case he doesn't realize how ridiculous that is. The falcon can catch only one duck at a time, and once he has eaten his fill I take the rest home. I freeze what's left for his summer meals, when the season is over and the waterfowl have abandoned the heat of the desert for cool northern climates.

I promise Draper that I will always clean up, never walk on the sod when there's a hard frost and stay out of the farm laborers' way. He waves his hand at me, dismissing my worries. "You're welcome here. Enjoy it while it lasts."

I wish I didn't know what he meant. But he suspects that the family that owns the land he's leasing may have sold a large plot a few blocks over. The falcon and I have lost four Coachella Valley ponds in the last year. They became three housing tracts and a golf course. I know that many farmers in Thermal own their land, and that property values are rising. It won't be long until the acreage is sold to make way for more homes. Farming in California is expensive and depends on the vagaries of weather. Even those who own the land they farm would be crazy not to sell and move to where farming is more affordable.

I don't want to lose the land where my falcon flies, but I have to admit that I would buy a house in Thermal. I have to admit that as much fist-shaking as I have done at the Wal-Mart down the street, I'll shop there. It's convenient. It's close. I like progress. I love California. I just don't know where my falcon is going to hunt 10 years from now.

WILLIAM STAFFORD

William Stafford was born in Hutchinson, Kansas, in 1914. During the Second World War, he was a conscientious objector and worked in civilian public service camps, an experience he recorded in the memoir Down in My Heart, *published in 1947. He received his PhD from the University of Iowa in 1954 and taught at Lewis and Clark College in Oregon until his retirement in 1980. His first major collection of poems,* Traveling through the Dark, *won the National Book Award for Poetry in 1963. After that he went on to publish more than sixty-five volumes of poetry and prose. He died in 1993.*

The following poem, "At the Bomb Testing Site," is taken from The Way It Is: New and Selected Poems *(1999), and depicts a lizard's view of 1950s-era nuclear bomb testing in the Mojave Desert.*

At the Bomb Testing Site

At noon in the desert a panting lizard
Waited for history, its elbows tense,
Watching the curve of a particular road,

As if something might happen.

It was looking at something farther off
than people could see, an important scene
acted in stone for little selves
at the flute end of consequences.

There was just a continent without much on it
under a sky that never cared less.
Ready for a change, the elbows waited.
The hands gripped hard on the desert.

EDYTHE HAENDEL SCHWARTZ

Edythe Haendel Schwartz was born and raised in New York and has spent most of her adult life in California. After retiring from her teaching career on the faculty of the Department of Child Development at California State University, Sacramento, she began to write and paint. Her poems and reviews have appeared in many journals and anthologies, and her chapbook, Exposure, *was a finalist in Finishing Line Press's New Women's Voices competition in 2007. She lives in Davis, California.*

The poem "Long Odds" was written after the author observed a desert tortoise digging a nest in the Anza-Borrego Desert State Park.

Long Odds

> It is moving its slow thighs, while all about it
> Reel shadows of the indignant desert birds.
> —William Butler Yeats

Where cholla meets Mojave aster,
stems raveled by hikers she ignores,
the tortoise shovels
a shallow basin—slow
work before she knows the hole will hold
what she has come to leave.

She settles in silence and starts to lay.
White orbs, wet and ripe, drop, pile
on each other like ping pong balls,
progeny curled in paper shells,
cells throbbing within. Empty,
she pushes sand over the nest, smooths it
with broad front paws and rests;
her eyes glaze over, seem not to see.

Some genetic map instructing, she turns
toward her burrow, escapes the task,
escapes the scent of rats, coyotes, skunks
who lurk to lick her eggs dry, their odds
no worse than ours—

313

SUSAN ZWINGER

Susan Zwinger was born in 1947 in Colorado and now lives on Whidbey Island, in Washington State. She is the author of several books focusing on environmental and cultural history, including Stalking the Ice Dragon: An Alaskan Journey *(1992) and* The Last Wild Edge: One Woman's Journey from the Arctic Circle to the Olympic Rain Forest *(1999). Her book* Still Wild, Always Wild: A Journey into the Desert Wilderness of California *(1997) highlights the landmark 1994 California Desert Protection Act.*

The following excerpt is from the chapter "With the Eyes of an Eagle: The California Desert Protection Act," in Still Wild, Always Wild.

from *Still Wild, Always Wild*

Somewhere out there is a land grand enough to hold all life's passions and contradictions, to expand one's soul with vast possibilities, to awaken all one's senses.

To find it, search north from the Mexican border up through the California desert via backroads and wilderness areas, through the lost and forgotten and least-explored outback. Find those places least written about. Ask locals. Be adventurous. Expand into infinity in a desert night. Take on the night vision of an elf owl, the radar of a leaf-nosed bat, the ears of a kit fox, the vibration sensors of a Gambel's quail. Turn off the radio. Turn off the stereo. Get out of the vehicle and walk. Gaze at the desert with the eyes of an eagle, then with the eyes of a snake. Sit for long periods of time in silence, listening. After a while, the desert will speak to you.

Time in the great deserts of the West is very different from city time. Because the earth's crust is sliced open, and so many objects from past history lie preserved in the dry, clean air, desert time is like layers of clear plate glass. You can drop down through them, intersecting lives separated by centuries. In the desert, antiquity is interlayered with the present—as when ancient rock art lies within military testing grounds. Sitting near a wall painted with a giant shaman figure created two thousand years ago, you may think you hear a foot thud on the hardpacked desert floor and turn swiftly to see no one. You may imagine the shaman dressing as, then actually becoming, a god in order to call the rain. You may hear the ladder creak as he emerges from a sacred cave, having performed a healing ritual.

Or you may hear the hoof beats of a seventeenth-century padre gingerly riding from one hardship to the next, seeking native people to convert, inserting Old World Spanish saints into the native pantheon.

Or you may overhear a voice from a century ago, an old white prospector having a talk with his mule, demanding from the desert too much, too soon, and too easy.

Simultaneously, you may hear contemporary Chemehuevi people chanting clan songs that will continue their tradition down through the ages, or hear a researcher mumbling at his Global Positioning Satellite navigator as he tracks desert tortoise.

In the desert, you will meet animals possessed with people-wisdom. You will glimpse Coyote, the Trickster, darting in and out, through the edge of cities, sewing an urban civilization uneasily onto the desert. The daredevil jackrabbit will dash two inches in front of your car bumper, making bets with his cronies about how close he can come to death. He will always win. The horned lizard, all pimpled and lumpy, will lie passively in your hand, too stuffed with sun and insects to object.

In your imagination, you will find yourself conducting interviews with the men who hunted with atl-atl twenty-six thousand years ago over near Barstow. You will find yourself chattering across time, across language, and across cultural gaps with the Ancient Ones as if they stood in your kitchen. Which, at your campfire, they will. They will teach you the rich variety of uses for the plants and animals. You will find your sensory perception far lacking compared to their own. Yet the longer you remain in wild places and outside your vehicle, the more your hearing, smelling, touching, electromagnetic radar, and vibration sensors will return to you.

Nowhere else on earth is geologic time so exposed as in a desert. There, you will travel swiftly through geological eras as you pass through highly eroded mountains, bizarre volcanic necks and cones, endless salt playas, towering sand dunes in sensuous curves of pink and tan. You will pass through brick red and black dorsal fins jutting eight hundred feet up out of the earth as if stegosaurs cavorted just under the ground, past long black snakes of glassy lava one hundred feet high, through badlands as colorful as an artist's palette, through narrow, contorted canyons, through limestone caverns, and through the world's largest Joshua tree forest. Each profusely flowering cactus, each hidden desert waterfall and plunge pool, will open senses you barely knew you had.

The fragile California desert stretches 100 miles from east to west and 240 miles from north to south. For ten long years, hundreds of American citizens worked to protect its remarkable wild lands, which include three of the great deserts of the world—the Mojave, the lower tip of the Great Basin, and the northern wedge of the Sonoran. On October 31, 1994, President Clinton signed the California Desert Protection Act into law, protecting another 3.57 million acres of land as wilderness. The sixty-nine new wilderness areas are as different from one another as topography can be—from high alpine snow fields discharging roaring streams to dry, barren salt playas—and they will take a lifetime to explore.

The Desert Protection Act created two new national parks, Death Valley and Joshua Tree. At over 3.3 million acres, Death Valley is now the largest park in the lower forty-eight states. Joshua Tree was increased to 793,000 acres by including natural ecosystems which had previously extended beyond the former monument boundaries. The act also established the dramatic landscape between Interstates 15 and 40 as the Mojave National Preserve, the newest addition to the National Park Service. In addition, wilderness habitat adjacent to Anza-Borrego and Red Rock Canyon State Parks has been greatly expanded.

Despite the passage of the act, debate continues as ferociously as ever: Why save so much desert? We have long disparaged it as "wasteland," "barren," "badland," "bleak," and "evil." Early writers and explorers claimed that it was unlivable, undesirable, unprofitable, and that every organism in it was at war with every other. Among European white explorers, the desert provoked feelings of anxiety and hopelessness. They expected to be attacked by fang, claw, and thorn.

Names hold power. And we have named the desert wilderness Death Valley, the Devil's Playground, Skull Valley, Last Chance Mountains, and the Badlands. In naming, we shape the wisdom of future generations. Nowhere are our land-use ethics and symbolic understanding of the desert more clearly revealed than in our naming process of desert phenomena. As in war, we devalue our "enemies" so that we can slaughter them with a clean conscience.

In the same manner, those who would like the "freedom" to continue to destroy the desert have been disseminating a dangerous and calculated pack of lies through newspapers, talk shows, and cyberspace. Here are a few:

1. All the roads to the new wilderness areas are closed. (Truth: Thousands of miles of road remain open, and "cherry stems" into wilderness areas provide access.)

2. Only rich, elite environmentalists can enjoy the desert because they can afford to be dropped in by helicopter. (Truth: The desert, except for military ranges and small research areas, is open to all citizens and is one of the cheapest forms of entertainment available. Citizens who care about the environment come from all walks of life and all income levels.)

316

3. The federal government is destroying your freedom. (Truth: Federal and state environmental protection laws reflect the wishes of a majority of citizens to preserve valuable life forms and topography for future generations.)

4. The California Desert Protection Act and related laws are un-American. (Truth: One of the greatest inventions that Americans have offered the world is our wilderness and park systems and our other efforts to protect the shrinking, remaining wild places on earth.)

We need the Desert Protection Act because the California desert environment is in grave danger—a fact that is supported by research in many branches of science: geology, botany, biology, chemistry, meteorology, sociology, and anthropology to name a few. In California and across the globe, man has been stripping the desert of vegetation through grazing, mining, gravel pits, ever-expanding off-road-vehicle trails, and an unsustainable increase in human population. Moreover, the California desert's proximity to some of the world's largest and fastest-growing urban areas—Los Angeles and Las Vegas—causes the fragile balances of nature to be overwhelmed by modern vehicles, roadways, resource extraction, housing expansion, and recreation. Finite fossil groundwater is recklessly depleted by golf courses and elegant resort expansions at the expense of the very ecosystem that attracts the growing population. For instance, in spite of the fact that the groundwater beneath Anza-Borrego State Park is sinking by two feet a year, the valley of Borrego Springs continues to build swimming pools, golf courses, and elaborate resorts, and agriculture continues to increase.

The loss of desert wilderness has serious consequences for all of us. Humans have much to learn from observing life at the critical edge of survival: its marvelous adaptations to blowing sand, high altitude, or extreme heat may someday prove invaluable to humans. For instance, within the volatile oils which protect desert plants from dessication, we have found many blessings for mankind, from the medicinal to the mechanical. The California desert also supports many plant and animal species that live nowhere else on earth. These species have evolved unique strategies to survive in extreme conditions, and within their DNA strands may be keys that can help us adapt to extremely limited resources. For example, we already know that the genes of a humble dune grass can revitalize our cereal grain species. Because we grow in giant monocultures, the grains we now depend on are subject to waves of disease, such as the fungus that destroyed Ireland's potatoes in the nineteenth century.

Yet we cannot claim that the only reason desert species should be saved is their usefulness to mankind. We must assume that each one is absolutely necessary in its own being. Our humanity depends on the farsighted act of saving the unique life forms of the desert, life forms which occur nowhere else on the planet. Complete natural phenomena such as bird and butterfly migrations are currently being destroyed because the wilderness links between South, Central, and North America are broken. When we lose one songbird or one butterfly species, we lose part of ourselves.

Desert wilderness is our American heritage. By protecting it, we give a gift not only to our nation but to the entire world. It increases our self-worth as a country and is a sign of our security and wisdom. It speaks well of us as a society that we are wealthy enough in both spiritual and natural resources that we can choose to stop before we destroy the legacy of our remaining wild lands.

I sit now on this stone shelf scribbling into my notebook, watching light drain from the landscape. A bat jerks by in flight, a tiny speck, squeaking like an uncoiled valve. Elf owls draw silken flight vectors across the sky, and Gambel's quail cross so close to my foot that their claws are audible. Coyotes begin mad chattering in cliffs like human babies crying.

The sky depends to cobalt, then to eggplant. Cliffs sponge into black. Black clumps of rabbit brush appear to float above the desert floor. Something large and catlike ripples by on silent paws. Time in the desert is so fluid that it seems to flow backward as well as forward, carrying me along.

LAWRENCE HOGUE

Lawrence Hogue was born in Eureka, California, in 1961 and is the author of
All the Wild and Lonely Places: Journeys in a Desert Landscape, *published*
in 2000. He lives in San Diego and works for the Desert Protective Council as
communications coordinator and editor of the organization's quarterly newsletter,
"El Paisano." A desert environmentalist, Hogue is involved in ongoing efforts to
save the endangered peninsular bighorn sheep, which is native to low-elevation
desert slopes, canyons, and washes from the mountains near Palm Springs to Baja
California, a region that includes Anza-Borrego Desert State Park.

In the following excerpt from "Third Grove: A Battle for the Bighorn," a chapter
in All the Wild and Lonely Places, *Hogue joins others in a summer sheep count.*

from *All the Wild and Lonely Places*

It's Sunday morning. We've spent three days waiting for the sheep, and so
far they've failed to show. Now, with just a few hours left, we're beginning to
lose hope. It seems we've examined every rock and cactus and brittlebush on
the slopes across from us and up and down the canyon for a mile in either
direction. We've spotted numerous "rock sheep" and "cactus sheep," those
shapes in the distance that at first glance look like bighorns but turn out to be
something else. We've squinted through binoculars and spotting scope for so
long that we both have headaches behind the eyes, as well as aching necks and
backs from craning our necks into odd positions. But still no sheep.

I'm not convinced that they've left the area—maybe we haven't been looking
in the right spots or haven't been looking persistently enough, or maybe
neither of us would recognize a sheep at this distance if we saw it. Last year, I
spent a day of the sheep count with Esther Rubin, a graduate student from the
University of California, Davis, who was studying the bighorns. We saw ten
or eleven sheep that day, and Rubin saw every one of them first. After years
of scanning distant slopes for bighorns, her eyes had become accustomed to
the way a sheep looks when viewed through a spotting scope at distances of
as much as a mile. I wish that Rubin and her trained eyes were here with us
now.

After two days of blistering heat, I'm beginning to lose enthusiasm for the
whole project. People will tell you that once the thermometer climbs above
100, it all feels the same. They are wrong. I've baked as I've hiked in this oven
of a canyon in early June, but now the desert is turned up to broil. During the
hottest part of the day, the shade of the palms and alders and cottonwoods

below us beckons like Circe. Only an act of will keeps me from descending to the water hole and splashing in.

The first day was the hardest. By now, I imagine I'm getting somewhat adapted. I keep my spritz bottle going most of the day and drink a liter of water mixed with electrolytes every couple of hours. I don't eat much—the heat saps my appetite. Mostly I just sit, moving occasionally to hug the shifting shade, to stretch, or to scan the slopes on the sunny side of the rock.

The best time is the first hour, before the sun rises over the canyon wall at eight o'clock, hitting us like a hammer. In that cool morning light, everything is clearly illuminated—all the rocks look just like rocks, the cacti just like cacti. Then, for an hour after the sun rises, staring at the slopes east of us is useless; all detail dissolves in the glare. Later, with the sun rising higher, the cacti begin imitating sheep—or maybe it's the sun warping our minds.

At five o'clock, when we can finally descend from our perch, even the tepid, algae-clotted water is refreshing as it hits my skin. Pumped through a water filter, it tastes fresh and cool compared with the hot liquid left in my bottle. With only two inches of rain having fallen this year and two and a half last year, the pool is only a couple of feet deep, the lowest Carl has seen it in his years of coming here. So far, Jorgensen seems to be right in his prediction of a low sheep count, at least here at Third Grove.

Despite the heat and the lack of sheep, there are a few rewards for spending time at a desert water hole in July. In the pool itself, I've noticed a form of life I've never seen in cooler months: water beetles diving into the pool and a walking stick, which climbs onto my boot as I pump water. And yesterday, from up at our count site, I heard a bird singing like a meadowlark. The sound was coming from one of the palms, and soon I spotted a yellow-and-black shape flashing among the green fronds. It turned out to be a Scott's oriole, a summer visitor here from Mexico and new to me.

To while away the rest of the time, Carl regales me with stories. He tells me about clever lawyer tricks he's pulled during lawsuits, about Burns night and drinking Scotch with members of the local House of Scotland. Mostly he tells me about his avocation, drafting a Civil War history based on old journals and letters he's found as part of his lawyering trade. He spends a few hours each day at his office working on it (part of the luxury of having his own practice), though he has little hope of publishing the manuscript.

As Carl continues the story this morning, I take a break from the binoculars. I'll admit it: I'm demoralized now, and I no longer believe that scanning the slopes rock by rock will do any good. I close my eyes for a moment, open them, and look across the canyon. There, perched on a rock out on the edge of the ridge, silhouetted against the sky in the perfect picture-postcard pose, stands a young ram. He's looking right at us. I gesture to Carl, and he sees him, too. Carl trains his spotting scope on the ram and I go back to my binoculars as the ram heads down and across the slope to the right. He seems to be a yearling (about a year and a half old now because the lambs are born around February). Whispering, we compare notes, making sure this really is a ram, given that young males resemble mature ewes.

The ram pauses to kick at a barrel cactus; then Carl notices that he's staring across the canyon at the slopes above us. I creep out from behind our rock to see what he's looking at. Another ram is walking across the slope above us, heading up-canyon. With half-curl horns, he's a teenager among sheep. Carl climbs to a slightly different spot and sees another one, this time with the three-quarters curl of a mature ram. The two rams continue up the canyon, walking a little faster now, aware of our presence.

The first young ram decides to follow. But he won't go down into the palm grove directly below us. Instead, he angles back to the left, down, and across the slope to a spot where it becomes sheer. This doesn't faze him. He perches on tiny protuberances in the cliff face, stepping nimbly from one to another. At one spot, he leaps down and across ten feet or so, landing with hooves together on a small shelf. Finally, the cliff ends where the side stream comes in, and he has to descend into the grove. We lose sight of him as he heads upstream through the trees to meet his newfound friends. Meanwhile, they pause just before a bend in the canyon, posing on large boulders as if waiting for him. He joins them, and they all disappear around the corner.

That's it—three days of waiting for sheep and about twenty minutes of actually seeing them. We feel fortunate to have seen that band of sheep on our hike in. We continue scanning the slopes for another hour or so, hoping to see these three coming back or maybe others heading up-canyon. But soon, it's one o'clock and time to leave. We pack our remaining gear quickly, douse ourselves in water from the pool, and head out. We hurry from pool to pool, taking long rests in the shade.

Checking with the rangers, we learn that Jorgensen's prediction was right—the count is low this year. Later, however, when all the numbers are in, Jorgensen will report that the lamb-to-ewe ratio is better than expected. Things look bleak for the bighorn, but there is a glimmer of hope.

On my first sheep count, in 1996, Esther Rubin picked out a site above Big Spring on a ridge separating two forks of Tubb Canyon, looking down-canyon to the east. In previous years, counters at this site had positioned themselves too close to the spring and been skunked—the spring is so overgrown that sheep no longer use it. Rubin planned to concentrate on the slopes of the canyon below us, where the terrain was rocky and open, dotted with brittlebush, grasses, barrel cactus, and agave. Everything behind and above us was covered in a thick stand of sugar bush. We were parked just below the line between desert scrub and chaparral.

To Rubin, that line was like a wall the sheep wouldn't cross. "We don't need to waste our time looking up that way," she told me. "That's good mountain lion cover, and the sheep don't usually go in there." In addition to food and water, the sheep need two things: steep, rocky slopes, known as "escape terrain," and open spaces where they can see predators approaching from far off. We spotted fourteen sheep that day—all of them out on the rocky slopes below us. Jorgensen agrees that bighorns won't go into chaparral. The brush has encroached on what used to be sheep habitat, even since 1967, when he began studying bighorns. Then, he saw forty-two sheep in one day at Fourth Grove. "Now I go there and it doesn't even look like sheep habitat," he said. That change is reflected in the annual counts at that site—zero for most of the 1990s and two in 1997.

If you've followed my argument so far, you may already be putting two and two together. The gears of my brain work slower. But over the years since that sheep count, I've begun putting together my own little grand theory, drawing from the area's environmental history as told by Florence Shipek and from Esther Rubin's and Mark Jorgensen's knowledge of sheep behavior. In a nutshell, it is this: One cause of the decline of the bighorn sheep is our own leave-it-alone-and-put-out-the-fires management of these public lands. The Indians used to burn off the chaparral, *creating* habitat for the bighorn as well as for deer and small game. By failing to continue that management, by viewing the chaparral as what should "naturally" be here and trying to preserve it through fire suppression, we have actually reduced sheep habitat, pushing the sheep farther down toward the desert floor and further toward extinction. In some places, the spread of chaparral may have divided two overlapping groups of bighorns, fragmenting the population even more.

An obvious solution is to start carrying out controlled burns of all the chaparral slopes on the desert side of the mountains, in effect reinstating the indigenous land management practices of the area. Jorgensen says that controlled burning could be one piece of the puzzle in bringing back the bighorn, but it also poses problems. First, of course, is the safety of homes near potential burn sites; controlled burns in this part of the country tend to get quickly out of control. But a greater problem could be unintended consequences for the sheep. Opening up chaparral areas to new growth of grasses and annuals could pull deer into the area and mountain lions along with them, in turn threatening the bighorn. Burning alone could make matters worse for the sheep.

If they were influencing the landscape to increase the numbers of game animals, the Cahuilla and the Kumeyaay must have had similar problems with predators. What was their solution? Lowell John Bean and Florence Shipek agree that both the Cahuilla and the Kumeyaay occasionally, though not often, killed a mountain lion. Like everything else in their physical environment, the mountain lion had a spirit, and it was regarded as a particularly powerful animal. For this reason, there was a general taboo associated with the killing of mountain lions. But there were exceptions if a particular lion seemed to pose a threat to people. When a hunter encountered a mountain lion, he would warn it to go away before it got hurt. Later, the hunter would come back with a hunting party. If the lion was still around, the hunters would kill it. Because mountain lions are powerful animals, only very powerful hunters could kill them, and they gained more power by virtue of the kill.

Jack Turner will hate me for proposing this, but maybe we need to shoot a few mountain lions to save the sheep. Turner believes that we shouldn't look to the Indians for examples of how to treat predators; rather, we should look to the Juwa Bushmen of Africa, who established what he calls a "peaceful covenant" with lions. But the Indians—at least those in southern California— weren't looking for a peaceful covenant with predators. Instead, they saw them as competitors for resources and as potential sources of danger. Fewer lions meant more game and less danger for people. Today, the presence of fewer mountain lions (or the removal of specific lions that prey on sheep) could improve the bighorn's chances for survival.

This is what park personnel and officials of the U.S. Fish and Wildlife Service are considering in their conservation plan for the bighorn sheep.

danger. Fewer lions meant more game and less danger for people. Today, the presence of fewer mountain lions (or the removal of specific lions that prey on sheep) could improve the bighorn's chances for survival.

This is what park personnel and officials of the U.S. Fish and Wildlife Service are considering in their conservation plan for the bighorn sheep.

ANN HAYMOND ZWINGER

Ann Haymond Zwinger was born in 1925 in Muncie, Indiana. She is the author of several books about the natural history of the southwestern United States, including Run, River, Run: A Naturalist's Journey Down One of the Great Rivers of the West, *based on a whitewater rafting excursion she took down the entire length of the Grand Canyon. She taught Southwest studies and English at Colorado College and is also the mother of environmental author and teacher Susan Zwinger, whose desert conservation writing is also featured in this anthology.*

In the following passage from her book The Mysterious Lands: A Naturalist Explores the Four Great Deserts of the Southwest *(1989), Zwinger takes a look at the unique landscape and animal life of the Coachella Valley Preserve, located adjacent to the urban sprawl of Palm Springs and its rapidly growing sister communities.*

from *The Mysterious Lands*

The Coachella Dunes pile up on the east side of a narrow valley extending some sixty miles from the Salton Sea on the southeast to the summit of San Gorgonio Pass to the northwest. The booming winds that pour out of San Gorgonio Pass and funnel down the Coachella Valley are legendary (attested to by the numerous wind-generator "farms" at the throat of the pass). Desert areas of lower barometric pressure draw relatively cool air in off the Pacific, creating an inversion at around three thousand feet over the pass. The narrow configuration of the pass creates a Venturi effect: air pressure drops and wind accelerates, unleashing high-velocity and extremely desiccating winds into the basin below, and whisking sand into low, rolling dunes against the Coachella Mountains.

The day has built to full blaze when I reach the dunes at noontime. I loathe hats but clamp one on my sweating forehead anyway, and check to see that my water bottle is full, knowing the water will soon be as unappetizing as warm dishwater. I roll down my sleeves and turn up my shirt collar to keep off the sun—loose clothing that does not hinder perspiring can halve the radiant heat load and cut water loss by two-thirds. I pick up my notebook, take a deep, reluctant breath, and push my way into a desert blanched with heat.

The dunes are so light in color that they reflect and focus the heat on any object above ground. Small dunes, the largest but twenty-five feet or so above the interdunal flats, face mostly south and east. I tromp through a rich growth of dune plants stabilizing small hummocks on the dunes' apron, laced with rabbit and fox tracks. Ladybugs clamber on desert yellow primrose and over desert four-o'clocks.

I bend over to admire a clump of yellow evening primroses and find myself contemplating instead a desert iguana tucked away neatly in the shadows. A stocky medium-size lizard with a five-inch body plus a five-inch tail, its buff-colored body is patterned with brown spots that form varied patterns across its back and coalesce to form lines around the tail, pied patterns that render it part and parcel of the broken shadows. A diagnostic row of beaded scales runs all the way down its back and onto its tail. It has a narrower range than that of creosote bush, which makes up almost all of its diet. It often climbs into the bushes to feed on the blossoms although it avoids the unpalatable leaves.

Desert iguanas are beautifully adapted to life on the desert. They may drink water if available, but can exist with only the water contained in their food. As they are primarily vegetarians, desert iguanas must have some extrarenal mechanism to get rid of the sodium and potassium salts they pick up in their diet. This they do in much the same way sea birds do, through a pair of glands opening into the nasal cavity that act as accessory kidneys and remove salt ions with a minimum loss of water.

In this afternoon heat the iguana operates very close to its tolerable limit. It regulates its body temperature within rather narrow limits by behavior, as it is doing now. It began its day in its burrow, waiting as long as two hours after sunup before emerging to forage, remaining nearby its burrow until its body temperature reached its optimum, a high 111 degrees F. Unlike most lizards, which tend to take over old rodent burrows, desert iguanas usually dig their own.

I lean closer to see it better and it bolts off across sand so hot I feel the heat through my soles. It runs with such a peculiar shuffling, swimming gait that I think it has a bad leg, before I realize it is brushing the hotter surface sand aside to bring its body into contact with cooler sand beneath. As the day cools—and I fervently wish it would—it will spend more time in the sun, orienting its body to receive the rays at a more perpendicular angle, until it finally tucks into its burrow for the night.

When I reach the dunes proper, I find them well used by ORVs, tire tracks superimposed on the chained imprints of lizards. Although it is windy now, it is nothing like what it can be, for which I am thankful. Fast-moving hot air quickly adds to heat stress. Because the winds from San Gorgonio Pass are so powerful, accumulations of larger grains and pebbles in the troughs emphasize the spectacular ripple patterns. Smaller-scale patterns crosshatch larger ripples going in another direction, as if the wind constantly wearied and shifted, quixotically rearranging the patterns to its liking every hour or so.

Small cylinders of damp sand dot the top of one dune, an inch or so high, as if tubes of sand had been piped up. Smaller ones are sealed. Several dozen stud the area, always several feet apart, probably belonging to large female wolf spiders, who find happy hunting on the dunes.

The distinctive and extraordinarily even track of a sidewinder loops across a sand flat. The impressions in the sand sit at an oblique angle to the rattlesnake's direction of travel, arcs with little walled ridges of heaped-up sand on the outer arc of the curve. Each impression has a hook made by the head and neck and a T by the tail, which is to say that this snake was heeding Horace Greeley's advice.

Sidewinders are specialized for living in dunes by this peculiar type of locomotion, not by any anatomical or physiological alterations. The principle of sidewinding is that of a sidewise rolling screw, touching ground with only two sections of its length at a time, efficient for fast travel in a loose substrate like sand because the force exerted is vertical, better than a horizontal force, which is inefficient in loose sand. I follow the tracks until they peter out, only to emerge a few feet farther, going in the other direction, looping around a bush, setting out straight again, and finally disappearing. They have no rhyme or reason and there are interruptions where I think this crazy snake must have been airborne. Where tamarisks have formed copses in the dunes, the hummocks are riddled with holes, oases for the sidewinder's favorite prey of small mammals and lizards.

What I have come here for, am enduring this fiery furnace for, is a glimpse of a fringe-toed lizard, one of the most highly specialized vertebrates of North America sand-dune habitats, a lizard endowed with distinctive anatomical adaptations to living in this flowing, unstable medium. Its sloping snout is wedge-shaped, with a countersunk lower jaw to slip easily into the sand. Its eyelids are thick and overlap and interlock to keep out sand. Scales form protective flaps over the ears, pressing back as the animal burrows. Elongated scales fringe its toes, which increase the surface area of the feet and therefore the traction, allowing greater purchase in loose sand by doubling back when its leg bends forward and flaring outward as its leg extends. Fringe-toed lizards are the only reptiles limited to and totally dependent upon dunes.

Fringe-toed lizards' ability to bury themselves on the instant is legendary. They wiggle their heads sideways and push strongly with muscular hind legs, holding their front legs close to their sides to reduce resistance as a scuba diver, arms held back, powers through the ocean with his legs. A U-shaped nasal passage prevents sand grains from penetrating its head. A cavity of air under its body keeps sand from falling in about it and constricting its next inhalation.

I follow endless tracks chaining the dune rims and skidding down over the edge into loose sand, where there are dark burrow holes on the slope face. But not a lizard is in sight. Possessing exceptionally keen eyesight, they may have spotted me long before I could see them. Or, being so closely matched to the sand that sometimes sophisticated optical equipment cannot differentiate between lizard and background, perhaps I couldn't see them even if they were out. I scan the dune crest for movement. Nothing.

Even though I know they bask only in early morning and late evening, the eternal optimist in me bids me wait. I sit down on the dune. Perspiration streams down my nose and drips on my notebook. I drain my water bottle. The sun notches another degree to the west, cleaning out the shadows, encouraging the wind, blow-torching my shoulders. If I don't get up I may be glued to the sand. After an hour, it takes great effort to heave myself to my feet and admit defeat. No lizard in its right mind would be out.

Disappointed, I shuffle my way out of the dunes, pondering the impossibility of fringe-toed lizards. I can see where one went to market, where they held the Olympic games, where one had another to tea, where the chorus line formed, but they themselves have sensibly retired beneath the surface, waiting for a respite from the heat. I envision dozens of them studding the lee side of the dunes, all neatly tucked beneath the sand, front legs folded close to their sides, hind limbs flexed for action, pineal gland shuttered, eyelids locked closed, all perversely, unreasonably, and inconsiderately keeping out of sight.

The 13,000-acre Coachella Valley Preserve protects not only fringe-toed lizards, but also a palm oasis, the anomaly of the Sonoran Desert. The flickering shade of the palm oasis is the complete antithesis of the unmitigated heat of the dunes. The shade is as much aural as visual. The fronds converse quietly, rustling like taffeta, from an air flow that moves through treetops thirty feet above but scarcely reaches me below. I can almost hear the whir of cameras as Rudolph Valentino gesticulated through *Son of the Sheik* or Cecil B. De Mille filmed *King of Kings* in this very palm oasis.

The only way in which early travelers survived crossing the desert was to know where there were tinajas, springs, or oases. On the first major crossing, when de Anza trekked the desert in the 1770s, he took with him an Indian as guide to direct him to water, and survived the trip. These oases are still marked on the map but now they are towns, among them Twentynine Palms, Thousand Palms, Palm Springs.

Native desert palms have been present in western North America for 100 million years, and at this time may be actively expanding their range. The genus developed in the Pliocene, probably in Baja California, and spread northward, retreating southward during the cooler, rainier times of the Pleistocene. When climate warmed after the Pleistocene, they again migrated northward, increasing their range, able to establish where there were springs and seeps. Every canyon with enough water to support a desert palm grove is on a fracture or fault that allows, or forces, water to rise to the surface.

The kind of soil in which the palms root is not as important as permanent water. If runoff and seepage increase, so do the number of desert palms; likewise, successive drought years decrease grove size. These groves look as if they are all the same height and the same age, but after reaching thirty feet the palms slow in growth, limited by the ability of the vascular system to pump water any higher; between seventy and eighty feet seems to be the limit. While close in height, they may actually be varied in age.

Since the seeds are so heavy, long-range dispersal depends on coyotes, which eat the whole fruit, passing the seeds through their digestive tracts unharmed—those seeds germinate better than uneaten ones. In the fall, coyote scat is filled with desert palm seeds, most of which are ready to sprout. Robins have been seen with seeds in their beaks, and possibly other birds may help in distributing palm seeds. Over eighty species of migratory birds frequent palm oases.

Fire is the most important factor, other than water, in maintaining desert palm oases. Palm oases are very flammable from the large skirts of dry fronds that thatch the trunks. Without burning, palms eventually decline in number, crowded out by plants with more extensive root systems. Fire eliminates the above-ground portion of competing plants so that palms can outgrow even the grasping tamarisk with its incredibly dense and shallow water-usurping root system. Although tamarisks will return by sprouting from rhizomes and roots, at least three years pass before they reach their former growth. Meanwhile, the industrious desert palms, growing one to two feet a year, will have produced multitudes of new seeds, which will have enjoyed ideal germinating conditions.

The cambium layer, which produces new growth, in deciduous trees is near the surface and thus vulnerable to fire; in palms, such transport tissue is scattered throughout the trunk. Even though the green leaves are killed when the dead hanging skirts catch fire, the palm generally survives and puts out new fronds in two to three weeks.

With no plants except desert palms transpiring, the soil actually becomes more moist, creating a perfect situation for palm seeds to germinate. In response, desert palms produce twice as much fruit after a fire. The survival rate of new palms is enhanced by arming the petioles of tender new fronds with spines, which discourage browsing. The large grove at Palm Springs burned in 1980; now, half a dozen years later, it has regained its lushness. Only blackened bark recalls the blaze. Nevertheless, each fire removes a little more wood, destroys a few more vascular bundles, eventually weakening the tree and making it vulnerable to boring beetles.

Nearly every tree has thumb-size exit holes, made by the two-inch larvae of the giant palm borer beetle on their way out. Once a desert palm is well-started, few insects attack it, but this beetle can literally reduce the core of a tree to sawdust. The fat, pale-yellow larvae spend up to six years eating their way through the trunk. Seventy to 90 percent of the palms are done in by these beetles, either directly or indirectly. The only control is fire, which decimates beetles close to the surface. The palm borer occurs in every palm oasis except the two remote ones in Arizona to which the beetles have not yet spread.

My cultural belief that fires are one of the evils of nature has required some massive rethinking on my part to discard, in order to accept that fire has always been a natural part of certain of the North American biomes, even a necessary part. The burning of the prairies fertilized and kept the tall grasses vigorous. The Indians who lived near the palm oases—and each oasis had its own group—often fired them to promote new growth. Plant and animal populations are adjusted to fire. Man is not, and has contrived to remove fire as much as possible from the ecosystem, frequently to its detriment.

But before there were careless campers there were, and still are, careless storms with careless lightning. Nature has its own cadence, in which there is little tolerance for houses built in floodplains or flammable chaparral, little consideration for cities built on fault zones, or irrigation works that turn the desert green. Nature acts without calling in a consultant or submitting an environmental impact statement.

BARRY LOPEZ

Barry Lopez was born in Port Chester, New York, in 1945 and lived in the Mojave Desert for part of his childhood. He writes fiction, essays, and poetry, often focused on the natural world. His Arctic Dreams won the National Book Award in 1986, and among his other books are Desert Notes/River Notes, published in 1990, and Crossing Open Ground, an essay collection published in 1989, both of which include pieces about the California deserts. Lopez now lives in Oregon.

In the following excerpt from the essay "The Stone Horse," first published in Antaeus magazine in 1986, Lopez writes of his efforts to locate a four-hundred-year-old stone horse intaglio near Blythe, California.

from "The Stone Horse"

A BLM archaeologist told me, with understandable reluctance, where to find the intaglio. I spread my Automobile Club of Southern California map of Imperial County out on his desk, and he traced the route with a pink felt-tip pen. The line crossed Interstate 8 and then turned west along the Mexican border.

"You can't drive any farther than about here," he said, marking a small X. "There's boulders in the wash. You walk up past them."

On a separate piece of paper, he drew a route in a smaller scale that would take me up the arroyo to a certain point where I was to cross back east, to another arroyo. At its head, on higher ground just to the north, I would find the horse.

"It's tough to spot unless you know it's there. Once you pick it up..." He shook his head slowly, in a gesture of wonder at its existence. I waited until I held his eye. I assured him I would not tell anyone else how to get there. He looked at me in stoical despair, like a man who had been robbed twice, whose belief in human beings was offered without conviction.

I did not go until the following day because I wanted to see it at dawn. I ate breakfast at four A.M. in El Centro and then drove south. The route was easy to follow, though the last section of road proved difficult, broken and drifted over with sand in some spots. I came to the barricade of boulders and parked. It was light enough by then to find my way over the ground with little trouble. The contours of the landscape were stark, without any masking vegetation. I worried only about rattlesnakes.

I traversed the stone plain as directed, but, in spite of the frankness of the land, I came on the horse unawares. In the first moment of recognition I was without feeling. I recalled later being startled, and that I held my breath. It was laid out on the ground with its head to the east, three times life-size. As I took in its outline I felt a growing concentration of all my senses, as though my attentiveness to the pale rose color of the morning sky and other peripheral images had now ceased to be important. I was aware that I was straining for sound in the windless air, and I felt the uneven pressure of the earth hard against my feet. The horse, outlined in a standing profile on the dark ground, was as vivid before me as a bed of tulips.

I've come upon animals suddenly before, and felt a similar tension, a precipitate heightening of the senses. And I have felt the inexplicable but sharply boosted intensity of a wild moment in the bush, where it is not until some minutes later that you discover the source of electricity—the warm remains of a grizzly bear kill, or the still moist tracks of a wolverine.

But this was slightly different. I felt I had stepped into an unoccupied corridor. I had no familiar sense of history, the temporal structure in which to think: this horse was made by Quechan people three hundred years ago. I felt instead a headlong rush of images: people hunting wild horses with spears on the Pleistocene veld of southern California; Cortés riding across the causeway into Montezuma's Tenochtitlán;[1] a short-legged Comanche, astride his horse like some sort of ferret, slashing through cavalry lines of young men who rode like farmers;[2] a hoof exploding past my face one morning in a corral in Wyoming. These images had the weight and silence of stone.

When I released my breath, the images softened. My initial feeling, of facing a wild animal in a remote region, was replaced with a calm sense of antiquity. It was then that I became conscious, like an ordinary tourist, of what was before me, and thought: this horse was probably laid out by Quechan people. *But when?* I wondered. The first horses they saw, I knew, might have been those that came north from Mexico in 1692 with Father Eusebio Kino.[3] But Cocopa people, I recalled, also came this far north on occasion, to fight with their neighbors, the Quechan. And *they could have seen horses with Melchior Diaz,*[4] at the mouth of the Colorado River in the fall of 1540. So, it could be four hundred years old. (No one in fact knows.)

---

[1] Hernando Cortés (c. 1485–1547), Spanish conqueror of Mexico, conquered in 1521 the Aztec capital city of Tenochtitlán, which was located on an island in Lake Texcoco, where Mexico City stands today. Montezuma II, ruler of the Aztecs, had welcomed Cortés to his capital on November 8, 1519, and the Spanish stayed there, holding him hostage.

[2] The Comanches, who often attacked settlers on the southern U.S. plains until the last of them were settled on a reservation in 1875, were known as the finest native American horsemen of the West.

[3] Kino (1644–1711) was the most famous of the seventeenth-century Spanish explorers and Jesuit missionaries in what became the U.S. Southwest.

[4] Diaz, one of Coronado's officers, explored the Sonoran Desert and the delta of the Colorado River in 1540.

I still had not moved. I took my eyes off the horse for a moment to look south over the desert plain into Mexico, to look east past its head at the brightening sunrise, to situate myself. Then, finally, I brought my trailing foot slowly forward and stood erect. Sunlight was running like a thin sheet of water over the stony ground and it threw the horse into relief. It looked as though no hand had ever disturbed the stones that gave it its form.

The horse had been brought to life on ground called desert pavement, a tight, flat matrix of small cobbles blasted smooth by sand-laden winds. The uniform, monochromatic blackness of the stones, a patina of iron and magnesium oxides called desert varnish, is caused by long-term exposure to the sun. To make this type of low-relief ground glyph, or intaglio, the artist either selectively turns individual stones over to their lighter side or removes them to expose the lighter soil underneath, creating a negative image. This horse, about eighteen feet from brow to rump and eight feet from wither to hoof, had been made in the latter way, and its outline was bermed at certain points with low ridges of stone a few inches high to enhance its three-dimensional qualities. (The left side of the horse was in full profile; each leg was extended at ninety degrees to the body and fully visible, as though seen in three-quarter profile.)

I was not eager to move. The moment I did I would be back in the flow of time, the horse no longer quivering in the same way before me. I did not want to feel again the sequence of quotidian events—to be drawn off into deliberation and analysis. A human being, a four-footed animal, the open land. That was all that was present—and a "thoughtless" understanding of the very old desires bearing on this particular animal: to hunt it, to render it, to fathom it, to subjugate it, to honor it, to take it as a companion.

What finally made me move was the light. The sun now filled the shallow basin of the horse's body. The weighted line of the stone berm created the illusion of a mane and the distinctive roundness of an equine belly. The change in definition impelled me. I moved to the left, circling past its rump, to see how the light might flesh the horse out from various points of view. I circled it completely before squatting on my haunches. Ten or fifteen minutes later I chose another view. The third time I moved, to a point near the rear hooves, I spotted a stone tool at my feet. I stared at it a long while, more in awe than disbelief, before reaching out to pick it up. I turned it over in my left palm and took it between my fingers to feel its cutting edge. It is always difficult, especially with something so portable, to rechannel the desire to steal.

I spent several hours with the horse. As I changed positions and as the angle of the light continued to change I noticed a number of things. The angle at which the pastern carried the hoof away from the ankle was perfect. Also, stones had been placed within the image to suggest at precisely the right spot the left shoulder above the foreleg. The line that joined thigh and hock was similarly accurate. The muzzle alone seemed distorted—but perhaps these stones had been moved by a later hand. It was an admirably accurate representation, but not what a breeder would call perfect conformation. There was the suggestion of a bowed neck and an undershot jaw, and the tail, as full as a winter coyote's, did not appear to be precisely to scale.

The more I thought about it, the more I felt I was looking at an individual horse, a unique combination of generic and specific detail. It was easy to imagine one of Kino's horses as a model, or a horse that ran off from one of Coronado's columns. *What kind of horses would these have been?* I wondered. In the sixteenth century the most sought-after horses in Europe were Spanish, the offspring of Arabian stock and Barbary horses that the Moors brought to Iberia and bred to the older, eastern European strains brought in by the Romans. The model for this horse, I speculated, could easily have been a palomino, or a descendant of horses trained for lion hunting in North Africa.

A few generations ago, cowboys, cavalry quartermasters, and draymen would have taken this horse before me under consideration and not let up their scrutiny until they had its heritage fixed to their satisfaction. Today, the distinction between draft and harness horses is arcane knowledge, and no image may come to mind for a blue roan or a claybank horse. The loss of such refinement in everyday conversation leaves me unsettled. People praise the Eskimo's ability to distinguish among forty types of snow but forget the skill of others who routinely differentiate between overo and tobiano pintos.[1] Such distinctions are made for the same reason. You have to do it to talk clearly about the world.

For parts of two years I worked as a horse wrangler and packer in Wyoming. It is dim knowledge now; I would have to think to remember if a buckskin was a kind of dun horse. And I couldn't throw a double-diamond hitch over a set of panniers—the packer's basic tie-down—without guidance. As I squatted there in the desert, however, these more personal memories seemed tenuous in comparison with the sweep of this animal in human time. My memories had no depth. I thought of the Hittite cavalry riding against the Syrians 3,500 years ago. And the first of the Chinese emperors, Ch'in Shih Huang, buried in Shensi Province in 210 B.C. with thousands of life-size horses and soldiers, a terra-cotta guardian army. What could I know

---

[1] The two color patterns for this breed of horse. The overo have white spreading irregularly up from the belly, mixed with a darker color, and the tobiano have white spreading down from the back in clear-cut patterns.

of what was in the mind of whoever made this horse? Was there some racial memory of it as an animal that had once fed the artist's ancestors and then disappeared from North America? And then returned in this strange alliance with another race of men?

Certainly, whoever it was, the artist had observed the animal very closely. Certainly the animal's speed had impressed him. Among the first things the Quechan would have learned from an encounter with Kino's horses was that their own long-distance runners—men who could run down mule deer—were no match for this animal.

From where I squatted I could look far out over the Mexican plain. Juan Bautista de Anza[1] passed this way in 1774, extending El Camino Real into Alta California from Sinaloa. He was followed by others, all of them astride the magical horse; *gente de razón*, the people of reason, coming into the country of *los primitivos*. The horse, like the stone animals of Egypt, urged these memories upon me. And as I drew them up from some forgotten corner of my mind—huge horses carved in the white chalk downs of southern England by an Iron Age people; Spanish horses rearing and wheeling in fear before alligators in Florida—the images seemed tethered before me. With this sense of proportion, a memory of my own—the morning I almost lost my face to a horse's hoof—now had somewhere to fit.

I rose up and began to walk slowly around the horse again. I had taken the first long measure of it and was now looking for a way to depart, a new angle of light, a fading of the image itself before the rising sun, that would break its hold on me. As I circled, feeling both heady and serene at the encounter, I realized again how strangely vivid it was. It had been created on a barren bajada[2] between two arroyos, as nondescript a place as one could imagine. The only plant life here was a few wands of ocotillo cactus. The ground beneath my shoes was so hard it wouldn't take the print of a heavy animal even after a rain. The only sounds I heard here were the voices of quail.

The archaeologist had been correct. For all its forcefulness, the horse is inconspicuous. If you don't care to see it you can walk right past it. That pleases him, I think. Unmarked on this bleak shoulder of the plain, the site signals to no one; so he wants no protective fences here, no informative plaque, to act as beacons. He would rather take a chance that no motorcyclist, no aimless wanderer with a flair for violence and a depth of ignorance, will ever find his way here.

The archaeologist had given me something before I left his office that now seemed peculiar—an aerial photograph of the horse. It is widely believed

---

[1] De Anza (1735–1788), later governor of new Mexico, explored the route to California, founding San Francisco in 1775.

[2] A broad slope of debris of rocks and gravel.

that an aerial view of an intaglio provides a fair and accurate depiction. It does not. In the photograph the horse looks somewhat crudely constructed; from the ground it appears far more deftly rendered. The photograph is of a single moment, and in that split second the horse seems vaguely impotent. I watched light pool in the intaglio at dawn; I imagine you could watch it withdraw at dusk and sense the same animation I did. In those prolonged moments its shape and so, too, its general character changed—noticeably. The living quality of the image, its immediacy to the eye, was brought out by the light-in-time, not, at least here, in the camera's frozen instant.

Intaglios, I thought, were never meant to be seen by gods in the sky above. They were meant to be seen by people on the ground, over a long period of shifting light. This could even be true of the huge figures on the Plain of Nazca in Peru, where people could walk for the length of a day beside them.[1] It is our own impatience that makes us think otherwise.

This process of abstraction, almost unintentional, drew me gradually away from the horse. I came to a position of attention at the edge of the sphere of its influence. With a slight bow I paid my respects to the horse, its maker, and the history of us all, and departed.

A short distance away I stopped the car in the middle of the road to make a few notes. I could not write down what I was thinking when I was with the horse. It would have seemed disrespectful, and it would have required another kind of attention. So now I patiently drained my memory of the details it had fastened itself upon. The road I'd stopped on was adjacent to the All American Canal, the major source of water for the Imperial and Coachella valleys. The water flowed west placidly. A disjointed flock of coots, small, dark birds with white bills, was paddling against the current, foraging in the rushes.

I was peripherally aware of the birds as I wrote, the only movement in the desert, and of a series of sounds from a village a half-mile away. The first sounds from this collection of ramshackle houses in a grove of cottonwoods were the distracted dawn voices of dogs. I heard them intermingled with the cries of a rooster. Later, the high-pitched voices of children calling out to each other came disembodied through the dry desert air. Now, a little after seven, I could hear someone practicing on the trumpet, the same rough phrases played over and over. I suddenly remembered how as children we had tried to get the rhythm of a galloping horse with hands against the thighs, or by fluttering our tongues against the roofs of our mouths.

[1] Gigantic lines from an unknown civilization are laid out geometrically on the thirty-seven-mile-long Plain of Nazca in southern Peru.

After the trumpet, the impatient calls of adults summoning children. Sunday morning. Wood smoke hung like a lens in the trees. The first car starts—a cold eight-cylinder engine, of Chrysler extraction perhaps, goosed to life, then throttled back to murmur through dual mufflers, the obbligato music of a shade-tree mechanic. The rote bark of mongrel dogs at dawn, the jagged outcries of men and women, an engine coming to life. Like a thousand villages from West Virginia to Guadalajara.

I finished my notes—where was I going to find a description of the horses that came north with the conquistadors? Did their manes come forward prominently over the brow, like this one's, like the forelocks of Blackfeet and Assiniboin men in nineteenth-century paintings? I set the notes on the seat beside me.

The road followed the canal for a while and then arced north, toward Interstate 8. It was slow driving and I fell to thinking how the desert had changed since Anza had come through. New plants and animals—the MacDougall cottonwood, the English house sparrow, the chukar from India[1]—have about them now the air of the native born. Of the native species, some—no one knows how many—are extinct. The populations of many others, especially the animals, have been sharply reduced. The idea of a desert impoverished by agricultural poisons and varmint hunters, by off-road vehicles and military operations, did not seem as disturbing to me, however, as this other horror, now that I had been those hours with the horse. The vandals, the few who crowbar rock art off the desert's walls, who dig up graves, who punish the ground that holds intaglios, are people who devour history. Their self-centered scorn, their disrespect for ideas and images beyond their ken, create the awful atmosphere of loose ends in which totalitarianism thrives, in which the past is merely curious or wrong.

I thought about the horse sitting out there on the unprotected plain. I enumerated its qualities in my mind until a sense of its vulnerability receded and it became an anchor for something else. I remembered that history, a history like this one, which ran deeper than Mexico, deeper than the Spanish, was a kind of medicine. It permitted the great breadth of human expression to reverberate, and it did not urge you to locate its apotheosis in the present.

Each of us, individuals and civilizations, has been held upside down like Achilles in the River Styx.[2] The artist mixing his colors in the dim light of Altamira;[3] an Egyptian ruler lying still now, wrapped in his bysses, stored

---

[1] A gray-and-black partridge introduced into dry parts of the western United States from India.

[2] Achilles' mother dipped him in the River Styx, holding him upside down by the heels, to make him invulnerable and hence immortal.

[3] A cave in northern Spain with Old Stone Age drawings of animals.

against time in a pyramid; the faded Dorset culture of the Arctic;[1] the Hmong and Samburu and Walbiri of historic time;[2] the modern nations. This great, imperfect stretch of human expression is the clarification and encouragement, the urging and the reminder, we call history. And it is inscribed everywhere in the face of the land, from the mountain passes of the Himalayas to a nameless bajada in the California desert.

Small birds rose up in the road ahead, startled, and flew off. I prayed no infidel would ever find that horse.

---

[1] A culture in Greenland and the Canadian eastern Arctic that flourished between approximately 800 B.C. and A.D. 1300; it is not certain exactly when or why Dorset culture disappeared, although the artifacts that remain give an idea of its nature and its daily life.

[2] The Hmong, also called Miao or Meo, are mountain-dwelling peoples of China and Southeast Asia; the Samburu are a tribe in Kenya; the Walbiri are aborigines of the desert in central Australia.

PERMISSIONS

Edward Abbey. Excerpt from "Death Valley" from *The Journey Home* by Edward Abbey. New York: Plume, 1991. © 1977 by Edward Abbey. Reprinted by permission of Dutton, a division of Penguin Group (USA) Inc.

Katherine Ainsworth. Excerpt from *The Man Who Captured Sunshine* by Katherine Ainsworth. Palm Springs, CA: ETC Publications, 1978. © 1978 by ETC Publications. Reprinted by permission of ETC Publications.

Anonymous. "Stampede for Jackrabbit Homesteads Deluges Los Angeles Land Office" from *Desert* magazine (Palm Desert, CA), April 1945.

Anonymous. Excerpt from a letter to the Alta California, July 1, 1866. Reprinted in *Mojave Road in Newspapers: Tales of the Mojave Road Number Six, Jan. 1976,* edited by Dennis G. Casebier. Norco, CA: King's Press, 1976.

Mary Austin. Excerpt from *The Land of Little Rain* by Mary Austin. Boston: Houghton Mifflin, 1903.

Pearl Bailey. Excerpt from *The Raw Pearl* by Pearl Bailey. New York: Harcourt, Brace, and World, 1968. © 1968 by Pearl Bailey. Reprinted by permission of Houghton Mifflin Harcourt Publishing Company.

Dick Barnes. "Bagdad Chase Road in July" and "Song: Mojave Narrows" from *A Word Like Fire* by Dick Barnes. New York: Other Press, 2005. © 2005 by Patricia Barnes. Reprinted by permission of Patricia Barnes Darby.

William Phipps Blake. Journal excerpts, Pacific Railroad Survey, 1853, reprinted in "USGS Survey of Railroad–Coachella Valley, 1853" by Franklin Hoyt, *The Periscope*. Indio, CA: Coachella Valley Historical Society, 1990.

Gayle Brandeis. "Climbing at Joshua Tree" by Gayle Brandeis. © 2009 by Gayle Brandeis. Reprinted by permission of Gayle Brandeis.

Elizabeth Crozer Campbell. Excerpt from *The Desert Was Home: Adventures and Tribulations of a Desert Homestead.* Los Angeles: Westernlore Press, 1961. © 1961, 1989 by Elizabeth Crozer Campbell. Reprinted by permission of Corless Wilson Eldred.

César E. Chávez. "Speech at Coachella, 1973" by César E. Chávez reprinted in *The Words of César Chávez,* edited by Richard J. Jensen and John C. Hammerback. College Station, TX: Texas A&M University Press, 2002. © 2009 by the César E. Chávez Foundation, www.chavezfoundation.org. Reprinted by permission of the César E. Chávez Foundation.

Craig Childs. Excerpt from *The Desert Cries: A Season of Flash Floods in a Dry Land.* Phoenix, AZ: Arizona Highways Books, 2002. © 2002 by the Arizona Department of Transportation, State of Arizona. Reprinted by permission of the Arizona Department of Transportation, State of Arizona.

Jeanette Clough. "Leaving Palm Springs" from *Island* by Jeanette Clough. Granada Hills, CA: Red Hen Press, 2007. © 2007 by Jeanette Clough. Reprinted by permission of Red Hen Press.

Editions, 1990. © 1990 by Susan Straight. Reprinted by permission of Milkweed Editions. (www.milkweed.org).

Hunter S. Thompson. Excerpt from *Fear and Loathing in Las Vegas* by Hunter S. Thompson. New York: Random House, Inc., 1972. © 1972 by Hunter S. Thompson. Reprinted by permission of Random House.

Clifford E. Trafzer. "Chemehuevi Indian Creation" by Clifford E. Trafzer. © 2009 by Clifford E. Trafzer. Portions of this essay are drawn from *Chemehuevi People of the Coachella Valley* by Clifford E. Trafzer, Luke Madrigal, and Anthony Madrigal. Coachella, CA: Chemehuevi Press, 1997. Reprinted by permission of Clifford E. Trafzer; the author wishes to thank the Twenty-Nine Palms Tribe.

John C. Van Dyke. Excerpt from *The Desert* by John C. Van Dyke. New York: Charles Scribner Sons, 1901.

Victor Villaseñor. Excerpt from *Macho!* by Victor Villaseñor. Houston, TX: Arte Publico Press, University of Houston. © 1991 by Arte Publico Press. Reprinted by permission of Arte Publico Press.

Wakako Yamauchi. "Songs My Mother Taught Me" from *Songs My Mother Taught Me: Stories, Plays, and a Memoir* by Wakako Yamauchi. New York: The Feminist Press, 1994. © 1994 by Wakako Yamauchi. Reprinted by permission of The Feminist Press at the City University of New York, www.feministpress.org. All rights reserved.

Sally Zanjani. Excerpt from *A Mine of Her Own: Women Prospectors in the American West, 1850–1950* by Sally Zanjani. Lincoln, NE: University of Nebraska Press, 1997. © 1997 by University of Nebraska Press. Reprinted by permission of University of Nebraska Press.

Ann Haymond Zwinger. Excerpt from *The Mysterious Lands: A Naturalist Explores the Four Great Deserts of the Southwest* by Ann Haymond Zwinger. Tucson, AZ: University of Arizona Press, 1989. © 1989 by Ann Haymond Zwinger. Reprinted by permission of Ann Haymond Zwinger.

Susan Zwinger. Excerpt from *Still Wild, Always Wild: A Journey into the Desert Wilderness of California* by Susan Zwinger. San Francisco: Sierra Club Books, 1997. © 1997 by Tehabi Books. Reprinted by permission of Susan Zwinger.

AUTHOR INDEX

A California Legacy Book

Santa Clara University and Heyday Books are pleased to publish the California Legacy series, vibrant and relevant writings drawn from California's past and present.

Santa Clara University—founded in 1851 on the site of the eighth of California's original twenty-one missions—is the oldest institution of higher learning in the state. A Jesuit institution, it is particularly aware of its contribution to California's cultural heritage and its responsibility to preserve and celebrate that heritage.

Heyday Books, founded in 1974, specializes in critically acclaimed books on California literature, history, natural history, and ethnic studies.

Books in the California Legacy series appear as anthologies, single author collections, reprints of important books, and original works. Taken together, these volumes bring readers a new perspective on California's cultural life, a perspective that honors diversity and finds great pleasure in the eloquence of human expression.

Series editor: Terry Beers

Publisher: Malcolm Margolin

Advisory committee: Stephen Becker, William Deverell, Charles Faulhaber, David Fine, Steven Gilbar, Ron Hansen, Gerald Haslam, Robert Hass, Jack Hicks, Timothy Hodson, Jeanne Wakatsuki Houston, Maxine Hong Kingston, Frank LaPena, Ursula K. Le Guin, Jeff Lustig, Ishmael Reed, Alan Rosenus, Robert Senkewicz, Gary Snyder, Kevin Starr, Richard Walker, Alice Waters, Jennifer Watts, Al Young.

Thanks to the English Department at Santa Clara University and to Regis McKenna for their support of the California Legacy series.

If you would like to be added to the California Legacy mailing list, please send your name, address, phone number, and email address to:

California Legacy Project
English Department
Santa Clara University
Santa Clara, CA 95053
For more on California Legacy titles, events, or other information, please visit www.californialegacy.org.

Other California Legacy Books

The Anza Trail and the Settling of California
Vladimir Guerrero

Califauna: A Literary Field Guide
Edited by Terry Beers and Emily Elrod

California Poetry: From the Gold Rush to the Present
Edited by Dana Gioia, Chryss Yost, and Jack Hicks

Death Valley in '49
William Lewis Manly

Gunfight at Mussel Slough: Evolution of a Western Myth
Edited by Terry Beers

Inlandia: A Literary Journey through California's Inland Empire
Edited by Gayle Wattawa with an Introduction by Susan Straight

The Journey of the Flame
Walter Nordhoff with a Foreword by Rebecca Solnit

Lands of Promise and Despair: Chronicles of Early California, 1535-1846
Edited by Rose Marie Beebe and Robert M. Senkewicz

Mountains and Molehills
Frank Maryatt, Foreword by Robert J. Chandler

A Separate Star: Selected Writings of Helen Hunt Jackson
Edited with an Introduction by Michelle Burnham

Spring Salmon, Hurry to Me!: The Seasons of Native California
Edited by Margaret Dubin and Kim Hogeland

Under the Fifth Sun: Latino Literature in California
Edited by Rick Heide

Unfolding Beauty: Celebrating California's Landscapes
Edited with an Introduction by Terry Beers

Wallace Stegner's West
Edited with an Introduction by Page Stegner

Inlandia Institute

Inlandia Institute is a lively center of literary activity located in Riverside, California. It grew out of the highly acclaimed anthology *Inlandia: A Literary Journey through California's Inland Empire*, published by Heyday Books in 2006.

Inlandia Institute strives to nurture the rich and ongoing literary traditions of inland southern California. Its mission is to recognize, support, and expand literary activity in the Inland Empire by publishing books and sponsoring programs that deepen people's awareness, understanding, and appreciation of this unique, complex, and creatively vibrant area.

For more information about Inlandia Institute titles and programs please visit www.inlandiainstitute.net and www.heydaybooks.com/imprints/inlandia-institute.

HEYDAY INSTITUTE

Since its founding in 1974, Heyday Books has occupied a unique niche in the publishing world, specializing in books that foster an understanding of the history, literature, art, environment, social issues, and culture of California and the West. We are a 501(c)(3) nonprofit organization based in Berkeley, California, serving a wide range of people and audiences.

We are grateful for the generous funding we've received for our publications and programs during the past year from foundations and more than three hundred and fifty individual donors. Major supporters include:

Anonymous; Audubon California; Judith and Phillip Auth; Barona Band of Mission Indians; B.C.W. Trust III; S. D. Bechtel, Jr. Foundation; Barbara and Fred Berensmeier; Berkeley Civic Arts Program and Civic Arts Commission; Joan Berman; Book Club of California; Peter and Mimi Buckley; Buena Vista Rancheria; Lewis and Sheana Butler; Butler Koshland Fund; California State Automobile Association; California State Coastal Conservancy; California State Library; Joanne Campbell; Candelaria Fund; John and Nancy Cassidy Family Foundation, through Silicon Valley Community Foundation; Creative Work Fund; Columbia Foundation; Colusa Indian Community Council; The Community Action Fund; Community Futures Collective; Compton Foundation, Inc.; Lawrence Crooks; Ida Rae Egli; Donald and Janice Elliott, in honor of David Elliott, through Silicon Valley Community Foundation; Evergreen Foundation; Federated Indians of Graton Rancheria; Mark and Tracy Ferron; George Gamble; Wallace Alexander Gerbode Foundation; Richard & Rhoda Goldman Fund; Ben Graber, in honor of Sandy Graber; Evelyn & Walter Haas, Jr. Fund; Walter & Elise Haas Fund; James and Coke Hallowell; Cheryl Hinton; James Irvine Foundation; Mehdi Kashef; Robert and Karen Kustel, in honor of Bruce Kelley; Marty and Pamela Krasney; Guy Lampard and Suzanne Badenhoop; LEF Foundation; Michael McCone; Morongo Band of Mission Indians; National Endowment for the Arts; National Park Service; Organize Training Center; Patagonia; Pease Family Fund, in honor of Bruce Kelley; Resources Legacy Fund; Robinson Rancheria Citizens Council; Alan Rosenus; San Francisco Foundation; San Manuel Band of Mission Indians; Deborah Sanchez; William Saroyan Foundation; Contee and Maggie Seely; Sandy Shapero; Jim Swinerton; Swinerton Family Fund; Taproot Foundation; Thendara Foundation; Marion Weber; Albert and Susan Wells; Peter Booth Wiley; Dean Witter Foundation; and Yocha Dehe Wintun Nation.

For more information about Heyday Institute, our publications and programs, please visit our website at www.heydaybooks.com.

About the Editor

Ruth Nolan is a poet and writer whose works have appeared in many literary magazines and several anthologies. A former wildland firefighter for the BLM California Desert District, she holds a BA in English from California State University, San Bernardino, and an MA in English/creative writing from Northern Arizona University. She is an associate professor of English at College of the Desert and lives in Palm Desert, California.